THE WAY TO GEOMETRY

Being Necessary and Usefull

for

Astronomers, Enginees, Geographers,. Architects, Land-meaters, Carpenters, Sea-men & Etc.

[ILLUSTRATED]

By

Peter Ramus

Translator: William Bedwell

ILLUSTRATED &

PUBLISHED BY

E-KİTAP PROJESİ & CHEAPEST BOOKS

www.cheapestboooks.com

Istanbul

ISBN:

978-625-3870-02-7

Contents

The Way To Geometry

OF GEOMETRY THE EİGHT BOOKE, of the diverse kindes of Triangles. 192

GEOMETRY, THE ONE AND TWENTİETH BOOK, Of Lines and Surfaces in solids……………………………………………………………………………………..405

THE TWENTY SECOND BOOKE, of P. Ramus Geometry, Of a Pyramis. **418**

Αβουτ Αυτηορ:

Petrus Ramus *(1515 – 1572) was an influential French humanist, logician, and educational reformer. A Protestant convert, he was killed during the St. Bartholomew's Day Massacre.*

Early life:

He was born at the village of Cuts in Picardy; his father was a farmer. He gained admission at age twelve, to the Collège de Navarre, working as a servant. A reaction against scholasticism was in full tide, at a transitional time for Aristotelianism. On the occasion of taking his degree (1536) Ramus allegedly took as his thesis Quaecumque ab Aristotele dicta essent, commentitia esse, *which Walter J. Ong paraphrases as follows:*

"All the things that Aristotle has said are inconsistent because they are poorly systematized and can be called to mind only by the use of arbitrary mnemonic devices."

According to Ong this kind of spectacular thesis was in fact routine at the time. Even so, Ong raises questions as to whether Ramus actually ever delivered this thesis.

Early academic career:

Ramus, as graduate of the university, started courses of lectures. At this period he was engaged in numerous separate controversies. One opponent in 1543 was the Benedictine Joachim Périon. He was accused, by Jacques Charpentier, professor of medicine, of undermining the foundations of philosophy and religion. Arnaud d'Ossat, a pupil and friend of Ramus, defended him against Charpentier. Ramus was made to debate Goveanus (Antonio de Gouveia), over two days. The matter was brought before the parlement *of Paris, and finally before Francis I. By him it was referred to a commission of five, who found Ramus guilty of having "acted rashly, arrogantly and impudently," and interdicted his lectures (1544).*

* * * * *

VIA REGIA

Ad

GEOMETRIAM.

THE WAY

TO

GEOMETRY.

Being necessary and usefull,

For

Astronomers. Engineres. Geographers. Architecks. Land-meaters. Carpenters. Sea-men. Paynters. Carvers, &c.

Written in Latine by Peter Ramvs, and now Translated and much enlarged by the Learned M^r^. William Bedwell.

LONDON, 1636.

TO THE:

WORSHIPFVL

M. Iohn Greaves, *Professor* of

Geometry in *Gresham Colledge* London;

All happinesse.

SIR,

Your acquaintance with the Author before his death was not long, which I have oft heard you say, you counted your great unhappinesse, but within a short time after, you knew not well whether to count your selfe more happie in that you once knew him, or unhappy in that upon your acquaintance you so suddenly lost him. This his worke then being to come forth to the censorious eye of the world, and as the manner usually is to have some Patronage, I have thought good to dedicate it to your selfe; and that for these two reasons especially.

First, in respect of the sympathy betwixt it, and your studies; Laboures of this nature being usually offered to such persons whose profession is that way setled.

Secondly, for the great love and respect you alwayes shewed to the Author, being indeed a man that would deserve no lesse, humble, void of pride, ever ready to impart his knowledge to others in what kind soever, loving and affecting those that affected learning.

For these respects then, I offer to you this Worke of your so much honoured friend. I my selfe also (as it is no lesse my duty) for his sake striving to make you hereby some part of a requitall, least I should be found guilty of ingratitude, which is a solecisme in manners, if having so fit an opportunity, I should not expresse to the world some Testimonie of love to you, who so much loved him. I desire then (good Sir) your kind acceptance of it, you knowing so well the ability of the Author, and being also able to judge of a Worke of this nature, and in that respect the better able to defend it from the furie of envious Detractours, of which there are not few. Thus with my best wishes to you, as to my much respected friend, I rest.

Yours to be commanded in any thing that he is able. Iohn Clerke.

To the Reader.

Friendly Reader, that which is here set forth to thy view, is a Translation out of *Ramus*. Formerly indeed Translated by one M[r]. *Thomas Hood,* but never before set forth with the Demonstrations and Diagrammes, which being cut before the Authors death, and the Worke it selfe finished, the Coppie I having in mine hands, never had thought for the promulgation of it, but that it should have died with its Author, considering no small prejudice usually attends the printing of dead mens Workes, and wee see the times, the world is now all eare and tongue, the most given with the *Athenians*, to little else than to heare and tell newes: And if *Apelles* that skilfull Artist alwayes found somewhat to be amended in those Pictures which he had most curiously drawne; surely much in this Worke might have beene amended if the Authour had lived to refine it, but in that it was onely the first draught, and that he was prevented by death of a second view, though perused by others before the Presse; I was ever unwilling to the publication, but that I was often and much solicited with iteration of strong importunity, and so in the end over-ruled: perswading me from time to time unto it, and that it being finished by the Authour, it was farre better to be published, though with some errours and escapes, than to be onely moths-meat, and so utterly lost. I would have thee, Courteous Reader know, that it is no conceit of the worth of the thing that I should expose the name and credit of the Authour to a publike censure; yet I durst be bold to say, had he lived to have fitted it, and corrected the Presse, the worke would have pointed out the workeman. For I may say, without vaine ostentation, he was a man of worth and note, and there was not that kinde of learning in which he had not some knowledge, but especially for the Easterne tongues, those deepe and profound Studies, in the judgement of the learned, which knew him well, he hath not left his fellow behind him; as his Workes also in Manuscript now extant in the publike Library of the famous Vniversity of Cambridge; do testifie no lesse; for him then being so grave and learned a Divine to meddle with a worke of this nature, he gives thee a reason in his owne following Preface for his principall end and intent of taking this Worke in hand, was not for the deepe and Iudiciall, but for the shallowest skull, the good and profit of the simpler sort, who as it was in the Latine, were able to get little or no benifite from it. Therefore considering the worth of the Authour, and his intent in the Worke. Reade it favourably, and if the faults be not too great, cover them with the mantle of love, and judge charitably offences unwillingly committed, and doe according to the termes of equitie, as thou wouldest be done unto, but it is a common saying, as *Printers* get Copies for their profit, so Readers often buy and reade for their pleasure; and there is no worke so exactly done that can escape the malevolous disposition of some detracting spirits, to whom I say, as one well, *Facilius est unicuivis nostrum aliena curiosè observare: quam proproia negotia rectè agere*. It is a great deale more easie to carpe at other mens doings, than to give better of his owne. And as *Arist.* τό πάσιν ἀρέσαι δυσχερέστατόν ἐστι; *omnibus placere difficilimum est*. But wherefore, Gentle Reader, should I make any doubt of thy curtesie, and favourable acceptance; for surely there can be nothing more contrary to equitie, than to speake evill of those that have taken paines to doe good, a Pagan would hardly doe this, much lesse I hope any good Christian. Read then, and if by reading, thou reapest any profit, I have my desire, if not, the fault shall be thine owne, reading haply more to judge and censure, than for any good and benefit which otherwise may be received from it; let but the same mind towards thine owne good possesse thee in reading it, as did the Author in writing it, and there shall be no neede to doubt of thy profit by it.

Thine in the common

***bond of love*, Iohn Clerke.**

The Authors Preface.

Two things, I feare me, will here be objected against me: The one concerneth my selfe, directly: The other mine Author, and the worke I have taken in hand the translating of him. Concerning my selfe, I suppose, some will aske, Why I being a Divine; should meddle or busie my selfe with these prophane studies? Geometry *may no way further Divinity, and therefore is no fit study for a Divine? This objection seemeth to smell of Brownisme, that is, of a ranke peevish humour overflowing the stomach of some, whereby they are caused to loath all manner of solid learning, yea of true Divinity it selfe, and therefore it doth not deserve an answer: And this we in our Title before signified. For we have not taken this paines for Turkes and others, who by the lawes of their profession are bound to abandon all manner of learning. But if any man shall propose it, as a question, with a desire of satisfaction, we are ready to answer him to the best of our abilitie. First, that* Theologia vera est ars artium & scientia scientiarum, *Divinity is the Art of Arts, and Science of Sciences; or Divinity is the Mistresse upon which all Arts and Sciences are to attend as servants and handmaides. And why then not* Geometry? *But in what place she should follow her, I dare not say: For I am no herald, and therefore I meddle not with precedencie: But if I were, she should be none of the hindermost of her traine.*

The Oratour saith, and very truly doubtlesse, That, Omnes artes, quæ ad humanitatē pertinent, habent commune quoddam vinculum, & cognatione quadam inter se continentur. *All Arts which pertaine unto humanity, they have a certaine common bond, and are knit together by a kinde of affinity. If then any Arts and Sciences may be thought necessary attendants upon this great Lady; Then surely* Geometry *amongst the rest must needes be one: For otherwise her traine will be but loose and shattered.*

Plato *saith* τὸν θεὸν ἀκεὶ γεωμετρεῖν, *That God doth alwayes worke by* Geometry, *that is, as the wiseman doth interprete it,* Sap. XI. 21. Omnia in mensura & numero & pondere disponere. *Dispose all things by measure, and number, and weight: Or, as the learned* Plutarch *speaketh; He adorneth and layeth out all the parts of the world according to rate, proportion, and similitude. Now who, I pray you, understandeth what these termes meane, but he which hath some meane skill in* Geometry? *Therefore none but such an one, may be able to declare and teach these things unto others.*

How many things are there in holy Scripture which may not well be understood without some meane skill in Geometry? *The Fabricke and bignesse of* Noah's *Arke: The Sciagraphy of the Temple set out by* Ezechiel, *Who may understand, but he that is skilfull in these Arts? I speake not of many and sundry words both in the New and Old Testaments, whose genuine and proper signification is merely Geometricall: And cannot well be conceived but of a Geometer.*

And here, that I may speake it without offence, I would have it observed, how many men, much magnified for learning, not onely in their speeches, which alwayes are not premeditated, but even in their writings, exposed to the view and censure of all men, doe often paralogizein, *speake much, and little to the purpose. This they could not so easily and often doe, if they had beene but meanely practised in these kinde of studies. Wherefore that Epigramme which was used to be written over their Philosophy Schoole doores,* οὐδῆις ἀγεωμέτρητος εἴσιτω, *No man ignorant of* Geometry *come within these doores: Now written over our Divinitie Schooles. And if any man shall thinke this an hard sentence, let him heare what Saint* Augustine *saith in the same case,* Nemo ad divinarum humanarumq; rerum cognitionem accedat, nisi prius annumerandi artem addiscat: *Let no man come neither within the Divinity nor Philosophy Schooles, except he have first learned Arithmeticke. Now that the one of these Arts cannot be learned without the other;* Euclide *our great Master, who made but one of both, hath sufficiently demonstrated.*

If I should alledge the like practise of famous Divines, greatly admired for their great skill in this profession, as T. Peckham *Arch-Bishop of Canterbury,* Maurolycus *Bishop of* Messana *in* Sicilia, Cusanus *Cardinall of* Rome, *and many others, before indifferent judges, I am sure I should not be condemned.*

Who doth not greatly magnifie the grave Seb. Munster, *the nimble* Ph. Melanchthon, *and the noble* Bernardino Baldo *Abbot of* Guastill, *and the painefull* Barth. Pitiscus *of* Grunberg, *for their knowledge and paines in these Arts and Sciences? And thus much shall at this time suffice, to have spoken unto the first Question: If any shall require further satisfaction, those I referre unto the forenamed Authors, whose authority peradventure may more prevaile with them, then my reasons may.*

The next is concerning mine Author, and the worke in hand Geometry, *it must needs be confest we are beholden to* Euclides *Elements for: And he that would be rich in that profession, may have, if he be not covetous, his fill there, if he will labour hard, and take paines for it, it is true. But in what time thinke yau, may a man learne all* Euclide, *and so by him be made skilfull in this Art? By himselfe I know not whether ever or never: And with the helpe of another, although very expert, I will not promise him that hee shall attaine to perfection in many yeares.*

Hippocrates *the Prince of Physicians hath, as they say, in his workes laid out the whole Art of Physicke; but I marvell how long a man should study him alone, and read him over and over, before he should be a good Physician? I feare mee all the friends that he hath, and neighbours round about him, yea, and himselfe too, would all die before he should be able to hele them, or per adventure ere he should be able to know what they ail'd; and after 30, or 40. yeeres of such his study, I would be very loath to commit my selfe unto him. How much therefore are the students of this noble Science beholding unto those men, who by their industry, practise, and painefull travells, have shewed them a ready and certaine way through this wildernesse?*

The Elements of Euclide *they do containe generally the whole art of* Geometry: *But if you will offer to travell thorow them alone, you shall finde them, I will warrant you, Elements indeed: for there you may walke through the spacious Aire, and over the great and wide sea, and in and about the vaste and arid wildernesse many a day and night, before you shall know where you are. This* Ramus, *my Authour in reading him found to be true; and confesseth him-*

selfe often to have beene at a stand: Often to have lost himselfe: Often to have hitte upon a rocke, when he had thought he had touch'd land.

Least therefore other men, in this journey doe not likewise loose themselves, for the benefit and safety, I meane, of others he hath prick'd them out a charde or chack'd out a way, which if thou shalt please to follow, it shall lead thee to thy wayes end, as directly, and in as short time, as conveniently may be. Yet in what time I cannot warrant thee: For all mens capacity, especially in these Arts, is not alike: All are not a like painefull, industrious, or diligent: All are not of the same ability of body, to be able to continue or sit at it: Or all not so free from other imployments or businesse calling them from their study, as some others are. For know this for certaine, Thou shalt here make no great progresse, except thou doe make it as it were a continued labour, Here you must observe that rule of the great Painter, Nulla dies sine linea, *Let no day passe over your head, in which you draw not some diagram or figure or other.*

One other thing let me also advise thee of, how capable soever thou art, refuse not, if thou maist have it, the helpe of a teacher; For except thou be another Hippocrates *or* Forcatelus, *whō our Authour mentioneth, thou canst not in these Arts and Sciences attaine unto any great perfection without infinite patience and great losse of most precious time, For they are therefore called* Μαθηματικόι, *Mathematicks, that is, doctrinal or disciplinary Arts, because they are not to be attained unto by our owne information and industry; but by the helpe and instruction of others.*

This Worke gentle Reader, was in part above 30. yeares since published by M. Thomas Hood, *a learned man, and loving friend of mine, who teaching these Arts, in the Staplers Chappell in Leadenhall London, for the benefit of his Schollers and Auditory, did set out the Elements apart by themselves. The whole at large, with the Diagrammes, and Demonstrations, hee promised, as appeareth in the Preface to that his Worke, at his convenient leysure to send out shortly, after them. This for ought we know or can learne, is not by him or any other performed: And yet are those alone, without these of small use or none to a learner, where a teacher is not alwayes at hand. Wherefore we are bold being (encouraged thereunto by some private friends, and especially by the learned M.* H. Brigges, *professour of* Geometry *in the famous Vniversity of* Oxford) *to publish this of ours long since finished and ended.*

The usuall termes, whether Latine or Greeke, commonly used by the Geometers, *we have set downe and expressed in English, as well as we could, as others, writing of this argument in our language, have done before us. These termes, I doubt not, may by some in English otherwise be expressed, but how harsh those termes, may unto Mathematicall eares, at the first appeare, I will not say; and use in short time will make these familiar, and as pleasing to the eare as those possibly may be.*

Our Authour, in the declaration of the Elements hath many passages, which in our judgement doe not make so much for the understanding of the matter in hand, as for the defence of the method here used, against Aristotle, Euclide, Proclus, *and others, which we have therfore wholly omitted. Some other things, which in our opinion, might in some respect illustrate any particular in this businesse, we have here and there inserted. Out of the learned* Finkius's Geometria Rotundi, *Wee have added to the fifth Booke certaine Propositions with their Consectaries out of* Ptolomi's *Almagest. The painfull and diligent* Rod. Snellius *out of the Lectures and Annotations of B.* Salignacus, I. Tho. Freigius, *and others, hath illustrated and altered here and there some few things.*

VIA REGIA AD GEOMETRIAM.

THE FIRST BOOKE OF Peter Ramus's Geometry, Which is of a Magnitude.

1. Geometry is the Art of measuring well.

The end or scope of Geometry is to measure well: Therefore it is defined of the end, as generally all other Arts are. *To measure well* therefore is to consider the nature and affections of every thing that is to be measured: To compare such like things one with another: And to understand their reason and proportion and similitude. For all that is to measure well, whether it bee that by Congruency and application of some assigned measure: Or by Multiplication of the termes or bounds: Or by Division of the product made by multiplication: Or by any other way whatsoever the affection of the thing to be measured be considered.

But this end of Geometry will appeare much more beautifull and glorious in the use and geometricall workes and practise then by precepts, when thou shalt observe Astronomers, Geographers, Land-meaters, Sea-men, Enginers, Architects, Carpenters, Painters, and Carvers, in the description and measuring of the Starres, Countries, Lands, Engins, Seas, Buildings, Pictures, and Statues or Images to use the helpe of no other art but of Geometry. Wherefore here the name of this art commeth farre short of the thing meant by it. (For *Geometria*, made of *Gè*, which in the Greeke language signifieth the Earth; and *Métron*, a measure, importeth no more, but as one would say *Land-measuring*. And *Geometra*, is but *Agrimensor*, A land-meter: or as *Tully* calleth him *Decempedator*, a Pole-man: or as *Plautus*, *Finitor*, a Marke-man.) when as this Art teacheth not only how to measure the Land or the Earth, but the Water, and the Aire, yea and the whole World too, and in it all Bodies, Surfaces, Lines, and whatsoever else is to bee measured.

Now *a Measure*, as *Aristotle* doth determine it, in every thing to be measured, is some small thing conceived and set out by the measurer; and of the Geometers it is called *Mensura famosa*, a knowne measure. Which kinde of measures, were at first, as *Vitruvius* and *Herodo* teache us, taken from mans body: whereupon *Protagoras* sayd, *That man was the measure of all things*, which speech of his, Saint *Iohn*, *Apoc.* 21. 17. doth seeme to approve. True it is, that beside those, there are some other sorts of measures, especially greater ones, taken from other things, yet all of them generally made and defined by those. And because the stature and bignesse of men is greater in some places, then it is ordinarily in others, therefore the measures taken from them are greater in some countries, then they are in others. Behold here a catalogue, and description of such as are commonly either used amongst us, or some times mentioned in our stories and other bookes translated into our English tongue.

Granum hordei, a Barley corne, like as a wheat corne in weights, is no kinde of measure, but is *quiddam minimum in mensura*, some least thing in a measure, whereof it is, as it were, made, and whereby it is rectified.

Digitus, a Finger breadth, conteineth 2. barly cornes length, or foure layd side to side:

Pollex, a Thumbe breadth; called otherwise *Vncia*, an ynch, 3. barley cornes in length:

Palmus, or *Palmus minor*, an Handbreadth, 4. fingers, or 3. ynches.

Spithama, or *Palmus major*, a Span, 3. hands breadth, or 9. ynches.

Cubitus, a Cubit, halfe a yard, from the elbow to the top of the middle finger, 6. hands breadth, or two spannes.

Ulna, from the top of the shoulder or arme-hole, to the top of the middle finger. It is two folde; A yard and an Elne. *A yard*, containeth 2. cubites, or 3. foote: *An Elne*, one yard and a quarter, or 2. cubites and ½.

Pes, a Foot, 4. hands breadth, or twelve ynches.

Gradus, or *Passus minor*, a Steppe, two foote and an halfe.

Passus, or *Passus major*, a Stride, two steppes, or five foote.

Pertica, a Pertch, Pole, Rod or Lugge, 5. yardes and an halfe.

Stadium, a Furlong; after the Romans, 125. pases: the English, 40. rod.

Milliare, or *Milliarium*, that is *mille passus*, 1000. passes, or 8. furlongs.

Leuca, a League, 2. miles: used by the French, spaniards, and seamen.

Parasanga, about 4. miles: a Persian, & common Dutch mile; 30. furlongs.

Schœnos, 40. furlongs: an Egyptian, or swedland mile.

Now for a confirmation of that which hath beene saide, heare the words of the Statute.

It is ordained, That 3. graines of Barley, dry and round, do make an Ynch: 12. *ynches do make a* Foote: 3. *foote do make a* Yard: *5. yardes and ½ doe make a Perch: And 40. perches in length, and 4. in breadth, doe make an Aker: 33. Edwar. 1. De terris mensurandis: & De compositione ulnarum & Perticarum.*

Item, *Bee it enacted by the authority aforesaid; That a* Mile *shall be taken and reckoned in this manner, and no otherwise; That is to say,* a Mile *to containe 8. furlongs: And every* Furlong *to containe 40. lugges or poles: And every* Lugge *or* Pole *to containe 16. foote and ½. 25. Eliza.* An Act for restraint of new building, &c.

These, as I said, are according to diverse countries, where they are used, much different one from another: which difference, in my judgment; ariseth especially out of the difference of the Foote, by which generally they are all made, whether they be greater of lesser. For the Hand being as before hath beene taught, the fourth part of the foot whether greater or lesser: And the Ynch, the third part of the hand, whether greater or lesser.

Item, the Yard, containing 3. foote, whether greater or lesser: And the Rodde 5. yardes and ½, whether greater or lesser, and so forth of the rest; It must needes follow, that the Foote beeing in some places greater then it is in other some, these measures, the Hand, I meane, the Ynch, the Yard, the Rod, must needes be greater or lesser in some places then they are in other. Of this diversity therefore, and difference of the foot, in forreine countries, as farre as mine intelligence will informe me, because the place doth invite me, I will here adde these few lines following. For of the rest, because they are of more speciall use, I will God willing, as just occasion shall be administred, speake more plentifully hereafter.

Of this argument divers men have written somewhat, more or lesse: But none to my knowledge, more copiously and curiously, then *Iames Capell*, a Frenchman, and the learned *Willebrand, Snellius*, of *Leiden* in Holland, for they have compared, and that very diligently, many and sundry kinds of these measures one with another. The first as you may see in his treatise *De mensuris intervallorum* describeth these eleven following: of which the greatest is *Pes Babylonius*, the Babylonian foote; the least, *Pes Toletanus*, the foote used about *Toledo* in Spaine: And the meane betweene both, *Pes Atticus*, that used about *Athens* in Greece. For they are one unto another as 20. 15. and 12. are one unto another. Therefore if the Spanish foote, being the least, be devided into 12. ynches, and every inch againe into 10. partes, and so the whole foote into 120. the *Atticke* foote shall containe of those parts 150. and the *Babylonian*, 200. To this *Atticke* foote, of all other, doth ours come the neerest: For our *English* foote comprehendeth almost 152. such parts.

The other, to witt the learned *Snellius*, in his *Eratosthenes Batavus*, a booke which hee hath written of the true quantity of the compasse of the Earth, describeth many more, and that after a farre more exact and curious manner.

Here observe, that besides those by us here set downe, there are certaine others by him mentioned, which as hee writeth are found wholly to agree with some one or other of these. For *Rheinlandicus*, that of *Rheinland* or *Leiden*, which hee maketh his base, is all one with *Romanus*, the *Italian* or *Roman* foote. *Lovaniensis*, that of *Lovane*, with that of *Antwerpe*: *Bremensis*, that of *Breme* in *Germany*, with that of *Hafnia*, in *Denmarke*. Onely his *Pes Arabicus*, the *Arabian* foote, or that mentioned in *Abulfada*, and *Nubiensis*: the Geographers I have overpassed, because hee dareth not, for certeine, affirme what it was.

Looke of what parts *Pes Tolitanus*, the spanish foote, or that of *Toledo* in Spaine, conteineth 120. of such is the *Pes*.

Heidelbergicus, that of Heidelberg, 137.

Hetruscus, that of Tuscan, in Italie, 138.

Sedanensis, of Sedan in France, 139.

Romanus, that of Rome in Italy, 144.

Atticus, of Athens in Greece, 150.

Anglicus, of England, 152.

Parisinus, of Paris in France, 160.

Syriacus, of Syria, 166.

Ægyptiacus, of Egypt, 171.

Hebraicus, that of Iudæa, 180.

Babylonius, that of Babylon, 200.

Looke of what parts *Pes Romanus*, the foote of Rome, (which is all one with the foote of *Rheinland*) is 1000. of such parts is the foote of

Toledo, in Spaine, 864.

Mechlin, in Brabant, 890.

Strausburgh, in Germany, 891.

Amsterdam, in Holland, 904.

Antwerpe, in Brabant, 909.

Bavaria, in Germany, 924.

Coppen-haun, in Denmarke, 934.

Goes, in Zeland, 954.

Middleburge, in Zeland, 960.

London, in England, 968.

Noremberge, in Germany, 974.

Ziriczee, in Zeland, 980.

The ancient *Greeke*, 1042.

Dort, in Holland, 1050.

Paris, in France, 1055.

Briel, in Holland, 1060.

Venice, in Italy, 1101.

Babylon, in Chaldæa, 1172.

Alexandria, in Egypt, 1200.

Antioch, in Syria, 1360.

Of all other therefore our English foote commeth neerest unto that used by the Greekes: And the learned Master *Ro. Hues*, was not much amisse, who in his booke or Treatise *De Globis*, thus writeth of it *Pedem nostrum Angli cum Græcorum pedi æqualem invenimus, comparatione facta cum Græcorum pede, quem Agricola & alij ex antiquis monumentis tradiderunt.*

Now by any one of these knowne and compared with ours, to all English men well knowne the rest may easily be proportioned out.

2. The thing proposed to bee measured is a Magnitude.

Magnitudo, a Magnitude or Bignesse is the subject about which Geometry is busied. For every Art hath a proper subject about which it doth employ al his rules and precepts: And by this especially they doe differ one from another. So the subject of Grammar was speech; of Logicke, reason; of Arithmeticke, numbers; and so now of Geometry it is a magnitude, all whose kindes, differences and affections, are hereafter to be declared.

3. A Magnitude is a continuall quantity.

A Magnitude is *quantitas continua*, a continued, or continuall quantity. A number is *quantitas discreta*, a disjoined quantity: As one, two, three, foure; doe consist of one, two, three, foure unities, which are disjoyned and severed parts: whereas the parts of a Line, Surface, and Body are contained and continued without any manner of disjunction, separation, or distinction at all, as by and by shall better and more plainely appeare. Therefore a Magnitude is here understood to be that whereby every thing to be measured is said to bee great: As a Line from hence is said to be long, a Surface broade, a Body solid: Wherefore Length, Breadth, and solidity are Magnitudes.

4. That is continuum, continuall, whose parts are contained or held together by some common bound.

This definition of it selfe is somewhat obscure, and to be understand onely in a geometricall sense: And it dependeth especially of the common bounde. For the parts (which here are so called) are nothing in the whole, but in a *potentia* or powre: Neither indeede may the whole magnitude bee conceived, but as it is compact of his parts, which notwithstanding wee may in all places assume or take as conteined and continued with a common bound, which Aristotle nameth a *Common limit*; but *Euclide* a *Common section*, as in a line, is a Point, in a surface, a Line: in a body, a Surface.

5. A bound is the outmost of a Magnitude.

Terminus, a Terme, or Bound is here understood to bee that which doth either bound, limite, or end *actu*, in deede; as in the beginning and end of a magnitude: Or *potentia*, in powre or ability, as when it is the common bound of the continuall magnitude. Neither is the Bound a parte of the bounded magnitude: For the thing bounding is one thing, and the thing bounded is another: For the Bound is one distance, dimension, or degree, inferiour to the thing bounded: A Point is the bound of a line, and it is lesse then a line by one degree, because it cannot bee divided, which a line may. A Line is the bound of a surface, and it is also lesse then a surface by one distance or dimension, because it is only length, wheras a surface hath both length and breadth. A Surface is the bound of a body, and it is lesse likewise then it is by one dimension, because it is onely length and breadth, whereas as a body hath both length, breadth, and thickenesse.

Now every Magnitude *actu*, in deede, is terminate, bounded and finite, yet the geometer doth desire some time to have an infinite line granted him, but no otherwise infinite or farther to bee drawane out then may serve his turne.

6. A Magnitude is both infinitely made, and continued, and cut or divided by those things wherewith it is bounded.

A line, a surface, and a body are made gemetrically by the motion of a point, line, and surface: Item, they are conteined, continued, and cut or divided by a point, line, and surface. But a Line is bounded by a point: a surface, by a line: And a Body by a surface, as afterward by their severall kindes shall be understood.

Now that all magnitudes are cut or divided by the same wherewith they are bounded, is conceived out of the definition of *Continuum*, e. 4. For if the common band to containe and couple together the parts of a Line, surface, & Body, be a Point, Line, and Surface, it must needes bee that a section or division shall be made by those common bandes: And that to bee dissolved which they did containe and knitt together.

7. A point is an undivisible signe in a magnitude.

A Point, as here it is defined, is not naturall and to bee perceived by sense; Because sense onely perceiveth that which is a body; And if there be any thing lesse then other to be perceived by sense, that is called a Point. Wherefore a Point is no Magnitude: But it is onely that which in a Magnitude is conceived and imagined to bee undivisible. And although it be voide of all bignesse or Magnitude, yet is it the beginning of all magnitudes, the beginning I meane *potentiâ*, in powre.

> *8. Magnitudes commensurable, are those which one and the same measure doth measure: Contrariwise, Magnitudes incommensurable are those, which the same measure cannot measure. 1, 2. d. X.*

Magnitudes compared betweene themselves in respect of numbers have Symmetry or commensurability, and Reason or rationality: Of themselves, Congruity and Adscription. But the measure of a magnitude is onely by supposition, and at the discretion of the Geometer, to take as pleaseth him, whether an ynch, an hand breadth, foote, or any other thing whatsoever, for a measure. Therefore two magnitudes, the one a foote long, the other two foote long, are commensurable; because the magnitude of one foote doth measure them both, the first once, the second twice. But some magnitudes there are which have no common measure, as the Diagony of a quadrate and his side, 116. p. X. *actu*, in deede, are *Asymmetra*, incommensurable: And yet they are *potentiâ*, by power, *symmetra*, commensurable, to witt by their quadrates: For the quadrate of the diagony is double to the quadrate of the side.

> *9. Rationall Magnitudes are those whose reason may bee expressed by a number of the measure given. Contrariwise they are irrationalls. 5. d. X.*

Ratio, Reason, Rate, or Rationality, what it is our Authour (and likewise *Salignacus*) have taught us in the first Chapter of the second booke of their Arithmetickes: Thither therefore I referre thee.

Data mensura, a Measure given or assigned, is of *Euclide* called *Rhetè*, that is spoken, (or which may be uttered) definite, certaine, to witt which may bee expressed by some number, which is no other then that, which as we said, was called *mensura famosa*, a knowne or famous measure.

Therefore Irrationall magnitudes, on the contrary, are understood to be such whose reason or rate may not bee expressed by a number or a measure assigned: As the side of the side of a quadrate of 20. foote unto a magnitude of two foote; of which kinde of magnitudes, thirteene sorts are mentioned in the tenth booke of *Euclides Elements*: such are the segments of a right line proportionally cutte, unto the whole line. The Diameter in a circle is rationall: But it is irrationall unto the side of an inscribed quinquangle: The Diagony of an Icosahedron and Dodecahedron is irrationall unto the side.

10. Congruall or agreeable magnitudes are those, whose parts beeing applyed or laid one upon another doe fill an equall place.

Symmetria, Symmetry or Commensurability and Rate were from numbers: The next affections of Magnitudes are altogether geometricall.

Congruentia, Congruency, Agreeablenesse is of two magnitudes, when the first parts of the one doe agree to the first parts of the other, the meane to the meane, the extreames or ends to the ends, and lastly the parts of the one, in all respects to the parts, of the other: so Lines are congruall or agreeable, when the bounding, points of the one, applyed to the bounding points of the other, and the whole lengths to the whole lengthes, doe occupie or fill the same place. So Surfaces doe agree, when the bounding lines, with the bounding lines: And the plots bounded, with the plots bounded doe occupie the same place. Now bodies if they do agree, they do seeme only to agree by their surfaces. And by this kind of congruency do we measure the bodies of all both liquid and dry things, to witt, by filling an equall place. Thus also doe the moniers judge the monies and coines to be equall, by the equall weight of the plates in filling up of an equall place. But here note, that there is nothing that is onely a line, or a surface onely, that is naturall and sensible to the touch, but whatsoever is naturall, and thus to be discerned is corporeall.

Therefore

> *11. Congruall or agreeable Magnitudes are equall. 8. ax. j.*

A lesser right line may agree to a part of a greater, but to so much of it, it is equall, with how much it doth agree: Neither is that axiome reciprocall or to be converted: For neither in deede are Congruity and Equality reciprocall or convertible. For a Triangle may bee equall to a Parallelogramme, yet it cannot in all points agree to it: And so to a Circle there is sometimes sought an equall quadrate, although incongruall or not agreeing with it: Because those things which are of the like kinde doe onely agree.

12. *Magnitudes are described betweene themselves, one with another, when the bounds of the one are bounded within the boundes of the other: That which is within, is called the inscript: and that which is without, the Circumscript.*

Now followeth Adscription, whose kindes are Inscription and Circumscription; That is when one figure is written or made within another: This when it is written or made about another figure.

Homogenea, Homogenealls or figures of the same kinde onely betweene themselves *rectitermina*, or right bounded, are properly adscribed betweene themselves, and with a round. Notwithstanding, at the 15. booke of *Euclides Elements* Heterogenea, Heterogenealls or figures of divers kindes are also adscribed, to witt the five ordinate plaine bodies betweene themselves: And a right line is inscribed within a periphery and a triangle.

But the use of adscription of a rectilineall and circle, shall hereafter manifest singular and notable mysteries by the reason and meanes of adscripts; which adscription shall be the key whereby a way is opened unto that most excellent doctrine taught by the subtenses or inscripts of a circle as *Ptolomey* speakes, or Sines, as the latter writers call them.

The second Booke of Geometry. Of a Line.

1. A Magnitude is either a Line or a Lineate.

The Common affections of a magnitude are hitherto declared: The *Species* or kindes doe follow: for other then this division our authour could not then meete withall.

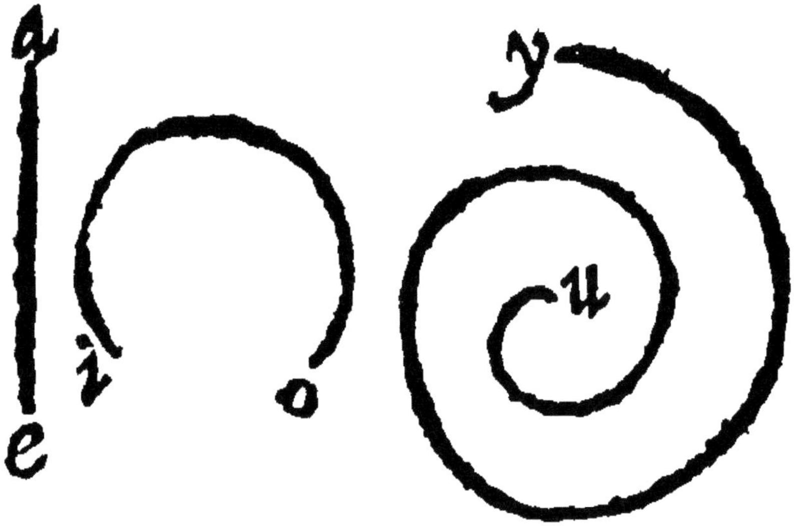

2. A Line is a Magnitude onely long.

As are *ae. io.* and *uy.* such a like Magnitude is conceived in the measuring of waies, or distance of one place from another: And by the difference of a lightsome place from a darke: *Euclide* at the 2 *d j.* defineth a line to be a length void of breadth: And indeede length is the proper difference of a line, as breadth is of a face, and solidity of a body.

3. The bound of a line is a point.

Euclide at the 3. *d j.* saith that the extremities or ends of a line are points. Now seeing that a Periphery or an hoope line hath neither beginning nor ending, it seemeth not to bee bounded with points: But when it is described or made it beginneth at a point, and it endeth at a pointe. Wherefore a Point is the bound of a line, sometime *actu*, in deed, as in a right line: sometime *potentiâ*, in a possibility, as in a perfect periphery. Yea in very deede, as before was taught in the definition of *continuum*, 4 *e.* all lines, whether they bee right lines, or crooked, are contained or continued with points. But a line is made by the motion of a point. For every magnitude generally is made by a geometricall motion, as was even now taught, and it shall afterward by the severall kindes appeare, how by one motion whole figures are made: How by a conversion, a Circle, Spheare, Cone, and Cylinder: How by multiplication of the base and heighth, rightangled parallelogrammes are made.

4. A Line is either Right or Crooked.

This division is taken out of the 4 d j. of *Euclide*, where rectitude or straightnes is attributed to a line, as if from it both surfaces and bodies were to have it. And even so the rectitude of a solid figure, here-after shall be understood by a right line perpendicular from the toppe unto the center of the base. Wherefore rectitude is propper unto a line: And therefore also obliquity or crookednesse, from whence a surface is judged to be right or oblique, and a body right or oblique.

5. A right line is that which lyeth equally betweene his owne bounds: A crooked line lieth contrariwise. 4. d. j.

Now a line lyeth equally betweene his owne bounds, when it is not here lower, nor there higher: But is equall to the space comprehended betweene the two bounds or ends: As here *ae.* is, so hee that maketh *rectum iter*, a journey in a straight line, commonly he is said to treade so much ground, as he needes must, and no more: He goeth *obliquum iter*, a crooked way, which goeth more then he needeth, as *Proclus* saith.

6. A right line is the shortest betweene the same bounds.

Linea recta, a straight or right line is that, as *Plato* defineth it, whose middle points do hinder us from seeing both the extremes at once; As in the eclipse of the Sunne, if a right line should be drawne from the Sunne, by the Moone, unto our eye, the body of the Moone beeing in the midst, would hinder our sight, and would take away the sight of the Sunne from us: which is taken from the Opticks, in which we are taught, that we see by straight beames or rayes. Therfore to lye equally betweene the boundes, that is by an equall distance: to bee the shortest betweene the same bounds; And that the middest doth hinder the sight of the extremes, is all one.

7. A crooked line is touch'd of a right or crooked line, when they both doe so meete, that being continued or drawne out farther they doe not cut one another.

Tactus, *Touching is propper to a crooked line, compared either with a right line or crooked, as is manifest out of the 2. and 3.* d 3. *A right line is said to touch a circle, which touching the circle and drawne out farther, doth not cut the circle, 2* d 3. *as here* ae, *the right line toucheth the periphery* iou. *And* ae. *doth touch the helix or spirall.*

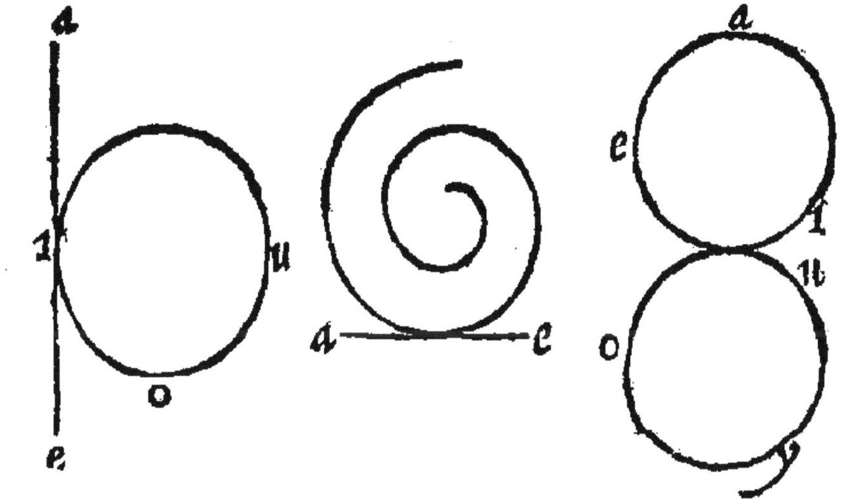

Circles are said to touch one another, when touching they doe not cutte one another, 3. *d* 3. as here the periphery doth *aej.* doth touch the periphery *ouy*.

Therefore

8. Touching is but in one point onely. è 13. p 3.

This Consectary is immediatly conceived out of the definition; for otherwise it were a cutting, not touching. So *Aristotle* in his *Mechanickes* saith; That a round is easiliest mou'd and most swift; Because it is least touch't of the plaine underneath it.

9. A crooked line is either a Periphery or an Helix. *This also is such a division, as our Authour could then hitte on.*

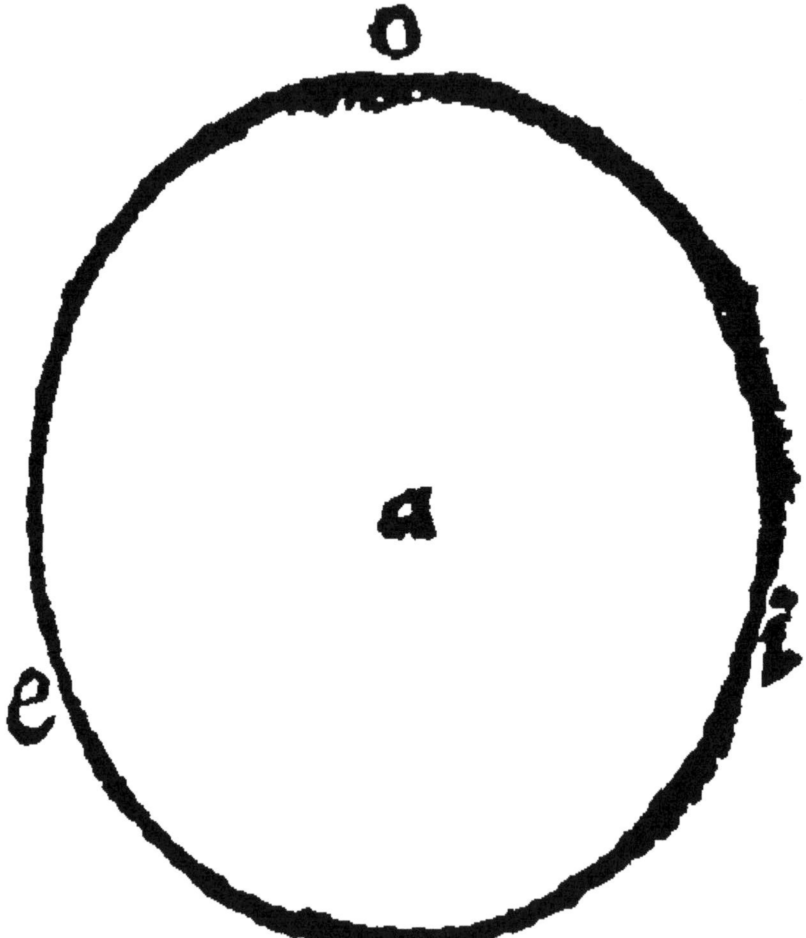

10. A Periphery is a crooked line, which is equally distant from the middest of the space comprehended.

Peripheria, a Periphery, or Circumference, as *eio.* doth stand equally distant from *a*, the middest of the space enclosed or conteined within it.

Therefore

11. A Periphery is made by the turning about of a line, the one end thereof standing still, and the other drawing the line.

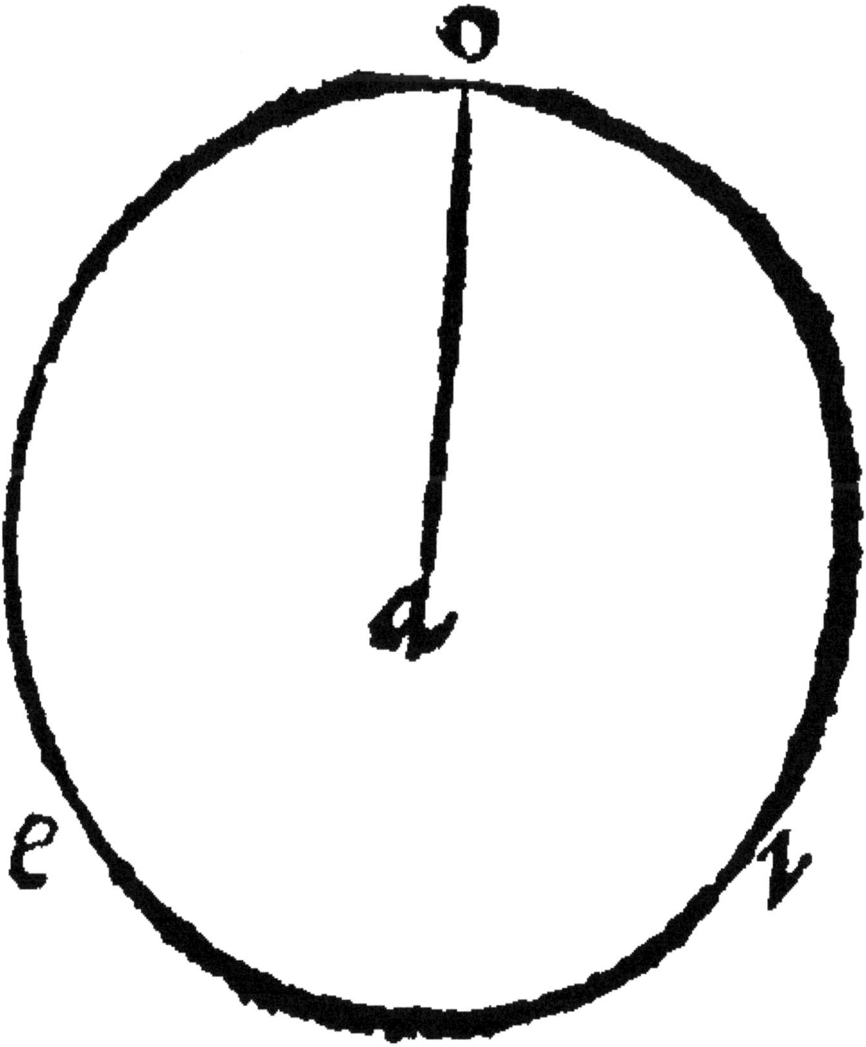

As in *eio.* let the point *a* stand still: And let the line *ao*, be turned about, so that the point *o* doe make a race, and it shall make the periphery *eoi*. Out of this fabricke doth *Euclide*, at the 15. d. j. frame the definition of a Periphery: And so doth hee afterwarde define a Cone, a Spheare, and a Cylinder.

Now the line that is turned about, may in a plaine, bee either a right line or a crooked line: In a sphericall it is onely a crooked line; But in a conicall or Cylindraceall it may bee a right line, as is the side of a Cone and Cylinder. Therefore in the conversion or turning about of a line making a periphery, there is considered onely the distance; yea two points, one in the center, the other in the toppe, which therefore Aristotle nameth *Rotundi principia*, the principles or beginnings of a round.

12. An Helix is a crooked line which is unequally distant from the middest of the space, howsoever inclosed.

Hæc tortuosa linea, This crankled line is of *Proclus* called *Helicoides*. But it may also be called *Helix*, a twist or wreath: The *Greekes* by this word do commonly either understand one of the kindes of Ivie which windeth it selfe about trees & other plants; or the strings of the vine, whereby it catcheth hold and twisteth it selfe about such things as are set for it to clime or run upon. Therfore it should properly signifie the spirall line. But as it is here taken it hath divers kindes; As is the *Arithmetica* which is Archimede'es Helix, as the *Conchois*, Cockleshell-like: as is the *Cittois*, Iuylike: The *Tetragonisousa*, the Circle squaring line, to witt that by whose meanes a circle may be brought into a square: The Admirable line, found out by *Menelaus*: The Conicall *Ellipsis*, the *Hyperbole*, the *Parabole*, such as these are, they attribute to *Menechmus*: All these *Apollonius* hath comprised in eight Bookes; but being mingled lines, and so not easie to bee all reckoned up and expressed, *Euclide* hath wholly omitted them, saith *Proclus*, at the 9. *p. j*.

13. Lines are right one unto another, whereof the one falling upon the other, lyeth equally: Contrariwise they are oblique. è 10. d j.

Hitherto straightnesse and crookednesse have beene the affections of one sole line onely: The affections of two lines compared one with another are *Perpendiculum*, Perpendicularity and *Parallelismus*, Parallell equality; Which affections are common both to right and crooked lines. Perpendicularity is first generally defined thus:

Lines are right betweene themselves, that is, perpendicular one unto another, when the one of them lighting upon the other, standeth upright and inclineth or leaneth neither way. So two right lines in a plaine may bee perpendicular; as are *ae.* and *io.* so two peripheries upon a sphearicall may be perpendiculars, when the one of them falling upon the other, standeth indifferently betweene, and doth not incline or leane either way. So a right line may be perpendicular unto a periphery, if falling upon it, it doe reele neither way, but doe ly indifferently betweene either side. And in deede in all respects lines right betweene themselves, and perpendicular lines are one and the same. And from the perpendicularity of lines, the perpendicularity of surfaces is taken, as hereafter shall appeare. Of the perpendicularity of bodies, *Euclide* speaketh not one word in his *Elements*, & yet a body is judged to be right, that is, plumme or perpendicular unto another body, by a perpendicular line.

Therefore,

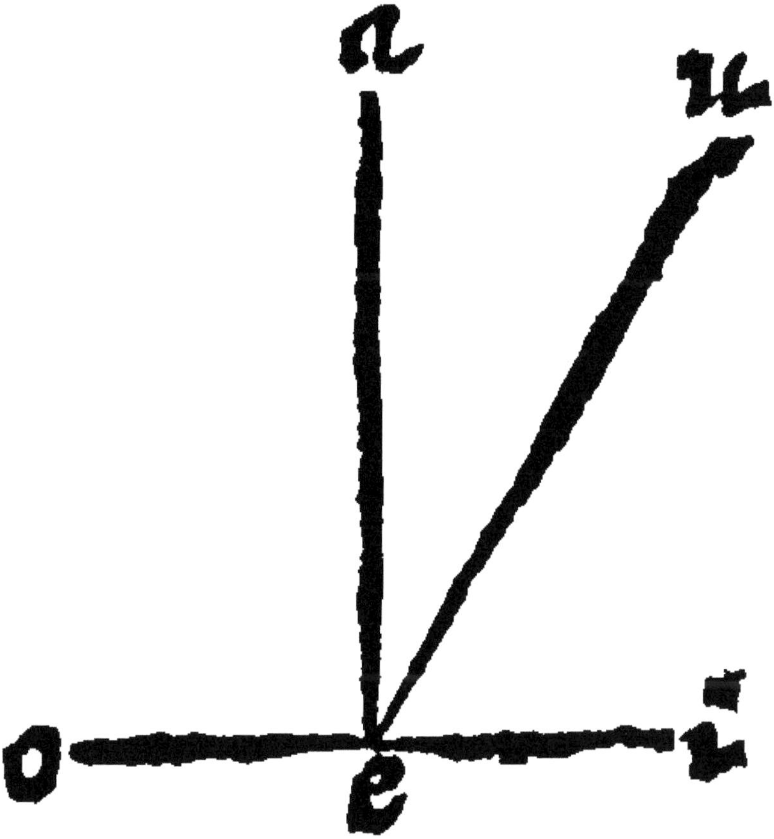

14. If a right line be perpendicular unto a right line, it is from the same bound, and on the same side, one onely. ê 13. p. xj.

Or, there can no more fall from the same point, and on the same side but that one. This consectary followeth immediately upon the former: For if there should any more fall unto the same point and on the same side, one must needes reele, and would not ly indifferently betweene the parts cut: as here thou seest in the right line *ae. io. eu.*

15. Parallell lines they are, which are everywhere equally distant. è 35. d j.

Parallelismus, Parallell-equality doth now follow: And this also is common to crooked lines and right lines: As heere thou seest in these examples following.

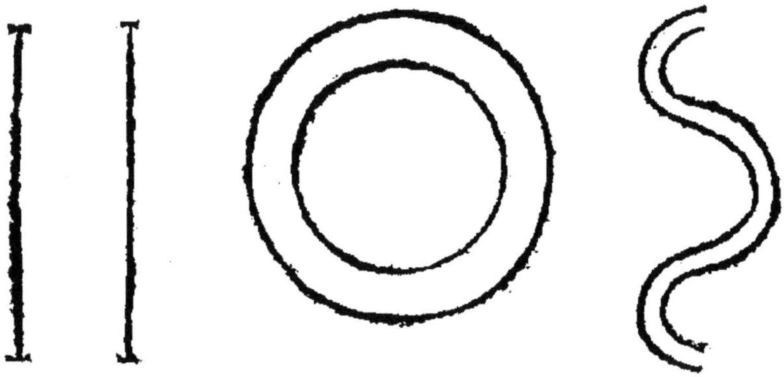

Parallell-equality is derived from perpendicularity, and is of neere affinity to it. Therefore Posidonius did define it by a common perpendicle or plumline: yea and in deed our definition intimateth asmuch. Parallell-equality of bodies is no where mentioned in *Euclides Elements*: and yet they may also bee parallells, and are often used in the Optickes, Mechanickes, Painting and Architecture.

Therefore,

16. Lines which are parallell to one and the same line, are also parallell one to another.

This element is specially propounded and spoken of right lines onely, and is demonstrated at the 30. *p. j.* But by an addition of equall distances, an equall distance is knowne, as here.

The third Booke of Geometry. *Of an Angle.*

1. A lineate is a Magnitude more then long.

A New forme of doctrine hath forced our Authour to use oft times new words, especially in dividing, that the logicall lawes and rules of more perfect division by a dichotomy, that is into two kindes, might bee held and observed. Therefore a Magnitude was divided into two kindes, to witt into a Line and a Lineate: And a Lineate is made the *genus* of a surface and a Body. Hitherto a Line, which of all bignesses is the first and most simple, hath been described: Now followeth a Lineate, the other kinde of magnitude opposed as you see to a line, followeth next in order. *Lineatum* therefore a Lineate, or *Lineamentum*, a Lineament, (as by the authority of our Authour himselfe, the learned *Bernhard Salignacus*, who was his Scholler, hath corrected it) is that Magnitude in which there are lines: Or which is made of lines, or as our Authour here, which is more then long: Therefore lines may be drawne in a surface, which is the proper soile or plots of lines; They may also be drawne in a body, as the Diameter in a Prisma: the axis in a spheare; and generally all lines falling from aloft: And therfore *Proclus* maketh some plaine, other solid lines. So Conicall lines, as the Ellipsis, Hyperbole, and Parabole, are called solid lines because they do arise from the cutting of a body.

2. To a Lineate belongeth an Angle and a Figure.

The common affections of a Magnitude were to be bounded, cutt, jointly measured, and adscribed: Then of a line to be right, crooked, touch'd, turn'd about, and wreathed: All which are in a lineate by meanes of a line. Now the common affections of a Lineate are to bee Angled and Figured. And surely an Angle and a figure in all Geometricall businesses doe fill almost both sides of the leafe. And therefore both of them are diligently to be considered.

3. An Angle is a lineate in the common section of the bounds.

So *Angulus Superficiarius*, a superficiall Angle, is a surface consisting in the common section of two lines: So *angulus solidus*, a solid angle, in the common section of three surfaces at the least.

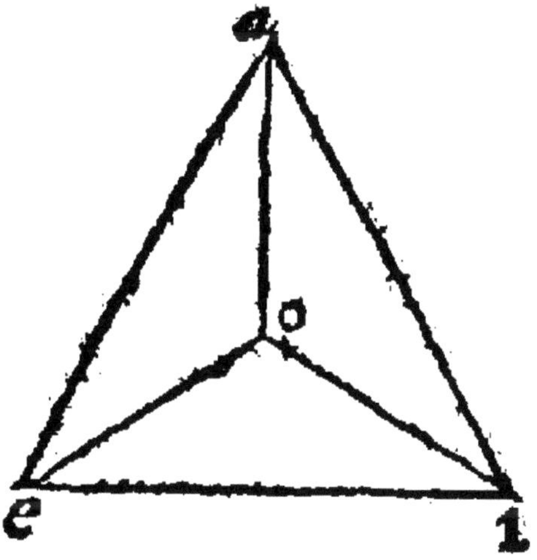

[But the learned B. *Salignacus* hath observed, that all angles doe not consist in the common section of the bounds, Because the touching of circles, either one another, or a rectilineal surface doth make an angle without any cutting of the bounds: And therefore he defineth it thus: *Angulus est terminorum inter se invicem inclinantium concursus: An angle is the meeting of bounds, one leaning towards another.*] So is *aei.* a superficiall angle: [And such also are the angles *ouy.* and *bcd.*] so is the angle *o.* a solid angle, to witt comprehended of the three surfaces *aoi. ioe.* and *aoe.* Neither may a surface, of 2. dimensions, be bounded with one right line: Nor a body, of three dimensions, bee bounded with two, at lest beeing plaine surfaces.

4. The shankes of an angle are the bounds compreding the angle.

Scèle or *Crura*, the Shankes, Legges, H. are the bounds insisting or standing upon the base of the angle, which in the Isosceles only or Equicrurall triangle are so named of *Euclide*, otherwise he nameth them *Latera*, sides. So in the examples aforesaid, *ea.* and *ei.* are the shankes of the superficiary angle *e*; And so are the three surfaces *aoi. ieo.* and *aeo.* the shankes of the said angle *o*. Therefore the shankes making the angle are either Lines or Surfaces: And the lineates formed or made into Angles, are either Surfaces or Bodies.

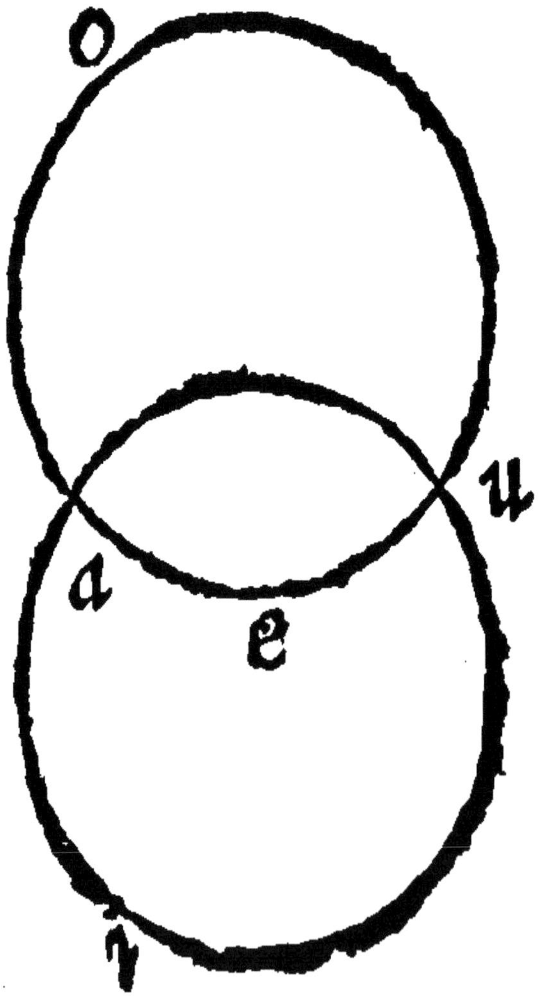

5. Angles homogeneall, are angles of the same kinde, both in respect of their shankes, as also in the maner of meeting of the same: [Heterogeneall, are those which differ one from another in one, or both these conditions.]

Therefore this *Homogenia*, or similitude of angles is twofolde, the first is of shanks; the other is of the manner of meeting of the shankes: so rectilineall right angles, are angles homogeneall betweene themselves. But right-lined right angles, and oblique-lined right angles between themselves, are heterogenealls. So are neither all obtusangles compared to all obtusangles: Nor all acutangles, to all acutangles, homogenealls, except both these conditions doe concurre, to witt the similitude both of shanke and manner of meeting. *Lunularis*, a Lunular, or Moonlike corner angle is homogeneall to a *Systroides* and *Pelecoides*, Hatchet formelike, in shankes: For each of these are comprehended of peripheries: The Lunular of one convexe; the other concave; as *iue*. The Systroides of both convex, as *iao*. The Pelecoides of both concave, as *eau*. And yet a lunular, in respect of the meeting of the shankes is both to the Systroides and Pelecoides heterogeneall: And therefore it is absolutely heterogeneall to it.

6. Angels congruall in shankes are equall.

This is drawne out of the 10. e j. For if twice two shanks doe agree, they are not foure, but two shankes, neither are they two equall angles, but one angle. And this is that which *Proclus* speaketh of, at the 4. p j. when hee saith, that a right lined angle is equall to a right lined angle, when one of the shankes of the one put upon one of the shankes of the other, the other two doe agree: when that other shanke fall without, the angle of the out-falling shanke is the greater: when it falleth within, it is lesser: For there is comprehendeth; here it is comprehended.

Notwithstanding although congruall or agreeable angles be equall: yet are not congruity and equality reciprocall or convertible: For a Lunular may bee equall to a right lined right angle, as here thou seest: For the angles of equall semicircles *ieo.* and *aeu.* are equall, as application doth shew. The angle *aeo.* is common both to the right angle *aei.* and to the lunar *aueo.* Let therefore the equall angle *aeo.* bee added to both: the right angle *aei.* shall be equall to the Lunular *aueo.*

The same Lunular also may bee equall to an obtusangle and Acutangle, as the same argument will demonstrate.

Therefore,

> *7. If an angle being equicrurall to an other angle, be also equall to it in base, it is equall: And if an angle having equall shankes with another, bee equall to it in the angle, it is also equall to it in the base. è 8. & 4. p j.*

For such angles shall be congruall or agreeable in shanks, and also congruall in bases. *Angulus isosceles*, or *Angulus æquicrurus*, is a triangle having equall shankes unto another.

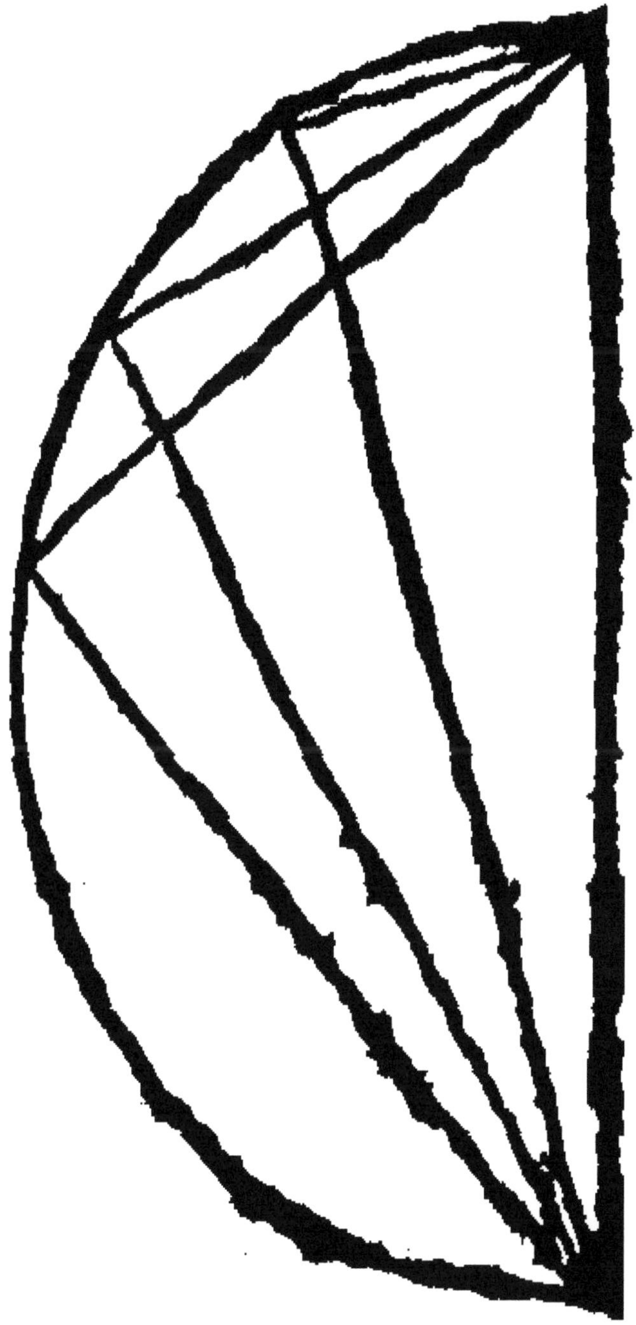

8. And if an angle equall in base to another, be also equall to it in shankes, it is equall to it.

For the congruency is the same: And yet if equall angles bee equall in base, they are not by and by equicrurall, as in the angles of the same section will appeare, as here. And so of two equalities, the first is reciprocall: The second is not. [And therefore is this Consectary, by the learned B. *Salignacus*, justly, according to the judgement of the worthy Rud. *Snellius*, here cancelled; or quite put out: For angles may be equall, although they bee unequall in shankes or in bases, as here, the angle *a.* is not greater then the angle *o*, although the angle *o* have both greater shankes and greater base then the angle *a*.]

And

9. If an angle equicrurall to another angle, be greater then it in base, it is greater: And if it be greater, it is greater in base: è 52 & 24. p j.

As here thou seest; [The angles *eai.* and *uoy.* are equicrurall, that is their shankes are equall one to another; But the base *ei* is greater then the base *uy*: Therefore the angle *eai*, is greater then the angle *uoy*. And contrary wise, they being equicrurall, and the angle *eai.* being greater then the angle *uoy.* The base *ei.* must needes be greater then the base *uy*.]

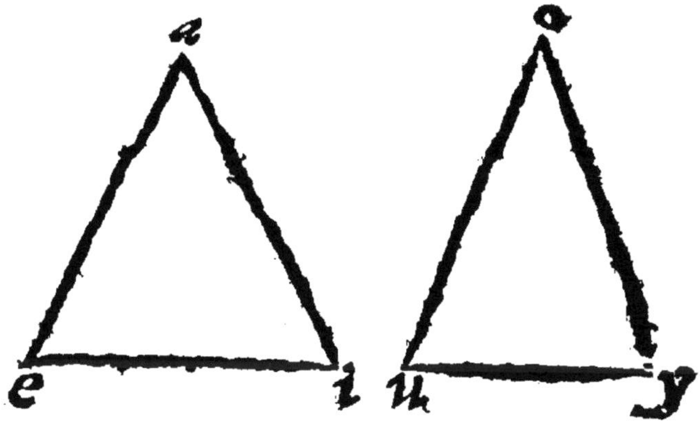

And

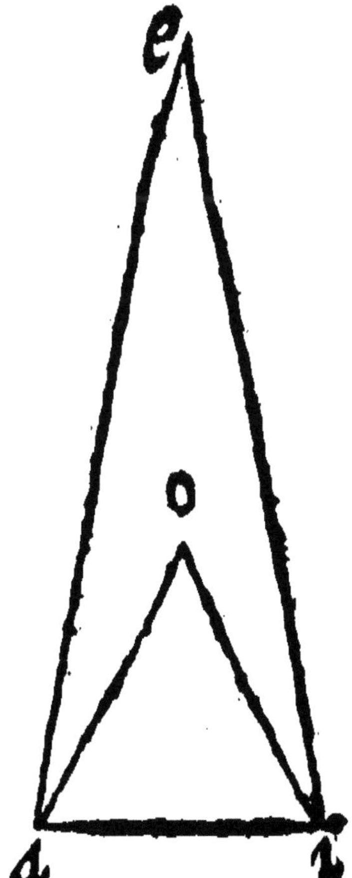

10. If an angle equall in base, be lesse in the inner shankes, it is greater.

Or as the learned Master *T. Hood* doth paraphrastically translate it. *If being equall in the base, it bee lesser in the feete (the feete being conteined within the feete of the other angle) it is the greater angle.* [That is, if one angle enscribed within another angle, be equall in base, the angle of the inscribed shall be greater then the angle of the circumscribed.]

As here the angle *aoi.* within the angle *aei.* And the bases are equall, to witt one and the same; Therefore *aoi.* the inner angle is greater then *aei.* the outter angle. *Inner* is added of necessity: For otherwise there will, in the section or cutting one of another, appeare a manifest errour. All these consectaries are drawne out of that same axiome of congruity, to witt out of the 10. e j. as *Proclus* doth plainely affirme and teach: It seemeth saith hee, that the equalities of shankes and bases, doth cause the equality of the verticall angles. For neither, if the bases be equall, doth the equality of the shankes leave the same or equall angles: But if the base bee lesser, the angle decreaseth: If greater, it increaseth. Neither if the bases bee equall, and the shankes unequall, doth the angle remaine the same: But when they are made lesse, it is increased: when they are made greater, it is diminished: For the contrary falleth out to the angles and shankes of the angles. For if thou shalt imagine the shankes to be in the same base thrust downeward, thou makest them lesse, but their angle greater: but if thou do againe conceive them to be pul'd up higher, thou makest them greater, but their angle lesser. For looke how much more neere they come one to another, so much farther off is the toppe removed from the base: wherefore you may boldly affirme, that the same base and equall shankes, doe define the equality of Angels. This *Poclus,*

Therefore,

11. If unto the shankes of an angle given, homogeneall shankes, from a point assigned, bee made equall upon an equall base, they shall comprehend an angle equall to the angle given. è 23. p j. & 26. p xj.

[This consectary teacheth how unto a point given, to make an angle equall to an Angle given. To the effecting and doing of each three things are required; First, that the shankes be homogeneall, that is in each place, either straight or crooked: Secondly, that the shankes bee made equall, that is of like or equall bignesse: Thirdly, that the bases be equall: which three conditions if they doe meete, it must needes be that both the angles shall bee equall: but if one of them be wanting, of necessity againe they must be unequall.]

This shall hereafter be declared and made plaine by many and sundry practises: and therefore here we bring no example of it.

12. An angle is either right or oblique.

Thus much of the Affections of an angle; the division into his kindes followeth. An angle is either Right or Oblique: as afore, at the 4 *e ij.* a line was right or straight, and oblique or crooked.

o
a
i
e

13. A right angle is an angle whose shankes are right (that is perpendicular) one unto another: An Oblique angle is contrary to this.

As here the angle *aio.* is a right angle, as is also *oie.* because the shanke *oi.* is right, that is, perpendicular to *ae.* [The instrument wherby they doe make triall which is a right angle, and which is oblique, that is greater or lesser then a right angle, is the square which carpenters and joyners do ordinarily use: For lengthes are tried, saith *Vitruvius*, by the Rular and Line: Heighths, by the Perpendicular or Plumbe: And Angles, by the square.] Contrariwise, an Oblique angle it is, when the one shanke standeth so upon another, that it inclineth, or leaneth more to one side, then it doth to the other: And one angle on the one side, is greater then that on the other.

Therefore,

14. All straight-shanked right angles are equall.

[That is, they are alike, and agreeable, or they doe fill the same place; as here are *aio.* and *eio.* And yet againe on the contrary: All straight shanked equall angles, are not right-angles.]

The axiomes of the equality of angles were three, as even now wee heard, one generall, and two Consectaries: Here moreover is there one speciall one of the equality of Right angles.

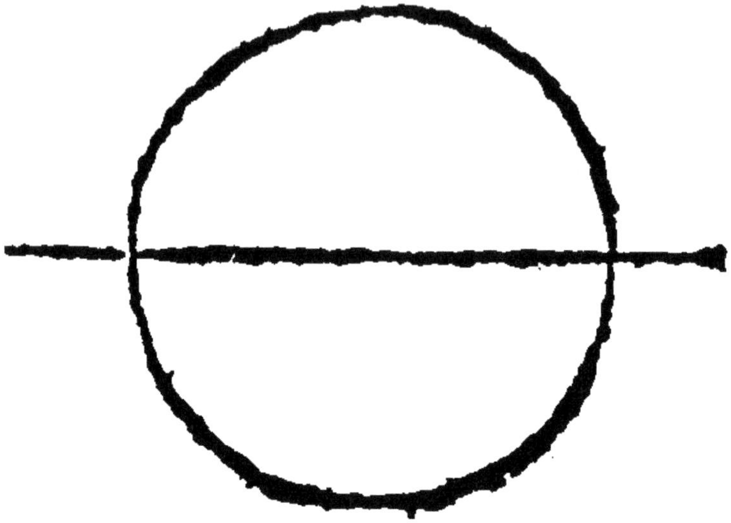

Angles therfore homogeneall and recticrurall, that is whose shankes are right, as are right lines, as plaine surfaces (For let us so take the word) are equall right angles. So are the above written rectilineall right angles equall: so are plaine solid right angles, as in a cube, equall. The axiome may therefore generally be spoken of solid angles, so they be recticruralls: Because all semicircular right angles are not equall to all semicircular right angles: As here, when the diameter is continued it is perpendicular, and maketh twice two angles, within and without, the outter equall betweene themselves, and inner equall betweene themselves: But the outer unequall to the inner: And the angle of a greater semicircle is greater, then the angle of a lesser. Neither is this affection any way reciprocall, That all equall angles should bee right angles. For oblique angles may bee equall betweene themselves: And an oblique angle may bee made equall to a right angle, as a Lunular to a rectilineall right angle, as was manifest, at the 6 e.

The definition of an oblique is understood by the obliquity of the shankes: whereupon also it appeareth; That an oblique angle is unequall to an homogeneall right angle: Neither indeed may oblique angles be made equall by any lawe or rule: Because obliquity may infinitly bee both increased and diminished.

15. An oblique angle is either Obtuse or Acute.

One difference of Obliquity wee had before at the 9 e ij. in a line, to witt of a periphery and an helix; Here there is another dichotomy of it into obtuse and acute: which difference is proper to angles, from whence it is translated or conferred upon other things and metaphorically used, as *Ingenium obtusum, acutum*; A dull, and quicke witte, and such like.

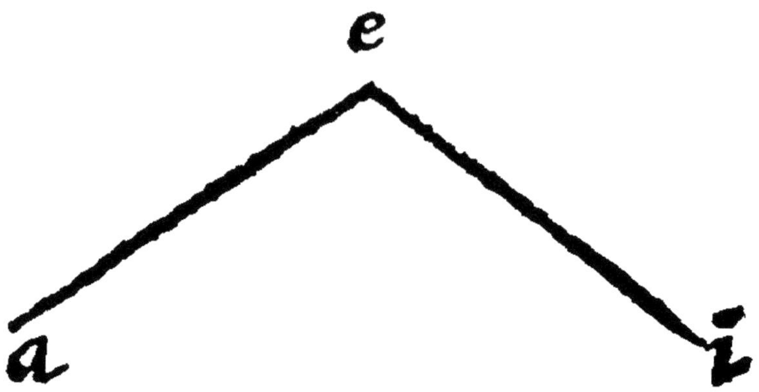

16. *An obtuse angle is an oblique angle greater then a right angle. 11. d j.*

Obtusus, Blunt or Dull; As here *aei.* In the definition the *genus* of both *Species* or kinds is to bee understood: For a right lined right angle is greater then a sphearicall right angle, and yet it is not an obtuse or blunt angle: And this greater inequality may infinitely be increased.

17. *An acutangle is an oblique angle lesser then a right angle. 12. d j.*

Acutus, Sharpe, Keene, as here *aei.* is. Here againe the same *genus* is to bee understood: because every angle which is lesse then any right angle is not an acute or sharp angle. For a semicircle and sphericall right angle, is lesse then a rectilineall right angle, and yet it is not an acute angle.

The fourth Booke, which is of a Figure.

1. A figure is a lineate bounded on all parts.

So the triangle *aei.* is a figure; Because it is a plaine bounded on all parts with three sides. So a circle is a figure: Because it is a plaine every way bounded with one periphery.

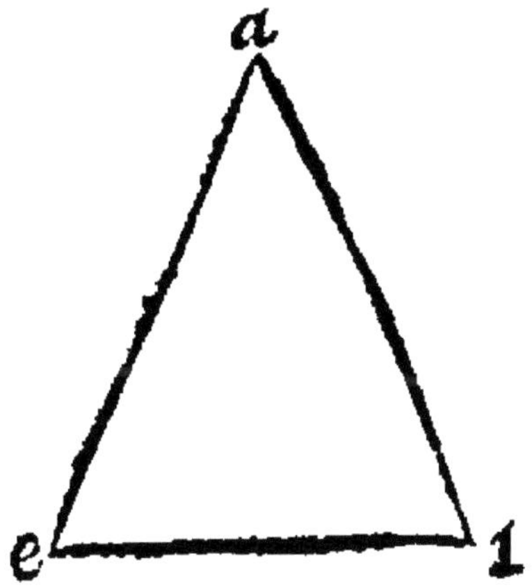

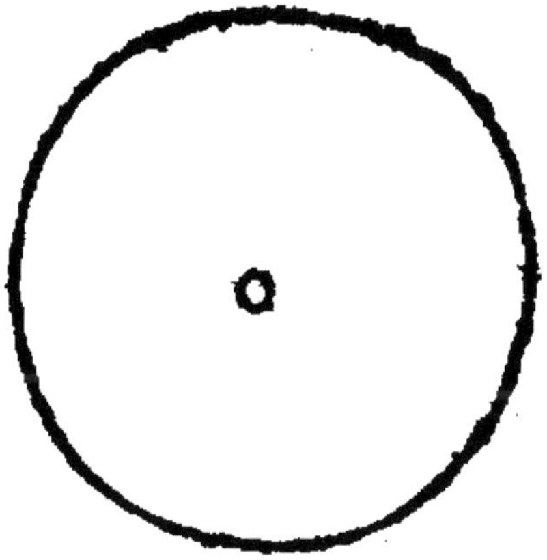

2. The center is the middle point in a figure.

In some part of a figure the Center, Perimeter, Radius, Diameter and Altitude are to be considered. The Center therefore is a point in the midst of the figure; so in the triangle, quadrate, and circle, the center is, *aei.*

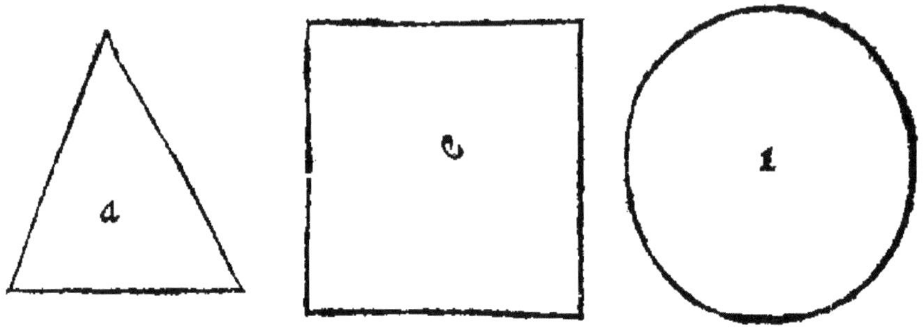

Centrum gravitatis, the center of weight, in every plaine magnitude is said to bee that, by the which it is handled or held up parallell to the horizon: Or it is that point whereby the weight being suspended doth rest, when it is caried. Therefore if any plate should in all places be alike heavie, the center of magnitude and weight would be one and the same.

3. The perimeter is the compasse of the figure.

Or, the perimeter is that which incloseth the figure. This definition is nothing else but the interpretation of the Greeke word. Therefore the perimeter of a Triangle is one line made or compounded of three lines. So the perimeter of the triangle *a*, is *eio.* So the perimeter of the circle *a* is a periphery, as in *eio.* So the perimeter of a Cube is a surface, compounded of sixe surfaces: And the perimeter of a spheare is one whole sphæricall surface, as hereafter shall appeare.

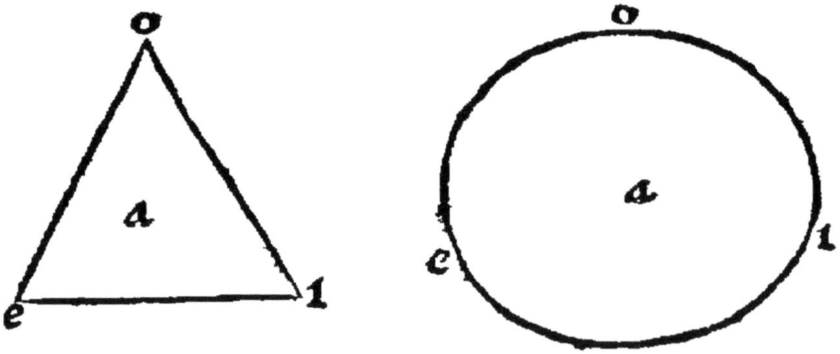

4. The Radius is a right line drawne from the center to the perimeter.

Radius, the Ray, Beame, or Spoake, as of the sunne, and cart wheele: As in the figures under written are *ae, ai, ao*. It is here taken for any distance from the center, whether they be equall or unequall.

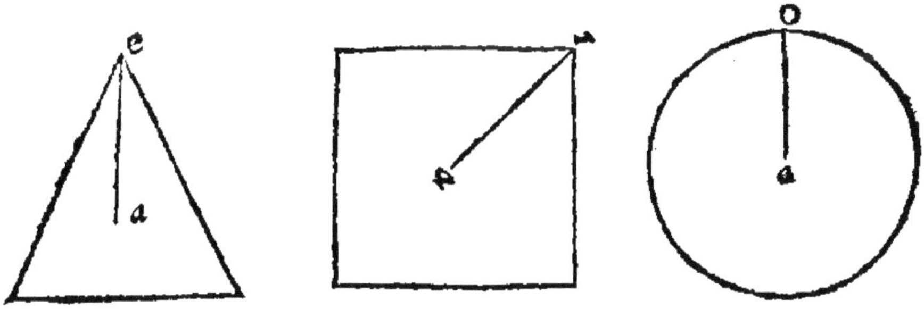

5. The Diameter is a right line inscribed within the figure by his center.

As in the figure underwritten are *ae, ai, ao*. It is called the *Diagonius*, when it passeth from corner to corner. In solids it is called the *Axis*, as hereafter we shall heare.

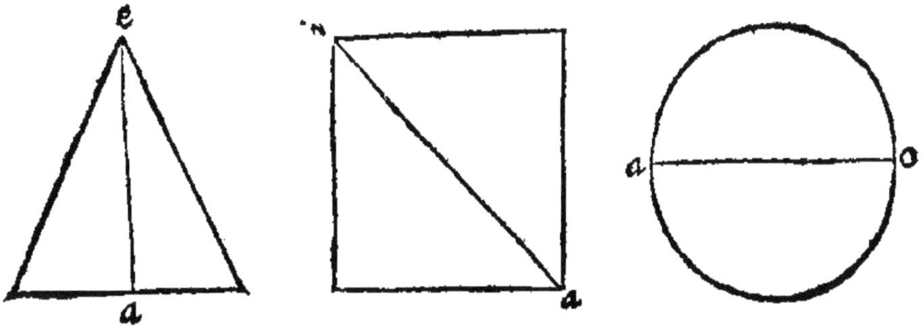

Therefore,

6. The diameters in the same figure are infinite.

Although of an infinite number of unequall lines that be only the diameter, which passeth by or through the center notwithstanding by the center there may be divers and sundry. In a circle the thing is most apparent: as in the Astrolabe the index may be put up and downe by all the points of the periphery. So in a speare and all rounds the thing is more easie to be conceived, where the diameters are equall: yet notwithstanding in other figures the thing is the same. Because the diameter is a right line inscribed by the center, whether from corner to corner, or side to side, the matter skilleth not. Therefore that there are in the same figure infinite diameters, it issueth out of the difinition of a diameter.

And

7. The center of the figure is in the diameter.

As here thou seest *a, e, i* this ariseth out of the definition of the diameter. For because the diameter is inscribed into the figure by the center: Therefore the Center of the figure must needes be in the diameter thereof: This is by *Archimedes* assumed especially at the 9, 10, 11, and 13 *Theoreme* of his *Isorropicks*, or *Æquiponderants*.

This consectary, saith the learned Rod. Snellius, is as it were a kinde of invention of the center. For where the diameters doe meete and cutt one another, there must the center needes bee. The cause of this is for that in every figure there is but one center only: And all the diameters, as before was said, must needes passe by that center.

And

8. It is in the meeting of the diameters.

As in the examples following. This also followeth out of the same definition of the diameter. For seeing that every diameter passeth by the center: The center must needes be common to all the diameters: and therefore it must also needs be in the meeting of them: Otherwise there should be divers centers of one and the same figure. This also doth the same *Archimedes* propound in the same words in the 8. and 12 theoremes of the same booke, speaking of Parallelogrammes and Triangles.

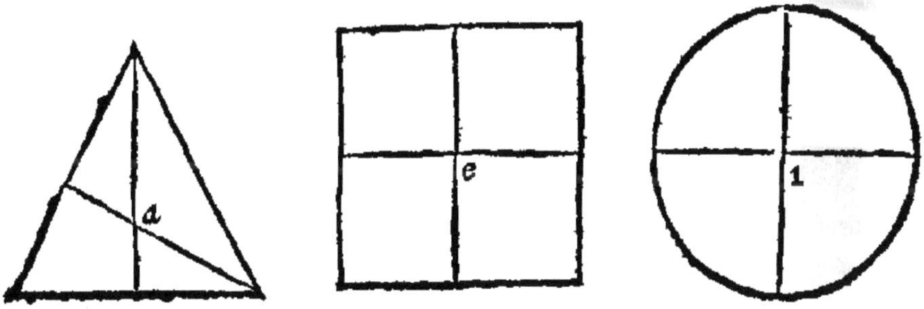

9. The Altitude is a perpendicular line falling from the toppe of the figure to the base.

Altitudo, the altitude, or heigth, or the depth: [For that, as hereafter shall bee taught, is but *Altitudo versa*, an heighth with the heeles upward.] As in the figures following are *ae*, *io*, *uy*, or *sr*. Neither is it any matter whether the base be the same with the figure, or be continued or drawne out longer, as in a blunt angled triangle, when the base is at the blunt corner, as here in the triangle, *aei*, is *ao*.

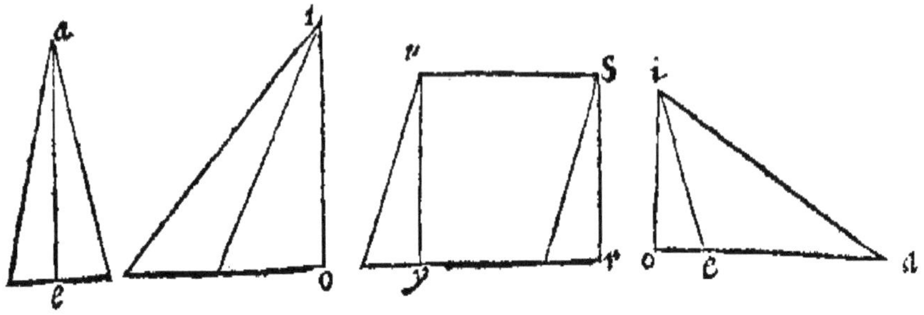

> *10. An ordinate figure, is a figure whose bounds are equall and angles equall.*

In plaines the Equilater triangle is onely an ordinate figure, the rest are all inordinate: In quadrangles, the Quadrate is ordinate, all other of that sort are inordinate: In every sort of Multangles, or many cornered figures one may be an ordinate. In crooked lined figures the Circle is ordinate, because it is conteined with equall bounds, (one bound alwaies equall to it selfe being taken for infinite many,) because it is equiangled, seeing (although in deede there be in it no angle) the inclination notwithstanding is every where alike and equall, and as it were the angle of the perphery be alwaies alike unto it selfe: whereupon of Plato and Plutarch a circle is said to be *Polygonia*, a multangle; and of Aristotle *Holegonia*, a totangle, nothing else but one whole angle. In mingled-lined figures there is nothing that is ordinate: In solid bodies, and pyramids the Tetrahedrum is ordinate: Of Prismas, the Cube: of Polyhedrum's, three onely are ordinate, the octahedrum, the Dodecahedrum, and the Icosahedrum. In oblique-lined bodies, the spheare is concluded to be ordinate, by the same argument that a circle was made to bee ordinate.

11. A prime or first figure, is a figure which cannot be divided into any other figures more simple then it selfe.

So in plaines the triangle is a prime figure, because it cannot be divided into any other more simple figure although it may be cut many waies: And in solids, the Pyramis is a first figure: Because it cannot be divided into a more simple solid figure, although it may be divided into an infinite sort of other figures: Of the Triangle all plaines are made; as of a Pyramis all bodies or solids are compounded; such are *aei.* and *aeio*.

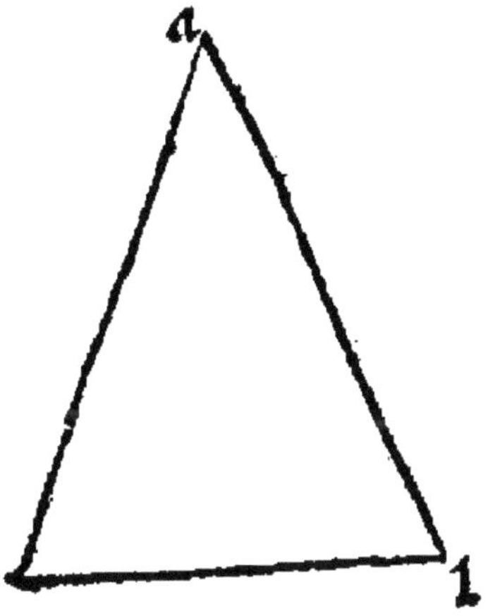

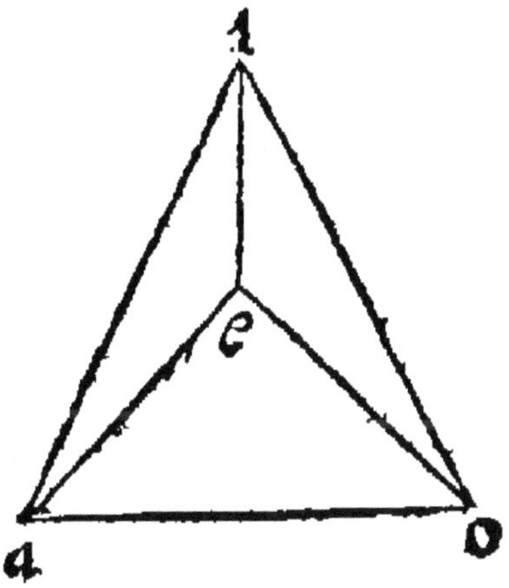

12. A rationall figure is that which is comprehended of a base and height rationall betweene themselves.

So *Euclide*, at the 1. d. ij. saith, that a rightangled parallelogramme is comprehended of two right lines perpendicular one to another, videlicet one multiplied by the other. For Geometricall comprehension is sometimes as it were in numbers a multiplication: Therefore if yee shall grant the base and height to bee rationalls betweene themselves, that their reason I meane may be expressed by a number of the assigned measure, then the numbers of their sides being multiplyed one by another, the bignesse of the figure shall be expressed. Therefore a Rationall figure is made by the multiplying of two rationall sides betweene themselves.

Therefore,

13. The number of a rationall figure, is called a Figurate number: And the numbers of which it is made, the Sides of the figurate.

As if a Right angled parallelogramme be comprehended of the base foure, and the height three, the Rationall made shall be 12. which wee here call the figurate: and 4. and 3. of which it was made, we name sides.

14. Isoperimetrall figures, are figures of equall perimeter.

This is nothing else but an interpretation of the Greeke word; So a triangle of 16. foote about, is a isoperimeter to a triangle 16. foote about, to a quadrate 16. foote about, and to a circle 16. foote about.

15. Of isoperimetralls homogenealls that which is most ordinate, is greatest: Of ordinate isoperimetralls heterogenealls, that is greatest, which hath most bounds.

So an equilater triangle shall bee greater then an isoperimeter inequilater triangle; and an equicrurall, greater then an unequicrurall: so in quadrangles, the quadrate is greater then that which is not a quadrate: so an oblong more ordinate, is greater then an oblong lesse ordinate. So of those figures which are heterogeneall ordinates, the quadrate is greater then the Triangle: And the Circle, then the Quadrate.

16. If prime figures be of equall height, they are in reason one unto another, as their bases are: And contrariwise.

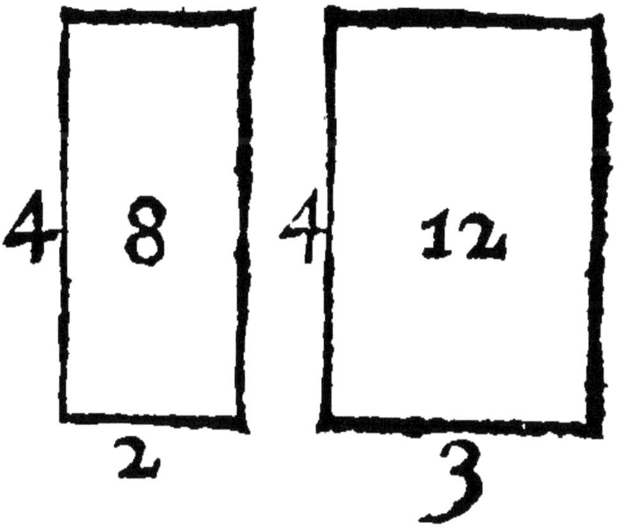

The proportion of first figures is here twofold; the first is direct in those which are of equall height. In Arithmeticke we learned; That if one number doe multiply many numbers, the products shall be proportionall unto the numbers, multiplyed. From hence in rationall figures the content of those which are of equall height is to bee expressed by a number. As in two right angled parallelogrammes, let 4. the same height, multiply 2. and 3. the bases: The products 8. and 12. the parallelogrammes made, are directly proportionall unto the bases 2. and 3. Therefore as 2. is unto 3. so is 8. unto 12. The same shall afterward appeare in right Prismes and Cylinders. In plaines, Parallelogramms are the doubles of triangles: In solids, Prismes are the triples of pyramides: Cylinders, the triples of Cones. The converse of this element is plaine out of the former also: First figures if they be in reason one to another as their bases are, then are they of equall height, to witt when their products are proportionall unto the multiplyed, the same number did multiply them.

Therefore,

17. If prime figures of equall heighth have also equall bases, they are equall.

[The reason is, because then those two figures compared, have equall sides, which doe make them equall betweene themselves; For the parts of the one applyed or laid unto the parts of the other, doe fill an equall place, as was taught at the 10. e. j. *Sn.*] So Triangles, so Parallelogrammes, and so other figures proposed are equalled upon an equall base.

18. If prime figures be reciprocall in base and height, they are equall: And contrariwise.

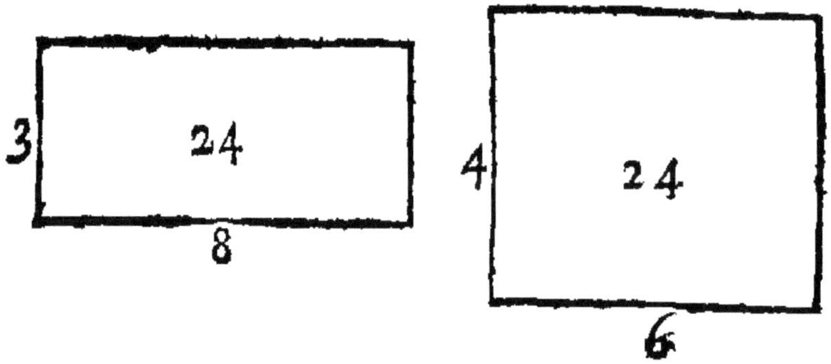

The second kind of proportion of first figures is reciprocall. This kinde of proportion rationall and expressible by a number, is not to be had in first figures themselves: but in those that are equally manifold to them, as was taught even now in direct proportion: As for example, Let these two right angled parallelogrammes, unequall in bases and heighths 3, 8, 4, 6, be as heere thou seest: The proportion reciprocall is thus, As 3 the base of the one, is unto 4, the base of the other: so is 6. the height of the one is to 8. the height of the other: And the parallelogrammes are equall, viz. 24. and 24. Againe, let two solids of unequall bases & heights (for here also the base is taken for the length and heighth) be 12, 2, 3, 6, 3, 4. The solids themselves shall be 72. and 72, as here thou seest; and the proportion of the bases and heights likewise is reciprocall: For as 24, is unto 18, so is 4, unto 3. The cause is out of the golden rule of proportion in Arithmeticke: Because twice two sides are proportionall: Therefore the plots made of them shall be equall. And againe, by the same rule, because the plots are equall: Therefore the bounds are proportionall; which is the converse of this present element.

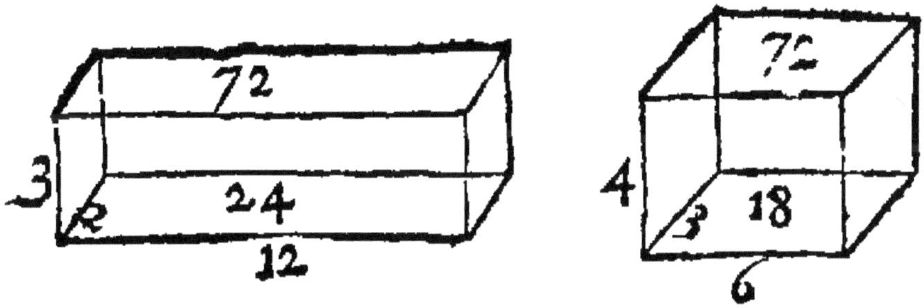

> *19. Like figures are equiangled figures, and proportionall in the shankes of the equall angles.*

First like figures are defined, then are they compared one with another, similitude of figures is not onely of prime figures, and of such as are compounded of prime figures, but generally of all other whatsoever. This similitude consisteth in two things, to witt in the equality of their angles, and proportion of their shankes.

Therefore,

> *20. Like figures have answerable bounds subtended against their equall angles: and equall if they themselves be equall.*

Or thus, They have their termes subtended to the equall angles correspondently proportionall: And equall if the figures themselves be equall; H. This is a consectary out of the former definition.

And

21. Like figures are situate alike, when the proportionall bounds doe answer one another in like situation.

The second consectary is of situation and place. And this like situation is then said to be when the upper parts of the one figure doe agree with the upper parts of the other, the lower, with the lower, and so the other differences of places. *Sn.*

And

22. Those figures that are like unto the same, are like betweene themselves.

This third consectary is manifest out of the definition of like figures. For the similitude of two figures doth conclude both the same equality in angles and proportion of sides betweene themselves.

And

23. If unto the parts of a figure given, like parts and alike situate, be placed upon a bound given, a like figure and likely situate unto the figure given, shall bee made accordingly.

This fourth consectary teacheth out of the said definition, the fabricke and manner of making of a figure alike and likely situate unto a figure given. *Sn.*

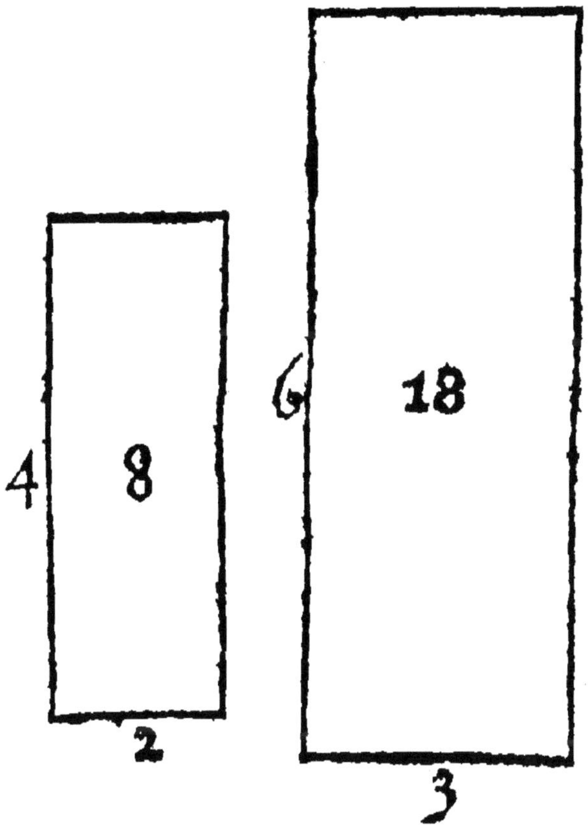

24. Like figures have a reason of their homologallor correspondent sides equally manifold unto their dimensions: and a meane proportionall lesse by one.

Plaine figures have but two dimensions, to witt Length, and Breadth: And therefore they have but a doubled reason of their homologall sides. Solids have three dimensions, videl. Length, Breadth, & thicknesse: therefore they shall have a treabled reason of their homologall or correspondent sides. In 8. and 18. the two plaines given, first the angles are equall: secondly, their homolegall side 2. and 4. and 3. and 6. are proportionall. Therefore the reason of 8. the first figure, unto 18. the second, is as the reason is of 2. unto 3. doubled. But the reason of 2. unto 3. doubled, by the 3. chap. ij. of Arithmeticke, is of 4. to 9. (for 2/3 2/3 is 4/9.) Therefore the reason of 8. unto 18, that is, of the first figure unto the second, is of 4. unto 9. In Triangles, which are the halfes of rightangled parallelogrammes, there is the same truth, and yet by it selfe not rationall and to be expressed by numbers.

Said numbers are alike in the trebled reason of their homologall sides; As for example, 60. and 480. are like solids; and the solids also comprehended in those numbers are like-solids, as here thou seest: Because their sides, 4. 3. 5. and 8. 6. 10. are proportionall betweene themselves. But the reason of 60. to 480. is the reason of 4. to 8. trebled, thus 4/8 4/8 4/8 = 64/512; that is of 1. unto 8. or *octupla*, which you shall finde in the dividing of 480. by 60.

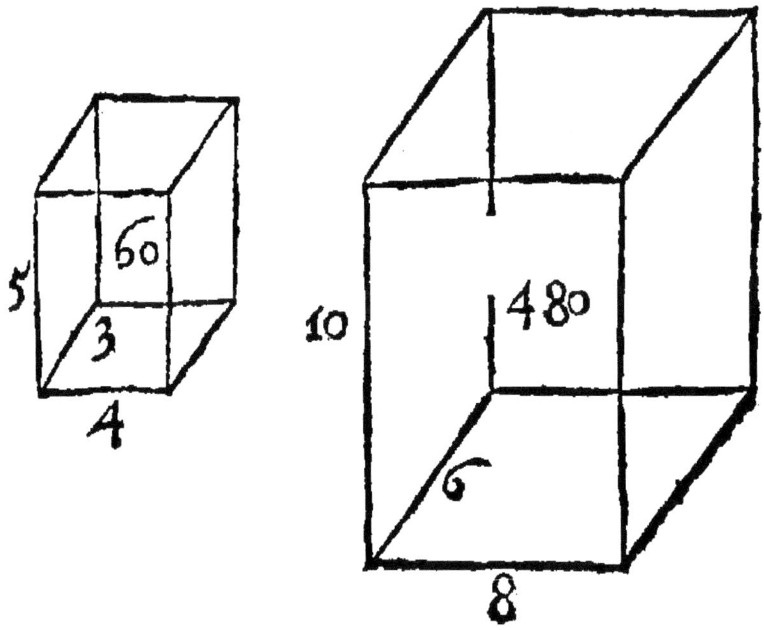

Thus farre of the first part of this element: The second, that like figurs have a meane, proportional lesse by one, then are their dimensions, shall be declared by few words. For plaines having but two dimensions, have but one meane proportionall, solids having three dimensions, have two meane proportionalls. The cause is onely Arithmeticall, as afore. For where the bounds are but 4. as they are in two plaines, there can be found no more but one meane proportionall, as in the former example of 8. and 18. where the homologall or correspondent sides are 2. 3. and 4. 6.

Therefore,

2	3		4	6
	3		4	
8		12		18

Againe by the same rule, where the bounds are 6. as they are in two solids, there may bee found no more but two meane proportionalls: as in the former solids 30. and 240. where the homologall or correspondent sides are 2. 4. 3. 6. 5. 10.

Therefore,

2	4		3	6	5		10
	4		3				
6		12		24			
				24	5		
	30		60	120		240	

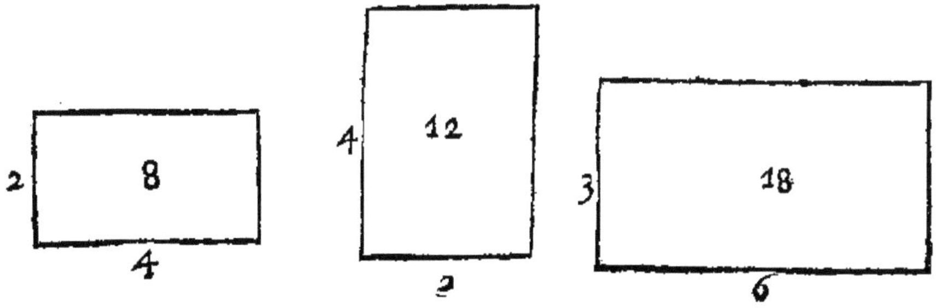

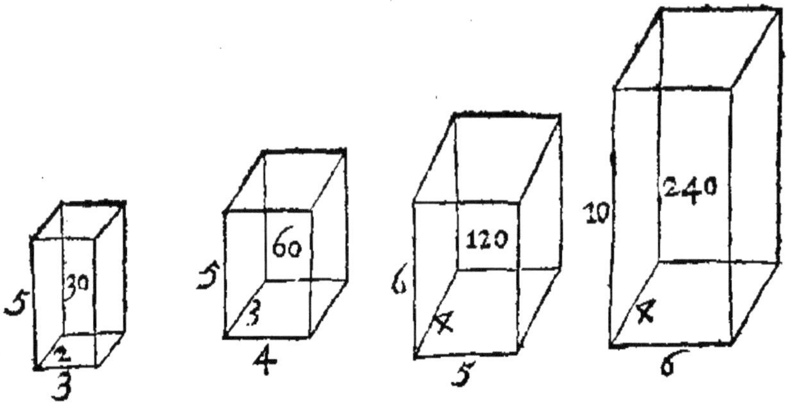

Therefore,

> *25. If right lines be continually proportionall, more by one then are the dimensions of like figures likelily situate unto the first and second, it shall be as the first right line is unto the last, so the first figure shall be unto the second: And contrariwise.*

Out of the similitude of figures two consectaries doe arise, in part only, as is their axiome, rationall and expressable by numbers. If three right lines be continually proportionall, it shall be as the first is unto the third: So the rectilineall figure made upon the first, shall be unto the rectilineall figure made upon the second, alike and likelily situate. This may in some part be conceived and understood by numbers. As for example, Let the lines given, be 2. foot, 4. foote, and 8 foote. And upon the first and second, let there be made like figures, of 6. foote and 24. foote; So I meane, that 2. and 4. be the bases of them. Here as 2. the first line, is unto 8. the third line: So is 6. the first figure, unto 24. the second figure, as here thou seest.

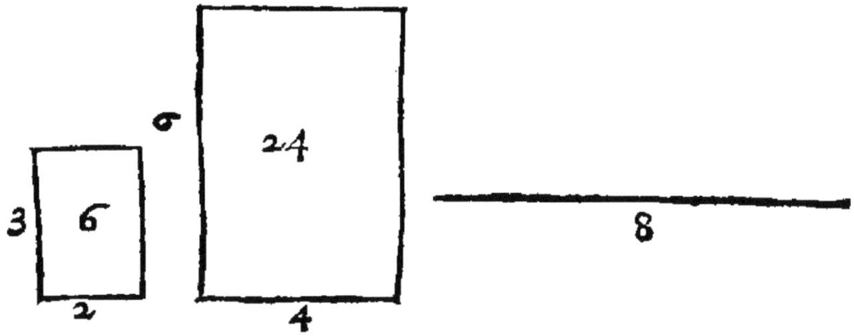

Againe, let foure lines continually proportionall, be 1. 2. 4. 8. And let there bee two like solids made upon the first and second: upon the first, of the sides 1. 3. and 2. let it be 6. Upon the second, of the sides 2. 6. and 4. let it be 48. As the first right line 1. is unto the fourth 8. So is the figure 6. unto the second 48. as is manifest by division. The examples are thus.

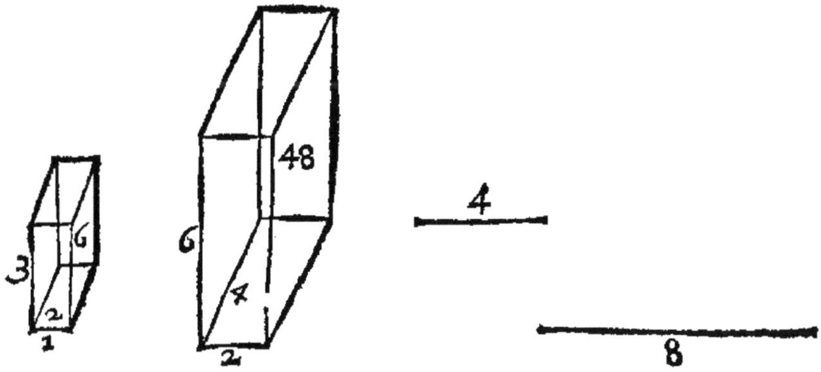

Moreover by this Consectary a way is laid open leading unto the reason of doubling, treabling, or after any manner way whatsoever assigned increasing of a figure given. For as the first right line shall be unto the last: so shall the first figure be unto the second.

And

26. If foure right lines bee proportionall betweene themselves: Like figures likelily situate upon them, shall be also proportionall betweene themselves: And contrariwise, out of the 22. p vj. and 37. p xj.

The proportion may also here in part bee expressed by numbers: And yet a continuall is not required, as it was in the former.

In Plaines let the first example be, as followeth.

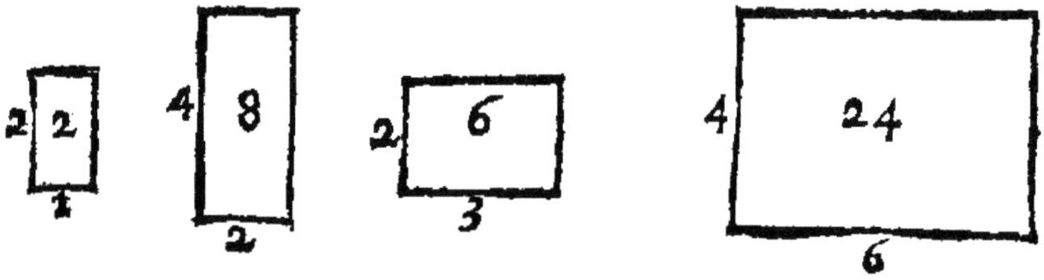

The cause of proportionall figures, for that twice two figures have the same reason doubled.

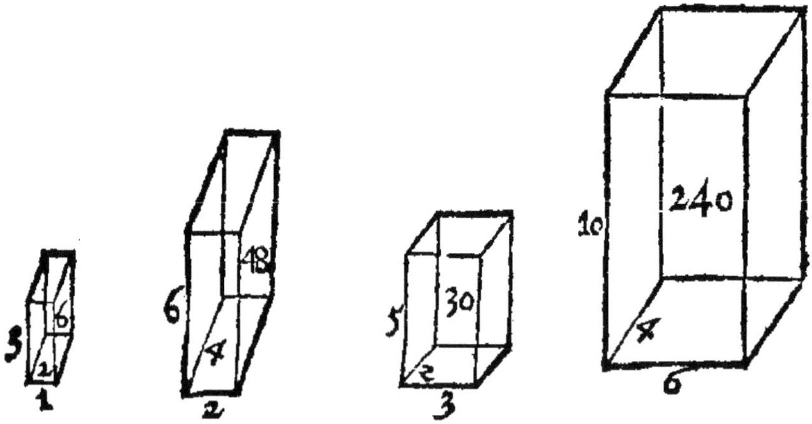

In Solids let this bee the second example. And yet here the figures are not proportionall unto the right lines, as before figures of equall heighth were unto their bases, but they themselves are proportionall one to another. And yet are they not proportionall in the same kinde of proportion.

The cause also is here the same, that was before: To witt, because twice two figures have the same reason trebled.

27. Figures filling a place, are those which being any way set about the same point, doe leave no voide roome.

This was the definition of the ancient Geometers, as appeareth out of *Simplicius*, in his commentaries upon the 8. chapter of *Aristotle's* iij. booke of Heaven: which kinde of figures *Aristotle* in the same place deemeth to bee onely ordinate, and yet not all of that kind. But only three among the Plaines, to witt a Triangle, a Quadrate, and a Sexangle: amongst Solids, two; the Pyramis, and the Cube. But if the filling of a place bee judged by right angles, 4. in a Plaine, and 8. in a Solid, the Oblong of plaines, and the Octahedrum of Solids shall (as shall appeare in their places) fill a place; And yet is not this Geometrie of *Aristotle* accurate enough. But right angles doe determine this sentence, and so doth *Euclide* out of the angles demonstrate, That there are onely five ordinate solids; And so doth *Potamon* the Geometer, as *Simplicus* testifieth, demonstrate the question of figures filling a place. Lastly, if figures, by laying of their corners together, doe make in a Plaine 4. right angles, or in a Solid 8. they doe fill a place.

Of this probleme the ancient geometers have written, as we heard even now: And of the latter writers, *Regiomontanus* is said to have written accurately; And of this argument *Maucolycus* hath promised a treatise, neither of which as yet it hath beene our good hap to see.

Neither of these are figures of this nature, as in their due places shall be proved and demonstrated.

28. A round figure is that, all whose raies are equall.

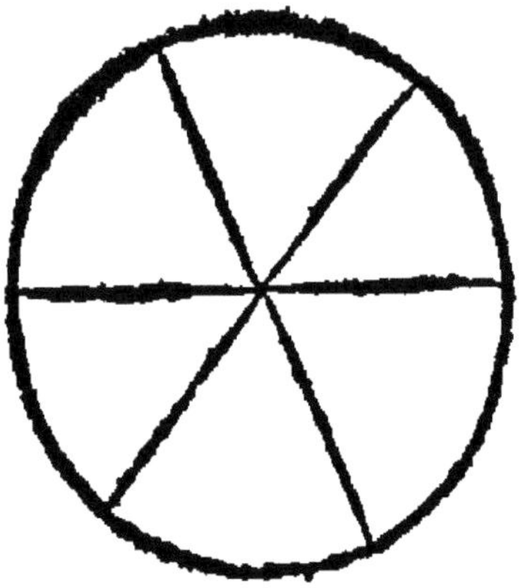

Such in plaines shall the Circle be, in Solids the Globe or Spheare. Now this figure, the Round, I meane, of all Isoperimeters is the greatest, as appeared before at the 15. e. For which cause *Plato*, in his *Timæus* or his Dialogue of the World said; That this figure is of all other the greatest. And therefore God, saith he, did make the world of a sphearicall forme, that within his compasse it might the better containe all things: And *Aristotle*, in his Mechanicall problems, saith; That this figure is the beginning, principle, and cause of all miracles. But those miracles shall in their time God willing, be manifested and showne.

Rotundum, a Roundle, let it be here used for *Rotunda figura*, a round figure. And in deede *Thomas Finkius* or *Finche*, as we would call him, a learned *Dane*, sequestring this argument from the rest of the body of Geometry, hath intituled that his worke *De Geometria rotundi*, Of the Geometry of the Round or roundle.

29. The diameters of a roundle are cut in two by equall raies.

The reason is, because the halfes of the diameters, are the raies. Or because the diameter is nothing else but a doubled ray: Therefore if thou shalt cut off from the diameter so much, as is the radius or ray, it followeth that so much shall still remaine, as thou hast cutte of, to witt one ray, which is the other halfe of the diameter. *Sn*.

And here observe, That *Bisecare*, doth here, and in other places following, signifie to cutte a thing into two equall parts or portions; And so *Bisegmentum*, to be one such portion; And *Bisectio*, such a like cutting or division.

30. Rounds of equall diameters are equall. Out of the 1. d. iij.

Circles and Spheares are equall, which have equall diameters. For the raies, which doe measure the space betweene the Center and Perimeter, are equall, of which, being doubled, the Diameter doth consist. *Sn*.

The fifth Booke, of Ramus *his Geometry, which is of Lines and Angles in a plaine Surface.*

1. A lineate is either a Surface or a Body.

Lineatum, (or *Lineamentum*) a magnitude made of lines, as was defined at 1. e. iij. is here divided into two kindes: which is easily conceived out of the said definition there, in which a line is excluded, and a Surface & a body are comprehended. And from hence arose the division of the arte Metriall into Geometry, of a surface, and Stereometry, of a body, after which maner *Plato* in his vij. booke of his Common-wealth, and *Aristotle* in the 7. chapter of the first booke of his *Posteriorums*, doe distinguish betweene Geometry and Stereometry: And yet the name of Geometry is used to signifie the whole arte of measuring in generall.

2. A Surface is a lineate only broade. 5. d j.

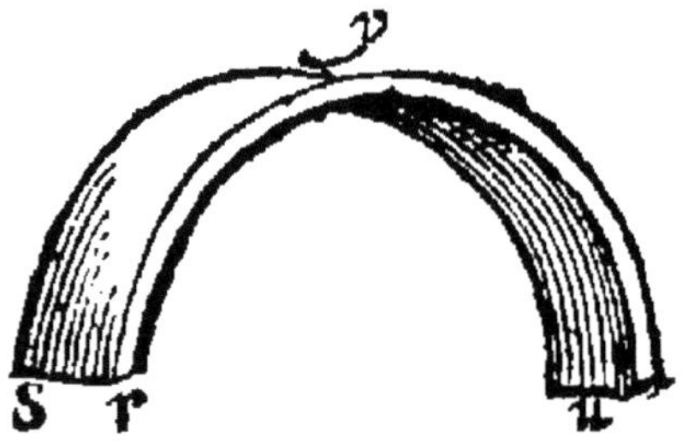

As here *aeio.* and *uysr.* The definition of a Surface doth comprehend the distance or dimension of a line, to witt Length: But it addeth another distance, that is Breadth. Therefore a Surface is defined by some, as *Proclus* saith, to be a magnitude of two dimensions. But two doe not so specially and so properly define it. Therefore a Surface is better defined, to bee a magnitude onely long and broad. Such, saith *Apollonius*, are the shadowes upon the earth, which doe farre and wide cover the ground and champion fields, and doe not enter into the earth, nor have any manner of thicknesse at all.

Epiphania, the Greeke word, which importeth onely the outter appearance of a thing, is here more significant, because of a Magnitude there is nothing visible or to bee seene, but the surface.

3. The bound of a surface is a line. 6. d j.

The matter in Plaines is manifest. For a three cornered surface is bounded with 3. lines: A foure cornered surface, with foure lines, and so forth: A Circle is bounded with one line. But in a Sphearicall surface the matter is not so plaine: For it being whole, seemeth not to be bounded with a line. Yet if the manner of making of a Sphearicall surface, by the conversiō or turning about of a semiperiphery, the beginning of it, as also the end, shalbe a line, to wit a semiperiphery: And as a point doth not only *actu*, or indeede bound and end a line: But is *potentia*, or in power, the middest of it: So also a line boundeth a Surface *actu*, and an innumerable company of lines may be taken or supposed to be throughout the whole surface. A Surface therefore is made by the motion of a line, as a Line was made by the motion of a point.

4. A Surface is either Plaine or Bowed.

The difference of a Surface, doth answer to the difference of a Line, in straightnesse and obliquity or crookednesse.

Obliquum, oblique, there signified crooked; Not right or straight: Here, uneven or bowed, either upward or downeward. *Sn.*

5. A plaine surface is a surface, which lyeth equally betweene his bounds, out of the 7. d j.

As here thou seest in *aeio.* That therefore a Right line doth looke two contrary waies, a Plaine surface doth looke all about every way, that a plaine surface should, of all surfaces within the same bounds, be the shortest: And that the middest thereof should hinder the sight of the extreames. Lastly, it is equall to the dimension betweene the lines: It may also by one right line every way applyed be tryed, as *Proclus* at this place doth intimate.

Planum, a Plaine, is taken and used for a plaine surface: as before *Rotundum*, a Round, was used for a round figure.

Therefore,

6. From a point unto a point we may, in a plaine surface, draw a right line, 1 and 2. post. j.

Three things are from the former ground begg'd: The first is of a Right line. A right line and a periphery were in the ij. booke defined: But the fabricke or making of them both, is here said to bee properly in a plaine.

The fabricke or construction of a right line is the 1. petition. And justly is it required that it may bee done onely upon a plaine: For in any other surface it were in vaine to aske it. For neither may wee possibly in a sphericall betweene two points draw a right line: Neither may wee possibly in a Conicall and Cylindraceall betweene any two points assigned draw a right line. For from the toppe unto the base that in these is only possible: And then is it the bounde of the plaine which cutteth the Cone and Cylinder. Therefore, as I said, of a right plaine it may onely justly bee demanded: That from any point assigned, unto any point assigned, a right line may be drawne, as here from *a* unto *e*.

Now the Geometricall instrument for the drawing of a right plaine is called *Amussis*, & by *Petolemey*, in the 2. chapter of his first booke of his Musicke, *Regula*, a Rular, such as heere thou seest.

And from a point unto a point is this justly demanded to be done, not unto points; For neither doe all points fall in a right line: But many doe fall out to be in a crooked line. And in a Spheare, a Cone & Cylinder, a Ruler may be applyed, but it must be a sphearicall, Conicall, or Cylindraceall. But by the example of a right line doth *Vitellio*, 2 *p j.* demaund that betweene two lines a surface may be extended: And so may it seeme in the Elements, of many figures both plaine and solids, by *Euclide* to be demanded; That a figure may be described, at the 7. and 8. e ij. Item that a figure may be made vp, at the 8. 14. 16. 23. 28. p. vj. which are of Plaines. Item at the 25. 31. 33. 34. 36. 49. p. xj. which are of Solids. Yet notwithstanding a plaine surface, and a plaine body doe measure their rectitude by a right line, so that *jus postulandi*, this right of begging to have a thing granted may seeme primarily to bee in a right plaine line.

Now the *Continuation* of a right line is nothing else, but the drawing out farther of a line now drawne, and that from a point unto a point, as we may continue the right line *ae.* unto *i.* wherefore the first and second Petitions of *Euclide* do agree in one.

And

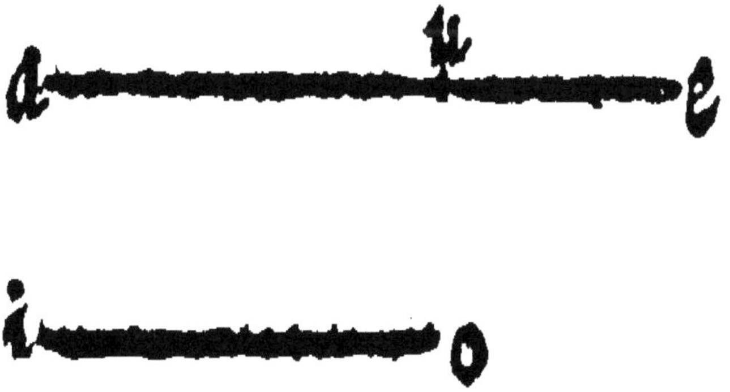

7. To set at a point assigned a Right line equall to another right line given: And from a greater, to cut off a part equall to a lesser. 2. and 3. p j.

As let the Right line given be *ae.* And to *i.* a point assigned, grant that *io.* equall to the same *ae.* may bee set. Item, in the second example, let *ae.* bee greater then *io.* And let there be cut off from the same *ae.* by applying of a rular made equall to *io.* the lesser, portion *au.* as here. For if any man shall thinke that this ought only to be don in the minde, hee also, as it were, beares a ruler in his minde, that he may doe it by the helpe of the ruler. Neither is the fabricke in deede, or making of one right equal to another: And the cutting off from greater Right line, a portion equall to a lesser, any whit harder, then it was, having a point and a distance given, to describe a circle: Then having a Triangle, Parallelogramme, and semicircle given, to describe or make a Cone, Cylinder, and spheare, all which notwithstanding *Euclide* did account as principles.

Therefore,

8. One right line, or two cutting one another, are in the same plaine, out of the 1. and 2. p xj.

One Right line may bee the common section of two plaines: yet all or the whole in the same plaine is one: And all the whole is in the same other: And so the whole is the same plaine. Two Right lines cutting one another, may bee in two plaines cutting one of another; But then a plaine may be drawne by them: Therefore both of them shall be in the same plaine. And this plaine is geometrically to be conceived: Because the same plaine is not alwaies made the ground whereupon one oblique line, or two cutting one another are drawne, when a periphery is in a sphearicall: Neither may all peripheries cutting one another be possibly in one plaine.

And

9. With a right line given to describe a peripherie.

This fabricke or construction is taken out of the 3. Petition which is thus. Having a center and a distance given to describe, make, or draw a circle. But here the terme or end of a circle is onely sought, which is better drawne out of the definition of a periphery, at the 10. e ij. And in a plaine onely may that conversion or turning about of a right line bee made:

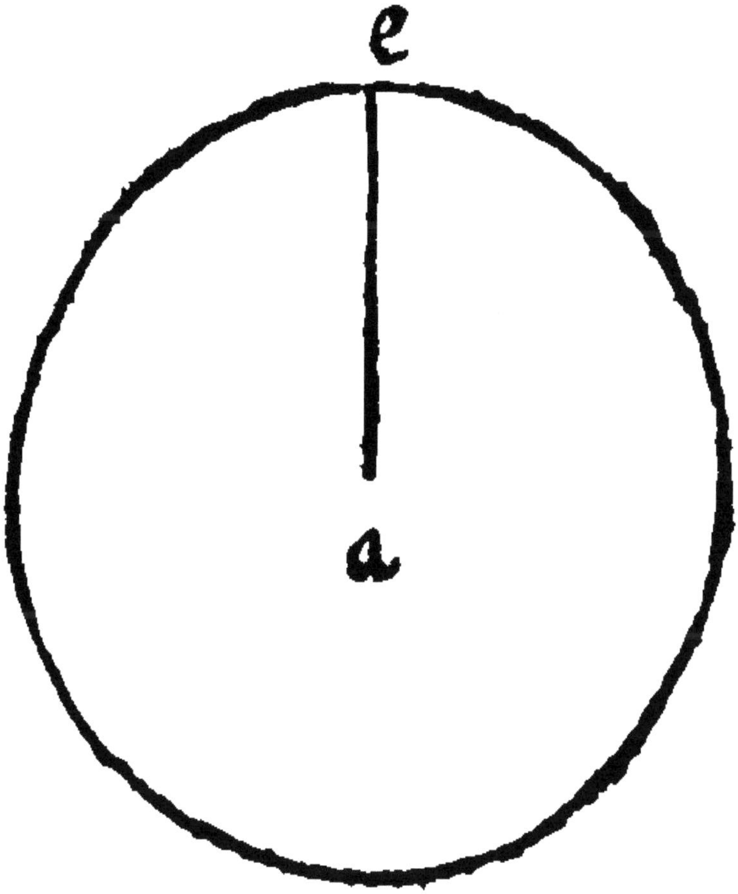

Not in a sphearicall, not in a Conicall, not in a Cylindraceall, except it be in top, where notwithstanding a periphery may bee described. Therefore before (to witt at the said 10. e ij.) was taught the generall fabricke or making of a Periphery: Here we are informed how to discribe a Plaine periphery, as here.

Now as the Rular was the instrument invented and used for the drawing of a right line: so also may the same *Rular*, used after another manner, be the instrument to describe or draw a periphery withall. And indeed such is that instrument used by the Coopers (and other like artists) for the rounding of their bottomes of their tubs, heads of barrells and otherlike vessells: But the

Compasses, whether straight shanked or bow-legg'd, such as here thou seest, it skilleth not, are for al purposes and practises, in this case the best and readiest. And in deed the Compasses, of all geometricall instruments, are the most excellent, and by whose help famous Geometers have taught: That all the problems of geometry may bee wrought and performed: And there is a booke extant, set out by *John Baptist*, an Italian, teaching, How by one opening of the Compasses all the problems of *Euclide* may be resolved: And *Jeronymus Cardanus*, a famous Mathematician, in the 15. booke of his Subtilties, writeth, that there was by the helpe of the Compasses a demonstration of all things demonstrated by *Euclide*, found out by him and one *Ferrarius*.

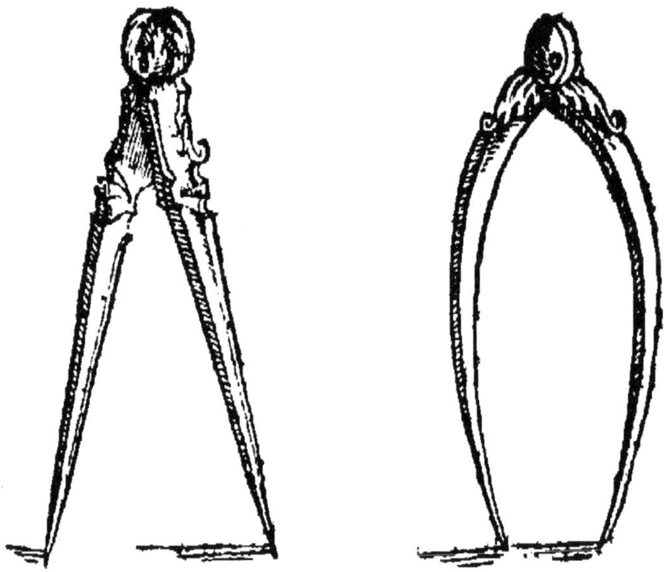

Talus, the nephew of *Dædalus* by his sister, is said in the viij. booke of *Ovids Metamorphosis*, to have beene the inventour of this instrument: For there he thus writeth of him and this matter:—*Et ex uno duo ferrea brachia nodo: Iunxit, ut æquali spatio distantibus ipsis: Altera pars staret, pars altera duceret orbem*.

Therfore

10. The raies of the same, or of an equall periphery, are equall.

The reason is, because the same right line is every where converted or turned about. But here by the Ray of the periphery, must bee understood the Ray the figure contained within the periphery.

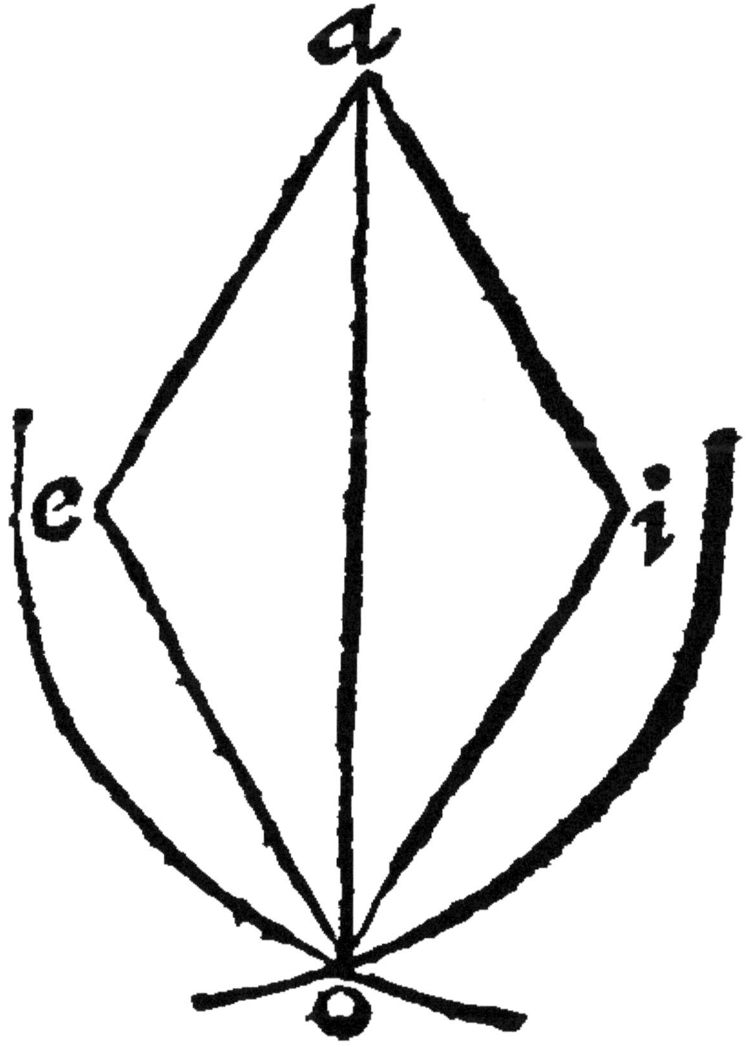

11. If two equall peripheries, from the ends of equall shankes of an assigned rectilineall angle, doe meete before it, a right line drawne from the meeting of them unto the toppe or point of the angle, shall cut it into two equall parts. 9. p j.

Hitherto we have spoken of plaine lines: Their affection followeth, and first in the Bisection or dividing of an Angle into two equall parts.

Let the right lined Angle to bee divided into two equall parts bee *eai.* whose equall shankes let them be *ae.* and *ai.* (or if they be unequall, let them be made equall, by the 7 e.) Then two equall peripheries from the ends *e* and *i.* meet before the Angle in *o*. Lastly, draw a line from *o.* unto *a*. I say the angle given is divided into two equall parts. For by drawing the right lines *oe.* and *oi.* the angles *oae.* and *oai.* equicrurall, by the grant, and by their common side *ao.* are equall in base *eo.* and *io.* by the 10 e (Because they are the raies of equall peripheries.) Therefore by the 7. e iij. the angles *oae.* and *oai.* are equall: And therefore the Angle *eai.* is equally divided into two parts.

12. If two equall peripheries from the ends of a right line given, doe meete on each side of the same, a right line drawne from those meetings, shall divide the right line given into two equall parts. 10. p j.

Let the right line given bee ae. *And let two equall peripheries from the ends* a. *and* e. *meete in* i. *and* o. *Then from those meetings let the right line* io. *be drawne. I say, That* ae. *is divided into two equall parts, by the said line thus drawne.*

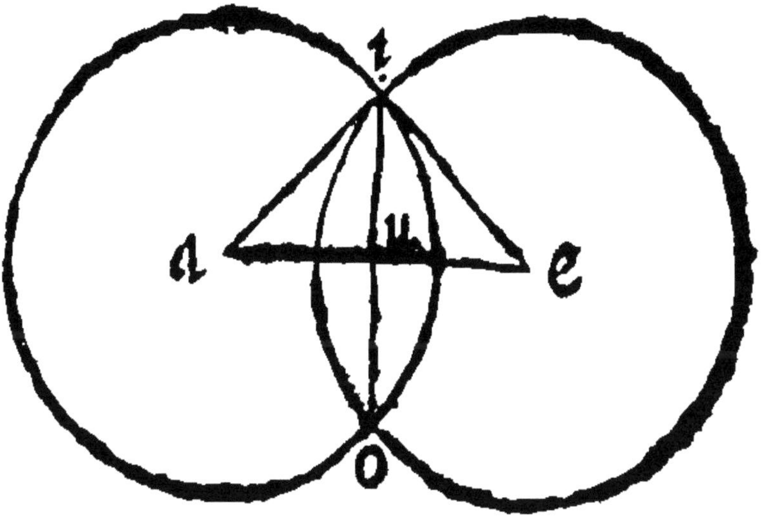

For by drawing the raies of the equall peripheries *ia.* and *ie.* the said *io.* doth cut the angle *aie.* into two equall parts, by the 11. e. Therefore the angles *aiu.* and *uie.* being equall and equicrurall (seeing the shankes are the raies of equall peripheries, by the grant.) have equall bases *au.* and *ue.* by the 7. e iij. Wherefore seeing the parts *au.* and *ue.* are equall, *ae.* the assigned right line is divided into two equall portions.

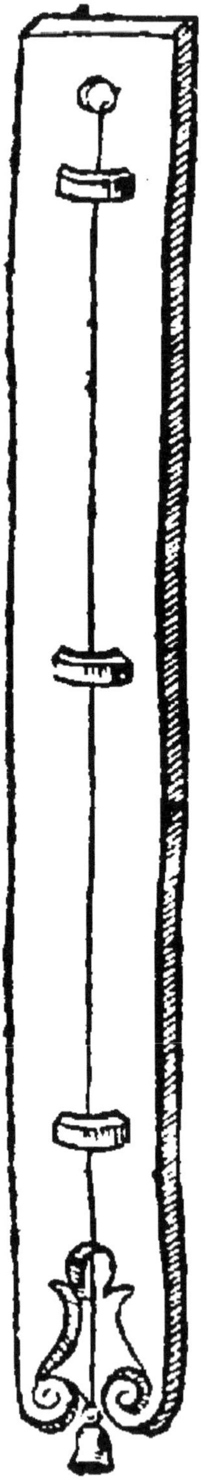

o
a
e
z

13. If a right line doe stand perpendicular upon another right line, it maketh on each side right angles: And contrary wise.

A right line standeth upon a right line, which cutteth, and is not cut againe. And the *Angles on each side*, are they which the falling line maketh with that underneath it, as is manifest out of *Proclus*, at the 15. pj. of *Euclide*; As here *ae.* the line cut: and *io.* the insisting line, let them be perpendicular; The angles on each side, to witt *aio.* and *eio.* shall bee right angles, by the 13. e iij.

The *Rular*, for the making of straight lines on a plaine, was the first Geometricall instrument: The *Compasses*, for the describing of a Circle, was the second: The *Norma* or *Square* for the true erecting of a right line in the same plaine upon another right line, and then of a surface and body, upon a surface or body, is the third. The figure therefore is thus.

Now *Perpendiculū*, an instrument with a line & a plummet of leade appendant upon it, used of Architects, Carpenters, and Masons, is meerely physicall: because heavie things naturally by their weight are in straight lines carried perpendicularly downeward. This instrument is of two sorts: The first, which they call a Plumbe-rule, is for the trying of an erect perpendicular, as whether a columne, pillar, or any other kinde of building bee right, that is plumbe unto the plaine of the horizont & doth not leane or reele any way. The second is for the trying or examining of a plaine or floore, whether it doe lye parallell to the horizont or not. Therefore when the line from the right angle, doth fall upon the middle of the base; it shall shew that the length is equally poysed. The Latines call it *Libra*, or *Libella*, a ballance: of the *Italians Livello*, and vel *Archipendolo, Achildulo*: of the *French, Nivelle*, or *Niueau*: of us a *Levill*.

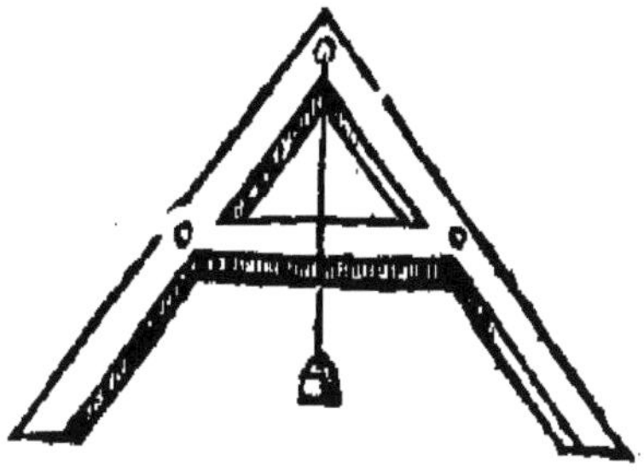

Therefore

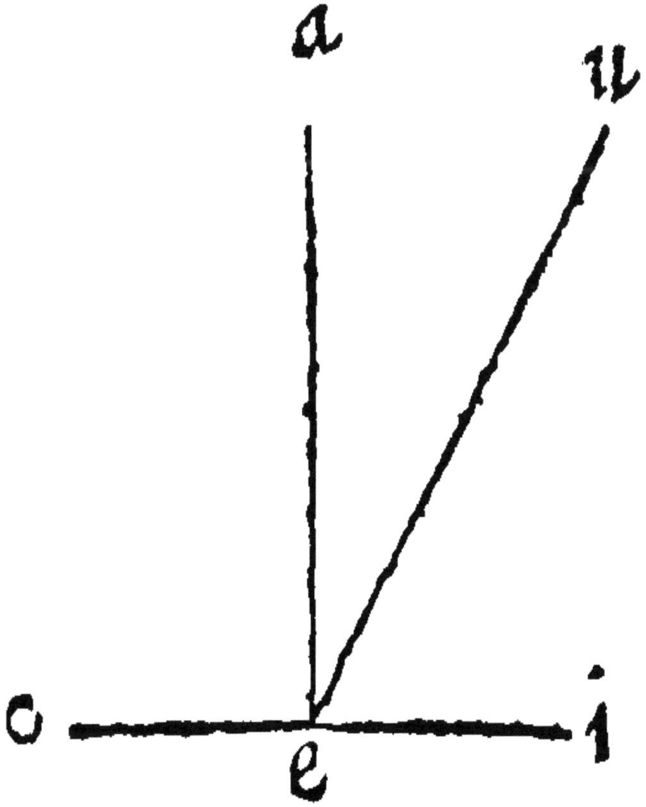

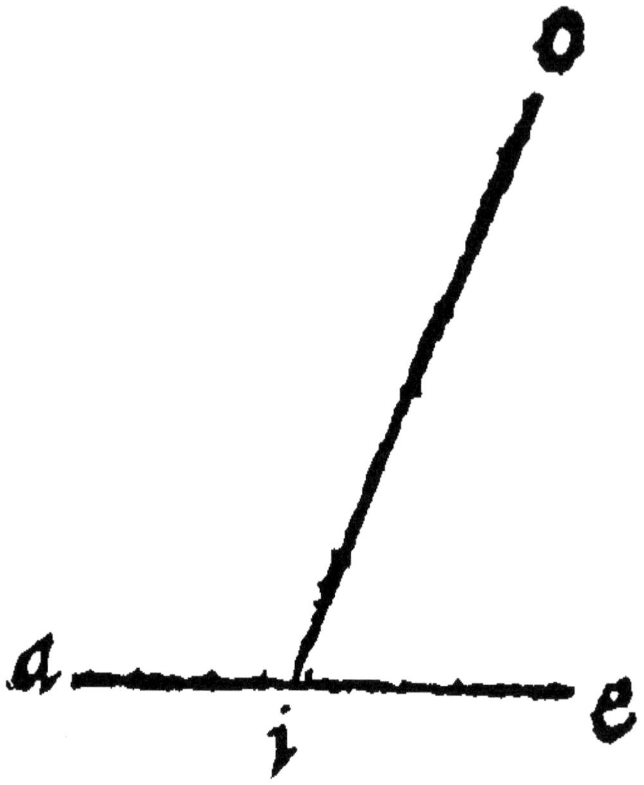

14. If a right line do stand upon a right line, it maketh the angles on each side equall to two right angles: and contrariwise out of the 13. and 14. p j.

For two such angles doe occupy or fill the same place that two right angles doe: Therefore they are equall to them by the 11. e j. If the insisting line be perpendicular unto that underneath it, it then shall make 2. right angles, by the 13. e. If it bee not perpendicular, & do make two oblique angles, as here *aio.* and *oie.* are yet shall they occupy the same place that two right angles doe: And therefore they are equall to two right angles, by the same.

The converse is forced by an argument *ab impossibli*, or *ab absurdo*, from the absurdity which otherwise would follow of it: For the part must otherwise needes bee equall to the whole. Let therefore the insisting or standing line which maketh two angles *aeo.* and *aeu.* on each side equall to two right angles, be *ae.* I say that *oe.* and *ei.* are but one right line. Otherwise let *oe.* bee continued unto *u.* by the 6. e. Now by the 14. e. or next former element, *aeo.* & *aeu.* are equall to two right angles; To which also *oea.* & *aei.* are equall by the grant: Let *aeo.* the common angle be taken away: then shall there be left *aeu.* equall to *aei.* the part to the whole, which is absurd and impossible. Herehence is it certaine that the two right lines *oe,* and *ei,* are in deede but one continuall right line.

And

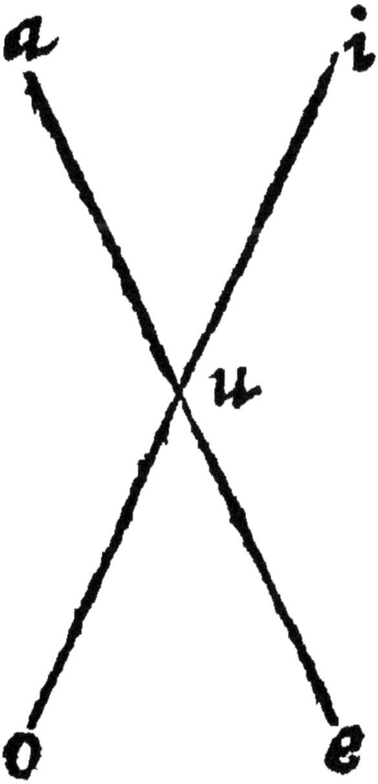

15. If two right lines doe cut one another, they doe make the angles at the top equall and all equall to foure right angles. 15. p j.

Anguli ad verticem, Angles at the top or head, are called Verticall angles which have their toppes meeting in the same point. The Demonstration is: Because the lines cutting one another, are either perpendiculars, and then all right angles are equall as heere: Or else they are oblique, and then also are the verticalls equall, as are *aui*, and *oue*: And againe, *auo*, and *iue*. Now *aui*, and *oue*, are equall, because by the 14. e. with *auo*, the common angle, they are equall to two right angles: And therefore they are equall betweene themselves. Wherefore *auo*, the said common angle beeing taken away, they are equall one to another.

And

16. If two right lines cut with one right line, doe make the inner angles on the same side greater then two right angles, those on the other side against them shall be lesser then two right angles.

As here, if *auy*, and *uyi*, bee greater then two right angles *euy*, and *uyo*, shall bee lesser then two right angles.

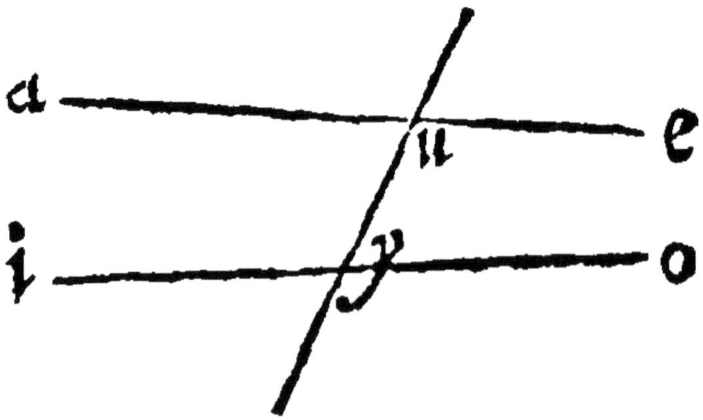

17. If from a point assigned of an infinite right line given, two equall parts be on each side cut off: and then from the points of those sections two equall circles doe meete, a right line drawne from their meeting unto the point assigned, shall bee perpendicular unto the line given. 11. p j.

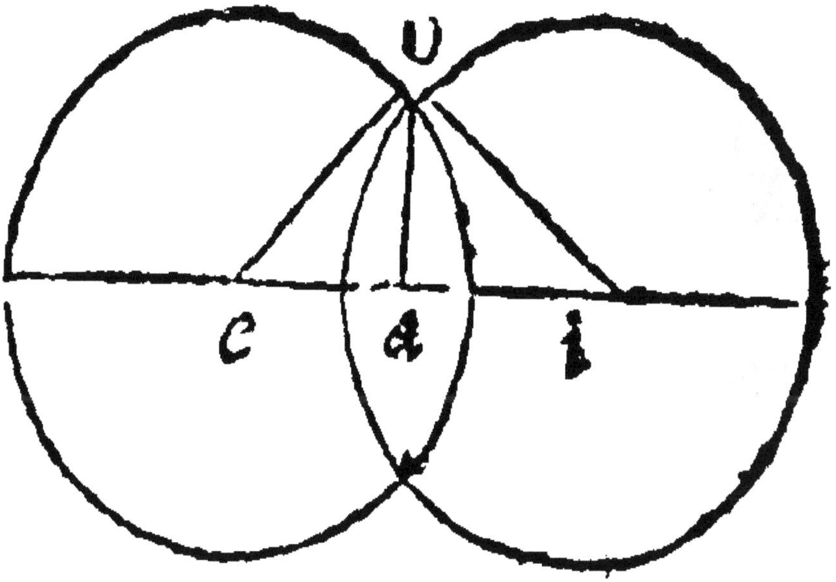

As let *a*, be the point assigned of the infinite line given: and from that on each side, by the 7. e. cut off equall portions *ae*, and *ai*, Then let two equall peripheries from the points *e*, and *i*, meete, as in *o*, I say that a right line drawne from *o*, the point of the meeting of the peripheries. unto *a*. the point given, shalbe perpendicular upon the line given. For drawing the right lines *oe*, & *oi*, the two angles *eao*, and *iao*, on each side, equicrurall by the construction of equall segments on each side, and *oa*, the common side, are equall in base by the 9. e. And therefore the angles themselves shall be equall, by the 7. e iij. and therefore againe, seeing that *ao*, doth lie equall betweene the parts *ea*, and *ia*, it is by the 13. e ij. perpendicular upon it.

18. If a part of an infinite right line, bee by a periphery for a point given without, cut off a right line from the said point, cutting in two the said part, shall bee perpendicular upon the line given. 12. p j.

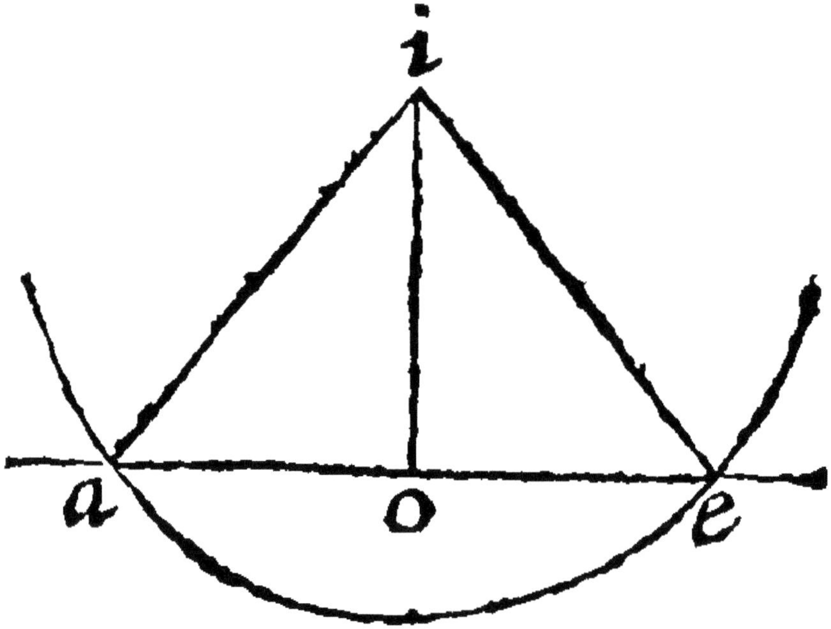

Of an infinite right line given, let the part cut off by a periphery of an externall center be *ae*: And then let *io*, cut the said part into two parts by the 12. e. I say that *io* is perpendicular unto the said infinite right line. For it standeth upright, and maketh *aoi*, and *eoi*, equall angles, for the same cause, whereby the next former perpendicular was demonstrated.

19. If two right lines drawne at length in the same plaine doe never meete, they are parallells. è 35. d j.

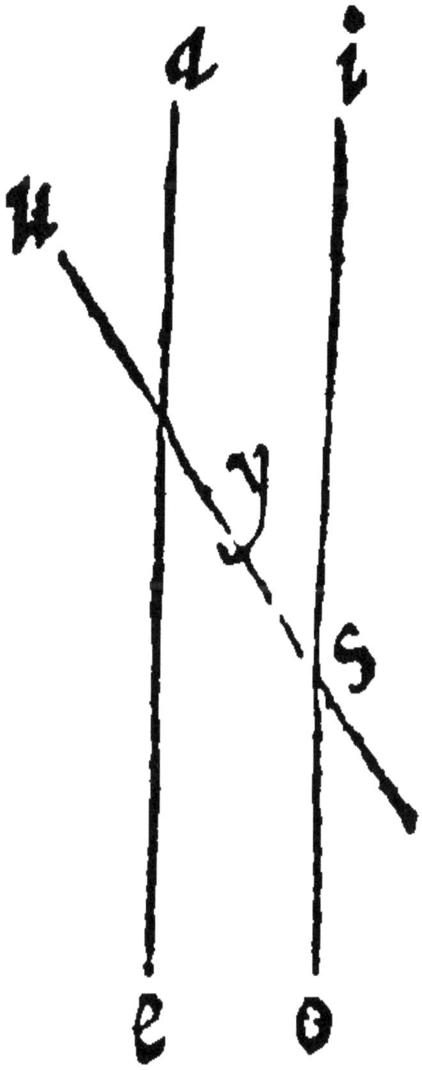

Thus much of the Perpendicularity of plaine right lines: *Parallelissmus*, or their parallell equality doth follow. *Euclid* did justly require these lines so drawne to be granted paralels: for then shall they be alwayes equally distant, as here *ae.* and *io*.

Therefore

20. If an infinite right line doe cut one of the infinite right parallell lines, it shall also cut the other.

As in the same example *uy.* cutting *ae.* it shall also cut *io.* Otherwise, if it should not cut it, it should be parallell unto it, by the 18 e. And that against the grant.

21. If right lines cut with a right line be pararellells, they doe make the inner angles on the same side equall to two right angles: And also the alterne angles equall betweene themselves: And the outter, to the inner opposite to it: And contrariwise, 29, 28, 27. p 1.

The paralillesme, or parallell-equality of right lines cut with a right line, concludeth a threefold equality of angles: And the same is againe of each of them concluded. Therefore in this one element there are sixe things taught; all which are manifest if a perpendicular, doe fall upon two parallell lines. The first sort of angles are in their owne words plainely enough expressed. But the word *Alternum*, alterne [or *alternate, H.*] here, as *Proclus* saith, signifieth situation, which in Arithmeticke signified proportion, when the antecedent was compared to the consequent; notwithstanding the metaphor answereth fitly. For as an acute angle is unto his successively following obtuse; So on the other part is the acute unto his successively following obtuse: Therefore alternly, As the acute unto the acute: so is the obtuse, unto the obtuse. But the outter and inner are opposite, of the which the one is without the parallels; the other is within on the same part not successively; but upon the same right line the third from the outer.

The cause of this threefold propriety is from the perpendicular or plumbline, which falling upon the parallells breedeth and discovereth all this variety: As here they are right angles which are the inner on the same part or side: Item, the alterne angles: Item the inner and the outter: And therefore they are equall, both, I meane, the two inner to two right angles: and the alterne angles between themselvs: And the outter to the inner opposite to it.

If so be that the cutting line be oblique, that is, fall not upon them plumbe or perpendicularly, the same shall on the contrary befall the parallels. For by that same obliquation or slanting, the right lines remaining and the angles unaltered, in like manner both one of the inner, to wit, *euy*, is made obtuse, the other, to wit, *uyo*, is made acute: And the alterne angles are made acute and obtuse: As also the outter and inner opposite are likewise made acute and obtuse.

If any man shall notwithstanding say, That the inner angles are unequall to two right angles: By the same argument may he say (saith *Ptolome* in *Proclus*) That on each side they be both greater than two right angles, and also lesser: As in the parallel right lines *ae* and *io*, cut with the right line *uy*, if thou shalt say that *auy* and *iyu*, are greater then two right angles, the angles on the other side, by the 16 e, shall be lesser then two right angles, which selfesame notwithstanding are also, by the gainesayers graunt, greater then two right angles, which is impossible.

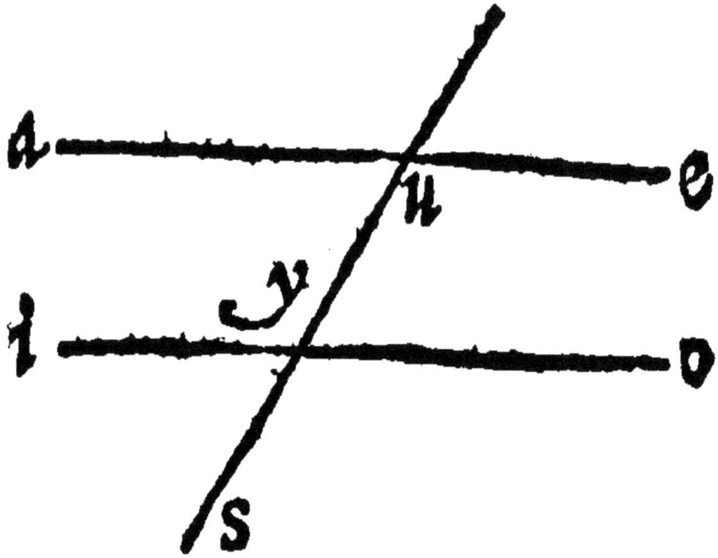

The same impossibility shall be concluded, if they shall be sayd, to be lesser than two right angles.

The second and third parts may be concluded out of the first. The second is thus: Twise two angles are equall to two right angles *oyu*, and *euy*, by the former part: Item, *auy*, and *euy*, by the 14 e. Therefore they are equall betweene themselves. Now from the equall, Take away *euy*, the common angle, And the remainders, the alterne angles, at *u*, and *y* shall be least equall.

The third is thus: The angles *euy*, and *oys*, are equall to the same *uyi*, by the second propriety, and by the 15 e. Therefore they are equall betweene themselves.

The converse of the first is here also the more manifest by that light of the common perpendicular, And if any man shall thinke, That although the two inner angles be equall to two right angles, yet the right may meete, as if those equall angles were right angles, as here; it must needes be that two right lines divided by a common perpendicular, should both leane, the one this way, the other that way, or at least one of them, contrary to the 13 e ij.

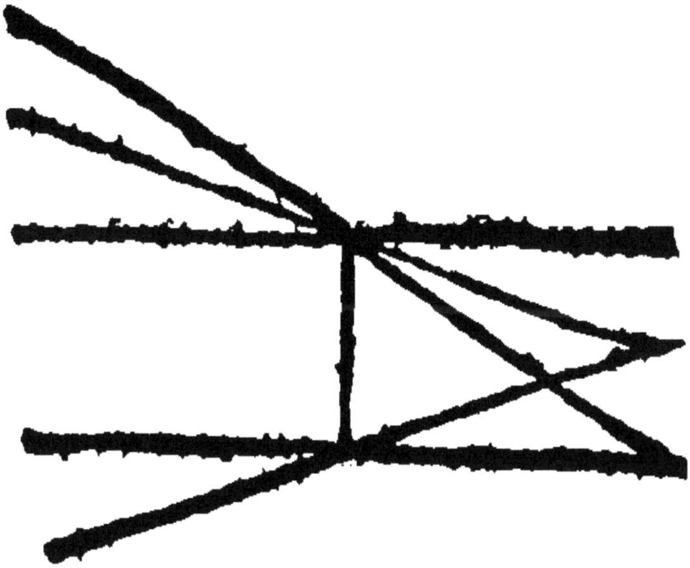

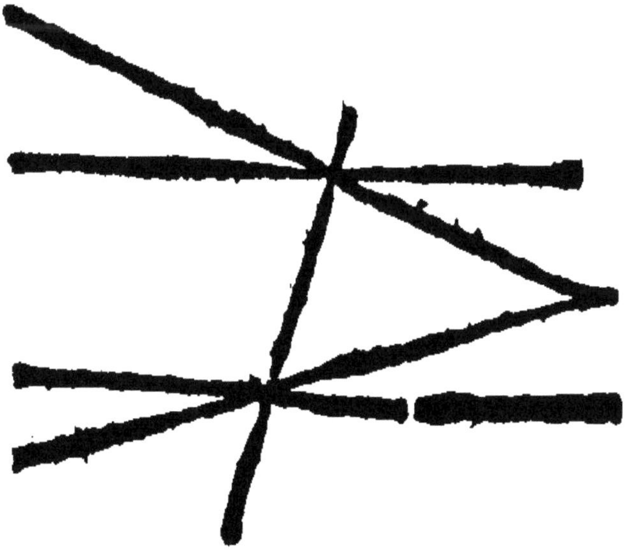

If they be oblique angles, as here, the lines one slanting or obliquely crossing one another, the angles on one side will grow lesse, on the other side greater. Therefore they would not be equall to two right angles, against the graunt.

From hence the second and third parts may be concluded. The second is thus: The alterne angles at *u* and *y*, are equall to the foresayd inner angles, by the 14 e: Because both of them are equall to the two right angles: And so by the first part the second is concluded.

The third is therefore by the second demonstrated, because the outter *oys*, is equall to the verticall or opposite angle at the top, by the 15 e. Therefore seeing the outter and inner opposite are equall, the alterne also are equall.

Wherefore as *Parallelismus*, parallell-equality argueth a three-fold equality of angles: So the threefold equality of angles doth argue the same parallel-equality.

Therefore,

22. If right lines knit together with a right line, doe make the inner angles on the same side lesser than two right Angles, they being on that side drawne out at length, will meete.

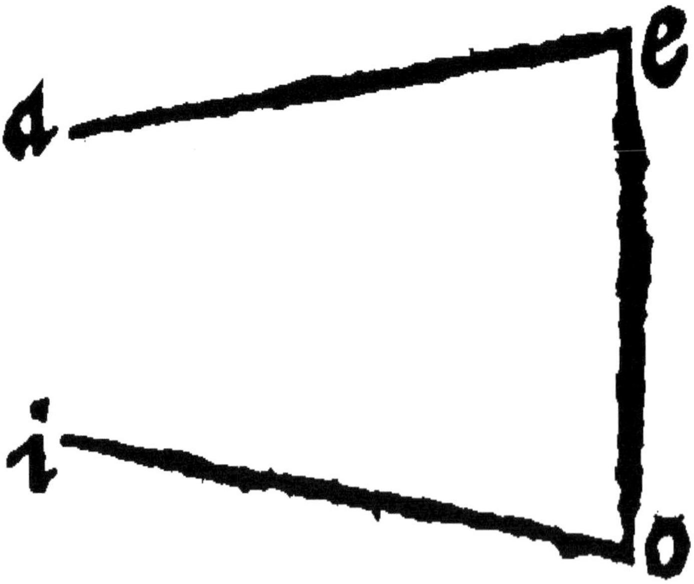

As here *ae*, and *io*, knit together with *eo*, doe make two angles *aeo*, and *ioe*, lesser than two right angles: They shall therefore, I say, meete if they be continued out that wayward. The assumption and complexion is out of the 21 e, of right lines in the same plaine. If right lines cut with a right line be parallels, they doe make the inner angles on the same part equall to two right-angles. Therefore if they doe not make them equall, but lesser, they shall not be parallel, but shall meete.

And

23. A right line knitting together parallell right lines, is in the same plaine with them. 7 p xj.

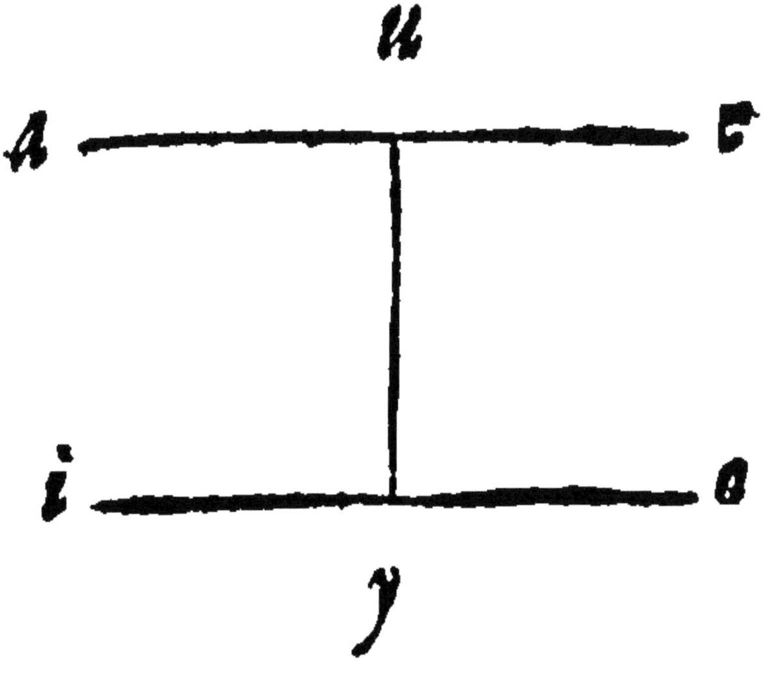

As here *uy*, knitting or joyning together the two parallels *ae*, and *io*, is in the same plaine with them as is manifest by the 8 e.

And

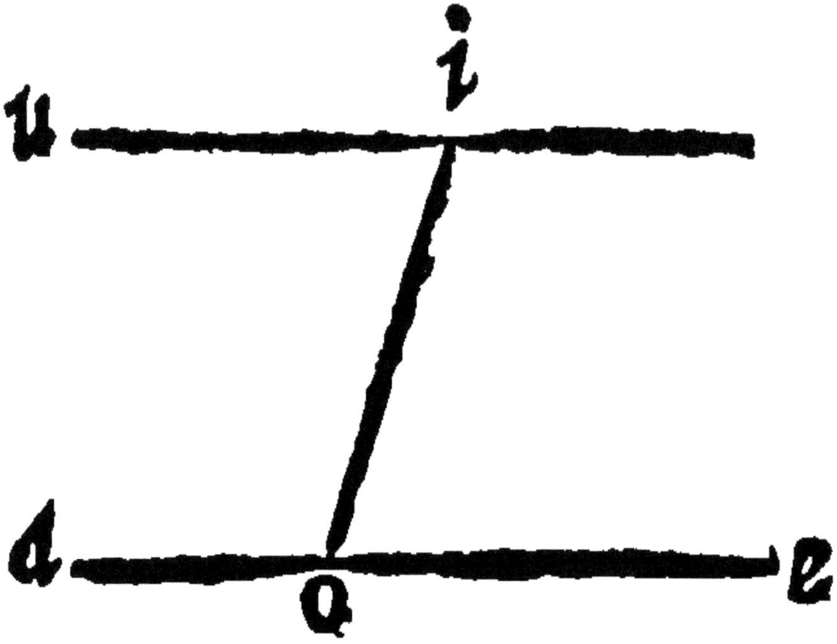

24. *If a right line from a point given doe with a right line given make an angle, the other shanke of the angle equalled and alterne to the angle made, shall be parallell unto the assigned right line. 31 p j.*

As let the assigned right line be *ae*: And the point given, let it be *i*. From which the right line, making with the assigned *ae*, the angle, *ioe*, let it be *io*: To the which at *i*, let the alterne angle *oiu,* be made equall: The right line *ui*, which is the other shanke, is parallel to the assigned *ae*.

An angle, I confesse, may bee made equall by the first propriety: And so indeed commonly the Architects and Carpenters doe make it, by erecting of a perpendicular. It may also againe in like manner be made by the outter angle: Any man may at his pleasure use which hee shall thinke good: But that here taught we take to be the best.

And

25. The angles of shanks alternly parallell, are equall. Or Thus, The angles whose alternate feete are parallells, are equall. H.

This consectary is drawne out of the third property of the 21 e. The thing manifest in the example following, by drawing out, or continuing the other shanke of the inner angle. But *Lazarus Schonerus* it seemeth doth thinke the adverbe *alterne*, (*alternely* or *alternately*) to be more then needeth: And therefore he delivereth it thus: The angles of parallel shankes are equall.

And

26. If parallels doe bound parallels, the opposite lines are equall è 34 p. j. Or thus: If parallels doe inclose parallels, the opposite parallels are equall. H.

Otherwise they should not be parallell. This is understood by the perpendiculars, knitting them together, which by the definition are equall betweene two parallells: And if of perpendiculars they bee made oblique, they shall notwithstanding remaine equall, onely the corners will be changed.

And

27. If right lines doe joyntly bound on the same side equall and parallell lines, they are also equall and parallell.

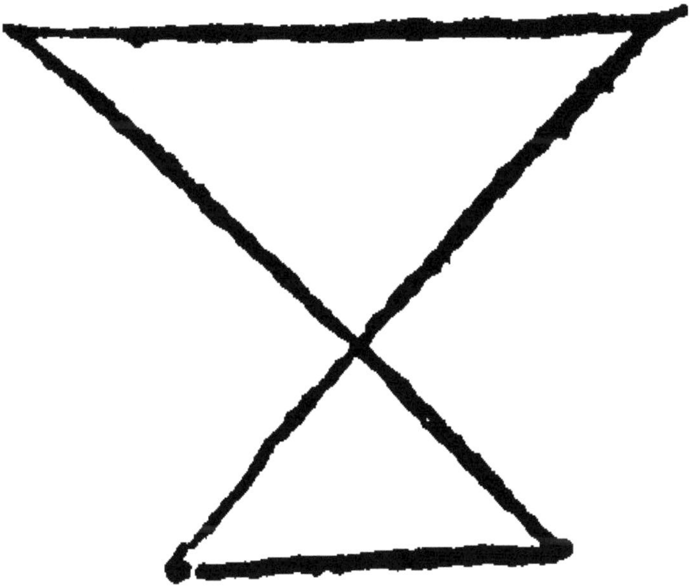

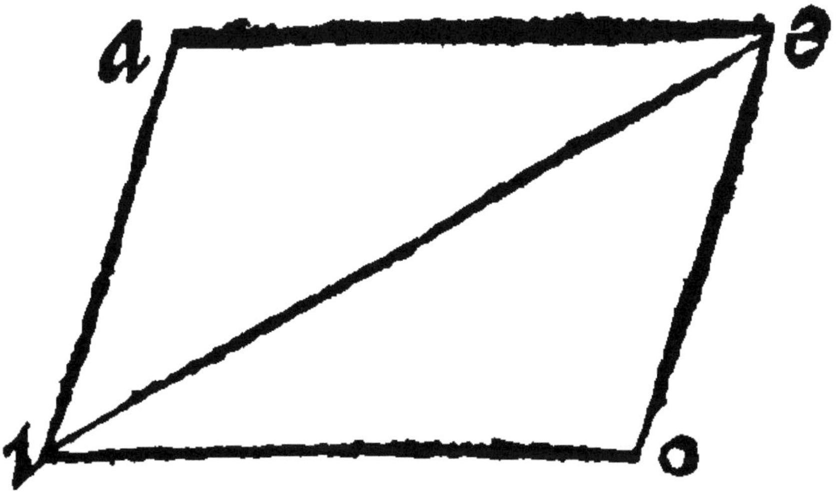

This element might have beene concluded out of the next precedent: But it may also be learned out of those which went before. As let *ae*, and *io*, equall parallels be bounded joyntly of *ai*, and *eo*: and let *ei* be drawn. Here because the right line *ei* falleth upon the parallels *ae*, and *io*, the alterne angles *aei* and *eio*, are equall, by the 21 e. And they are equall in shankes *ae*, and *io*, by the grants, and *ei*, is the common shanke: Therefore they are also equall in base *ai*, and *eo*, by the 7 e iij. This is the first: Then by 21 e, the alterne angles *eia*, and *ieo*, are equall betweene themselves: And those are made by *ai* and *eo*, cut by the right line *ei*: Therefore they are parallell; which was the second.

On the same part or side it is sayd, least any man might understand right lines knit together by opposite bounds as here.

28. If right lines be cut joyntly by many parallell right lines, the segments betweene those lines shall bee proportionall one to another, out of the 2 p vj and 17 p xj.

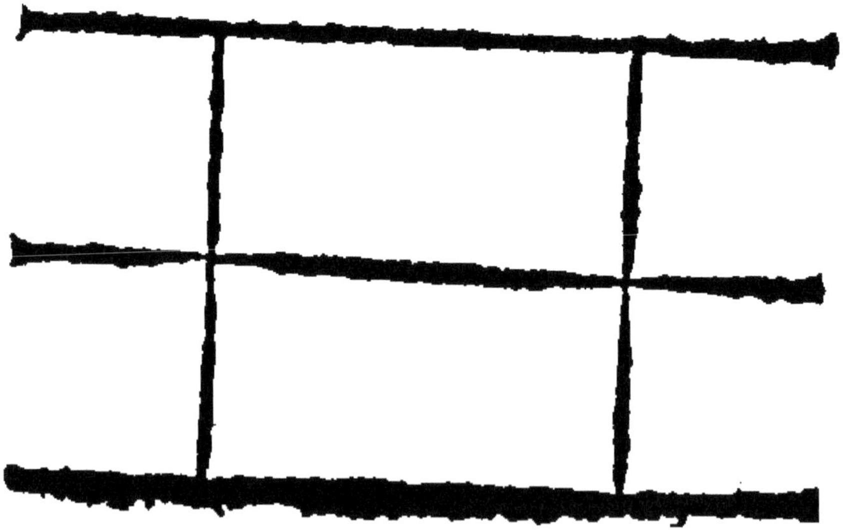

Thus much of the Perpendicle, and parallell equality of plaine right lines: Their Proportion is the last thing to be considered of them.

The truth of this element dependeth upon the nature of the parallells: And that throughout all kindes of equality and inequality, both greater and lesser. For if the lines thus cut be perpendiculars, the portions intercepted betweene the two parallels shall be equall: for common perpendiculars doe make parallell equality, as before hath beene taught, and here thou seest.

If the lines cut be not parallels, but doe leane one toward another, the portions cut or intercepted betweene them will not be equall, yet shall they be proportionall one to another. And looke how much greater the line thus cut is: so much greater shall the intersegments or portions intercepted be. And contrariwise, Looke how much lesse: so much lesser shall they be.

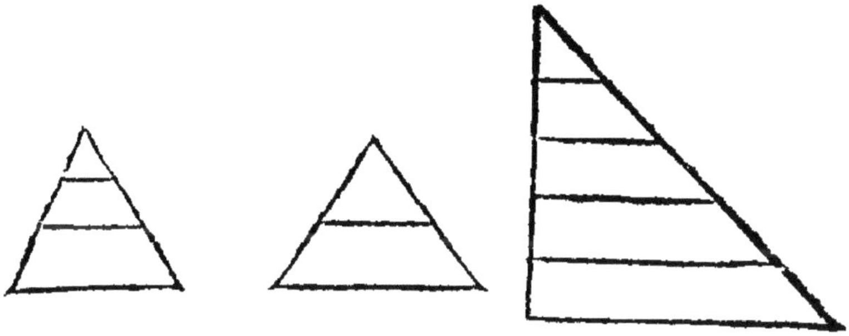

The third parallell in the toppe is not expressed, yet must it be understood.

This element is very fruitfull: For from hence doe arise and issue, First the manner of cutting a line according to any rate or proportion assigned: And then the invention or way to finde out both the third and fourth proportionalls.

29. If a right line making an angle with another right line, be cut according to any reason [or proportion] assigned, parallels drawne from the ends of the segments, unto the end of the sayd right line given and unto some contingent point in the same, shall cut the line given according to the reason given.

Schoner hath altered this Consectary, and delivereth it thus: *If a right line making an angle with a right line given, and knit unto it with a base, be cut according to any rate assigned, a parallell to the base from the ends of the segments, shall cut the line given according to the rate assigned.* 9 and 10 p vj.

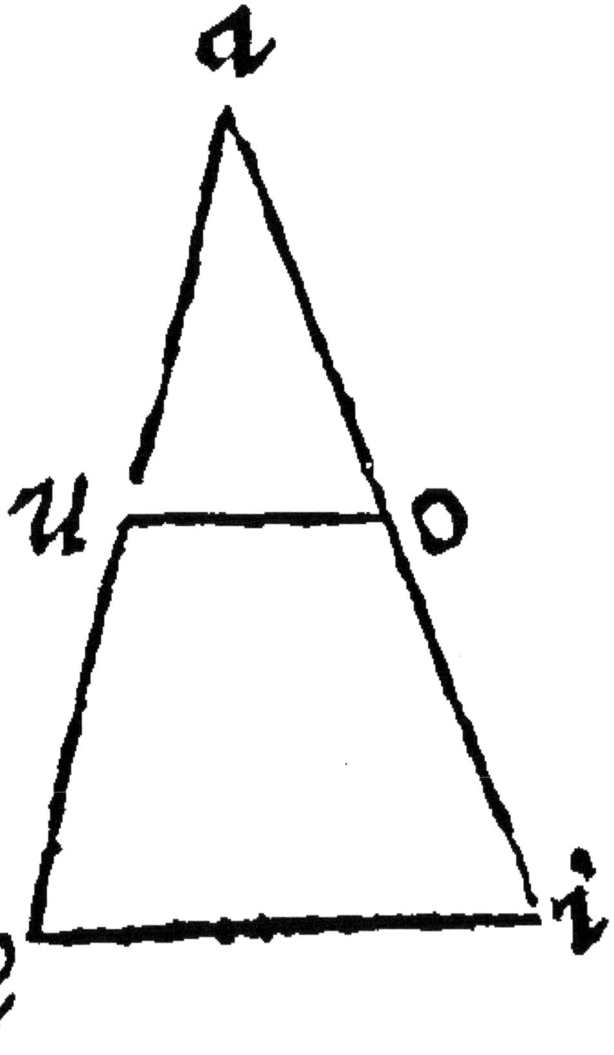

Punctum contingens, A contingent point, that is falling or lighting in some place at al adventurs, not given or assigned.

This is a marvelous generall consectary, serving indifferently for any manner of section of a right line, whether it be to be cut into two parts, or three parts, or into as many parts, as you shall thinke good, or generally after what manner of way soever thou shalt command or desire a line to be cut or divided.

Let the assigned Right line to be cut into two equall parts be *ae*. And the right line making an angle with it, let it be the infinite right line *ai.* Let *ao*, one portion thereof be cut off. And then by the 7 e, let *oi*, another part thereof be taken equall to it. And lastly, by the 24 e, draw parallels from the points *i*, and *o*, unto *e*, the end of the line given, and to *u*; a contingent point therein. Now the third parallell is understood by the point *a*, neither is it necessary that it should be expressed. Therefore the line *ae*, by the 28, is cut into two equall portions: And as *ao*, is to *oi*: So is *au*, to *ue*. But *ao*, and *oi*, are halfe parts. Therefore *au*, and *ue*, are also halfe parts.

And here also is the 12 e comprehended, although not in the same kinde of argument, yet in effect the same. But that argument was indeed shorter, although this be more generall.

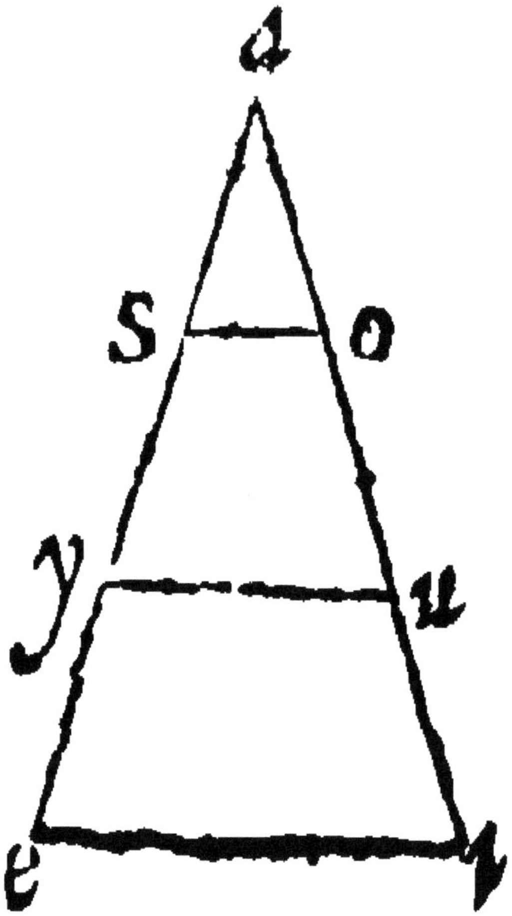

Now let *ae* be cut into three parts, of which the first let it bee the halfe of the second: And the second, the halfe of the third: And the conterminall or right line making an angle with the sayd assigned line, let it be cut one part *ao*: Then double this in *ou*: Lastly let *ui* be taken double to *ou*, and let the whole diagramme be made up with three parallels *ie*, *uy*, and *os*, The fourth parallell in the toppe, as afore-sayd, shall be understood. Therefore that section which was made in the conterminall line, by the 28 e, shall be in the assigned line: Because the segments or portions intercepted are betweene the parallels.

And

30. If two right lines given, making an angle, be continued, the first equally to the second, the second infinitly, parallels drawne from the ends of the first continuation, unto the beginning of the second, and some contingent point in the same, shall intercept betweene them the third proportionall. 11. p vj.

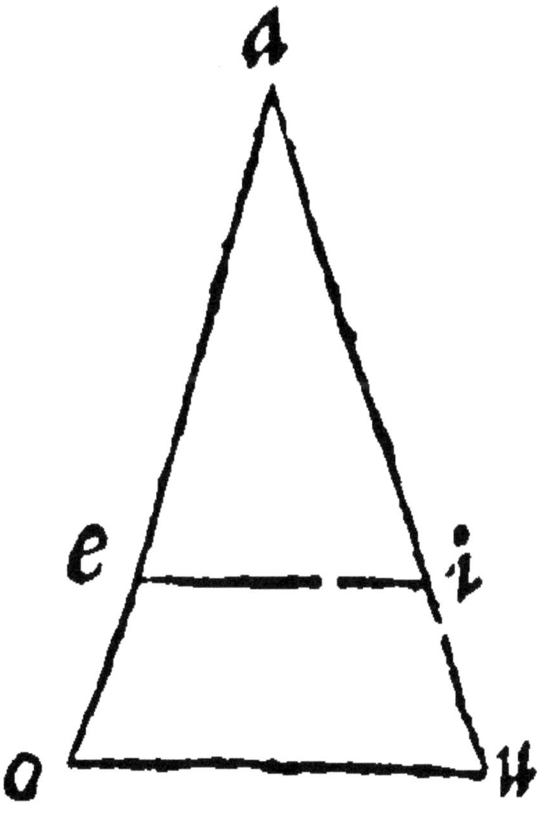

Let the right lines given, making an angle, be *ae*, and *ai*: and *ae*, the first, let it be continued equally to the same *ai*, and the same *ai*, let it be drawne out infinitly: Then the parallels *ei*, and *ou*, drawne from the ends of the first continuation, unto *i*, the beginning of the second: and *u*, a contingent point in the second, doe cut off *iu*, the third proportionall sought. For by the 28 e, as *ae*, is unto *eo*, so is *ai*, unto *iu*.

And

31. If of three right lines given, the first and the third making an angle be continued, the first equally to the second, and the third infinitly; parallels drawne from the ends of the first continuation, unto the beginning of the second, and some contingent point, the same shall intercept betweene them the fourth proportionall. 12. p vj.

Let the lines given be these: The first *ae*, the second *ei*, the third *ao*, and let the whole diagramme be made up according to the prescript of the consectary. Here by 28. e, as *ae*, is to *ei* so is *ao*, to *ou*. Thus farre *Ramus*.

Lazarus Schonerus, who, about some 25. yeares since, did revise and augment this worke of our Authour, hath not onely altered the forme of these two next precedent consectaries: but he hath also changed their order, and that which is here the second, is in his edition the third: and the third here, is in him the second. And to the former declaration of them, hee addeth these words: From hence, having three lines given, is the invention of the fourth proportionall; and out of that, having two lines given, ariseth the invention of the third proportionall.

2 *Having three right lines given, if the first and the third making an angle, and knit together with a base, be continued, the first equally to the second; the third infinitly; a parallel from the end of the second, unto the continuation of the third, shall intercept the fourth proportionall. 12. p vj.*

The Diagramme, and demonstration is the same with our 31. e or 3 c of *Ramus*.

3 *If two right lines given making an angle, and knit together with a base, be continued, the first equally to the second, the second infinitly; a parallell to the base from the end of the first continuation unto the second, shall intercept the third proportionall. 11. p vj.*

The Diagramme here also, and demonstration is in all respects the same with our 30 e, or 2 c of *Ramus*.

Thus farre *Ramus*: And here by the judgement of the learned *Finkius*, two elements of *Ptolomey* are to be adjoyned.

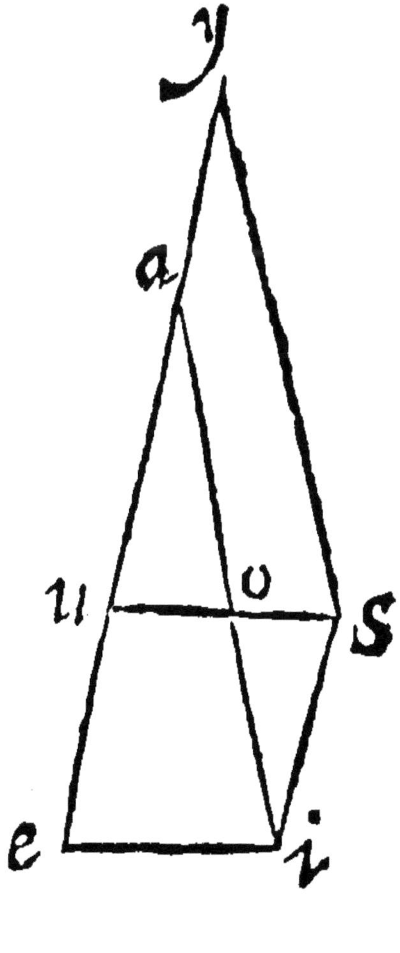

32 If two right lines cutting one another, be againe cut with many parallels, the parallels are proportionall unto their next segments.

It is a consectary out of the 28 e. For let the right lines *ae.* and *ai*, cut one another at *a*, and let two parallell lines *uo*, and *ei*, cut them; I say, as *au*, is to *uo*, so *ae*, is to *ei*. For from the end *i*, let *is*, be erected parallell to *ae*, and let *uo*, be drawne out untill it doe meete with it. Then from the end *s*, let *sy*, be made parallell to *ai*: and lastly, let *ea*, be drawne out, untill it doe meete with it. Here now *ay*, shall be equall to the right line *is*, that is, by the 26. e, to *ue*: and at length, by the 28. e, as *ua*, is to *uo*; so is *ay*, that is, *ue*, to *os*. Therefore, by composition or addition of proportions, as *ua*, is unto *uo*, so *ua*, and *ue*, shall be unto *uo*, and *os*, that is, *ei*, by the 27. e.

The same demonstration shall serve, if the lines do crosse one another, or doe vertically cut one another, as in the same diagramme appeareth. For if the assigned *ai*, and *us*, doe cut one another vertically in *o*, let them be cut with the parallels *au*, and *si*: the precedent fabricke or figure being made up, it shall be by 28. e. as *au*, is unto *ao*, the segment next unto it: so *ay*, that is, *is*, shall be unto *oi*, his next segment.

The 28. e teacheth how to finde out the third and fourth proportionall: This affordeth us a meanes how to find out the continually meane proportionall single or double.

Therefore;

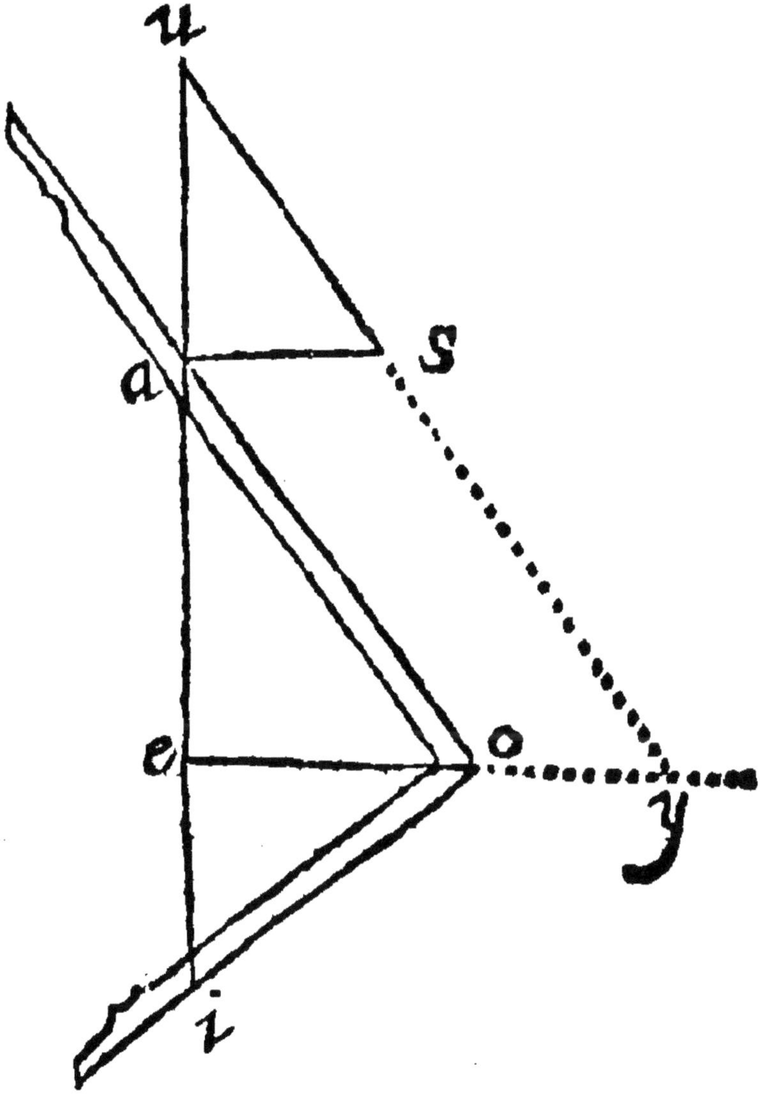

33. If two right lines given be continued into one, a perpendicular from the point of continuation unto the angle of the squire, including the continued line with the continuation, is the meane proportionall betweene the two right lines given.

A squire (*Norma, Gnomon,* or *Canon*) is an instrument consisting of two shankes, including a right angle. Of this we heard before at the 13. e. By the meanes of this a meane proportionall unto two lines given is easily found: whereupon it may also be called a *Mesolabium*, or *Mesographus simplex*, or single meane finder.

Let the two right lines given, be *ae*, and *ei*. The meane proportional between these two is desired. For the finding of which, let it be granted that as *ae*, is to *eo*, so *eo*, is to *ei*: therefore let *ae*, be continued or drawne out unto *i*, so that *ei*, be equall to the other given. Then from *e*, the point of the continuation, let *eo*, an infinite perpendicular be erected. Now about this perpendicular, up and downe, this way and that way, let the squire *ao*, be moved, so that with his angle it may comprehend at *eo*, and with his shanks it may include the whole right line *ai*. I say that *eo*, the segment of the perpendicular, is the meane proportionall between *ae*, and *ei*, the two lines given. For let *ea*, be continued or drawne out into *u*, so that the continuation *au*, be equall unto *eo*: and unto *a*, the point of the continuation, let the angle *uas*, be made equall, and equicrurall to the angle *oei*, that is, let the shanke *as*, be made equall to the shanke *ei*. Wherefore knitting *u*, and *s*, together, the right lines *us*, and *oi*, shall be equall; and the angles *eoi*, *aus*, by the 7. e iij. And by the 21. e, the lines *sa*, and *oe*, are parallell: and the angle *sao*, is equall to the angle *aoe*. But the angles *sae*, and *aoi*, are right angles by the Fabricke and by the grant; and therefore they are equall, by the 14. e iij. Wherefore the other angles *oae*, and *eoi*, that is, *sua*, are equall. And therefore by the 21. e. *us*, and *ao* are parallell; and *us*, and *eo*, continued shall meete, as here in *y*: and by the 26. e. *oy*, and *as* are equall. Now, by the 32. e. as *ue*, is to *ua*, so is *ey*, to *as*. Therefore by subduction or subtraction of proportions, as *ea*, is to *ua*, so is *eo*, that is, *ua*, to *oy*, that is *as*.

And

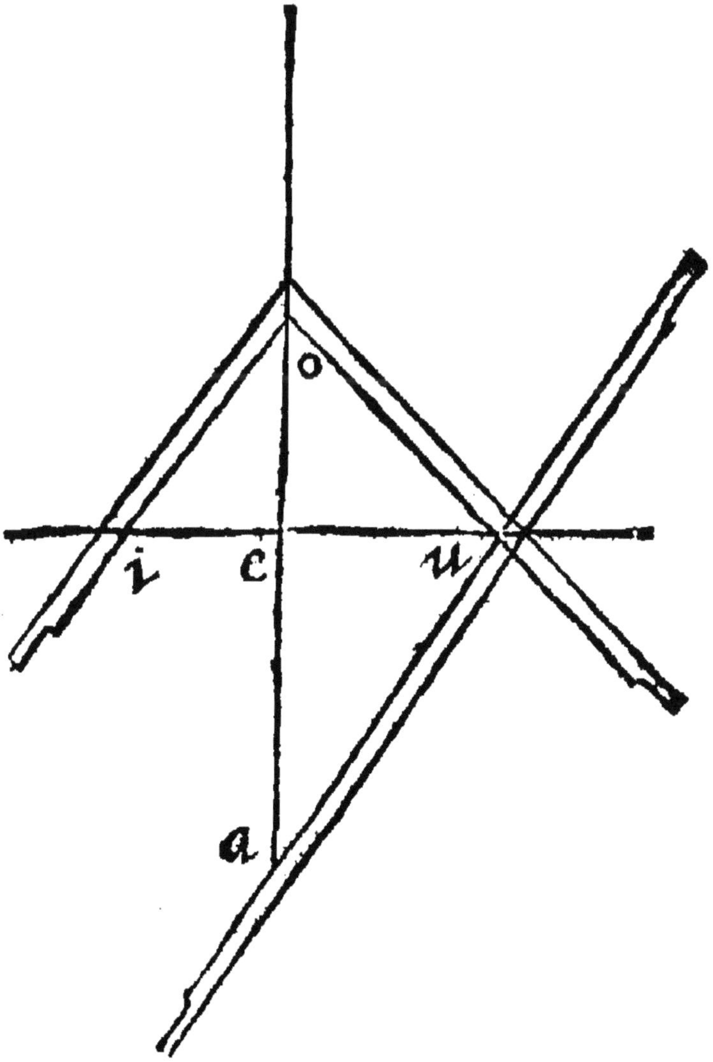

> *34 If two assigned right lines joyned together by their ends rightanglewise, be continued vertically; a square falling with one of his shankes, and another to it parallell and moveable upon the ends of the assigned, with the angles upon the continued lines, shall cut betweene them from the continued two meanes continually proportionall to the assigned.*

The former consectary was of a single mesolabium; this is of a double, whose use in making of solids, to this or that bignesse desired is notable.

Let the two lines assigned be *ae*, and *ei*; and let there be two meane right lines, continually proportionall betweene them sought, to wit, that may be as *ae*, is unto one of the lines found; so the same may be unto the second line found. And as that is unto this, so this may be unto *ei*. Let therefore *ae*, and *ei*, be joyned rightanglewise by their ends at *e*; and let them be infinite continued, but vertically, that is, from that their meeting from the lines ward, from *ei*, towards *u*, but *ae*, towards *o*. Now for the rest, the construction; it was *Plato's Mesographus*; to wit, a squire with the opposits parallell. One of his sides *au*, moueable, or to be done up and downe, by an hollow riglet in the side adjoyning. Therefore thou shalt make thee a Mesographus, if unto the squire thou doe adde one moveable side, but so that how so ever it be moved, it be still parallell unto the opposite side [which is nothing else, but as it were a double squire, if this squire be applied unto it; and indeed what is done by this instrument, may also be done by two squires, as hereafter shall be shewed.] And so long and oft must the moveable side be moved up and downe, untill with the opposite side it containe or touch the ends of the assigned, but the angles must fall precisely upon the continued lines: The right lines from the point of the continuation, unto the corners of the squire, are the two meane proportionals sought.

As if of the Mesographus *auoi*, the moveable side be *au*; thus thou shalt move up and downe, untill the angles *u*, and *o*, doe hit just upon the infinite lines; and joyntly at the same instant *ua*, and *oi*, may touch the ends of the assigned *a*, and *i*. By the former consectary it shall be as *ei*, is to *eo*, so *eo*, shall be unto *eu*: and as *eo*, is to *eu*, so shall *eu*, be unto *ea*.

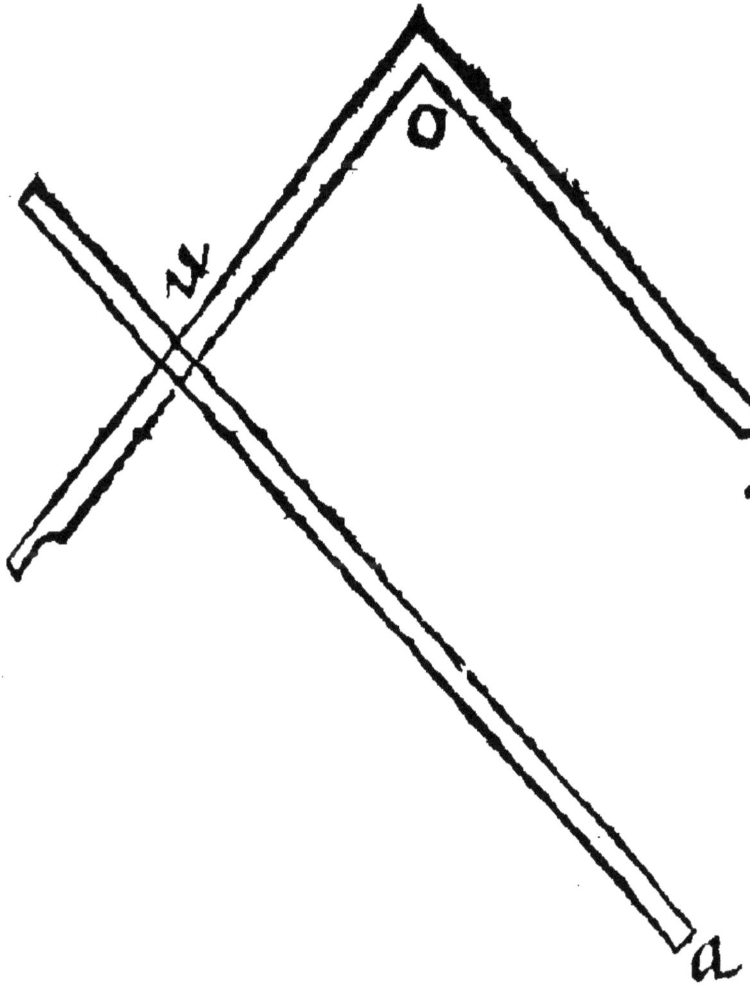

And thus wee have the composition and use, both of the single and double Mesolabium.

35. If of foure right lines, two doe make an angle, the other reflected or turned backe upon themselves, from the ends of these, doe cut the former; the reason of the one unto his owne segment, or of the segments betweene themselves, is made of the reason of the so joyntly bounded, that the first of the makers be joyntly bounded with the beginning of the antecedent made; the second of this consequent joyntly bounded with the end; doe end in the end of the consequent made.

Ptolomey hath two speciall examples of this *Theorem*: to those *Theon* addeth other foure.

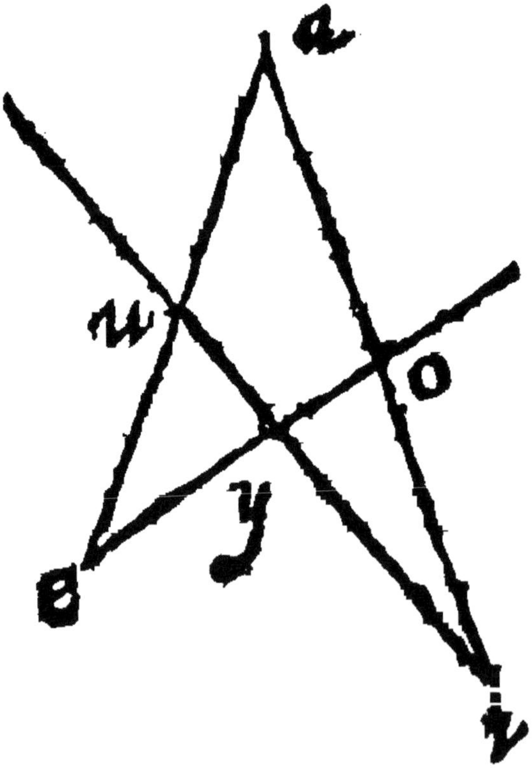

Let therefore the two right lines be *ae*, and *ai*: and from the ends of these other two reflected, be *iu*, and *eo*, cutting themselves in *y*; and the two former in *u*, and *o*. The reason of the particular right lines made shall be as the draught following doth manifest. In which the antecedents of the makers are in the upper place: the consequents are set under neathe their owne antecedents.

The I. is *Ptolemeys* and *Theons* I.			
The makers:		*The reason made.*	
iu,	*ye.*		
uy,	*eo,*	*ia,*	*ao.*
The II. is *Theons* VI.			
au,	*ey.*		
ue,	*yo,*	*ai,*	*io.*
The III. is *Theons* III.			
ea,	*ui.*		
au,	*iy,*	*eo,*	*oy.*
The IIII. is *Theons* II.			
oa,	*iu.*		
ai,	*uy,*	*oe,*	*ey.*
The V. is *Ptolemys*, II. *Theons* IIII.			
iy,	*ue.*		
yu,	*ea,*	*io,*	*ao.*
The VI. is *Theons* V.			
eu,	*ai.*		
ua,	*io,*	*ey,*	*yo.*

The businesse is the same in the two other, whether you doe crosse the bounds or invert them.

Here for demonstrations sake we crave no more, but that from the beginning of an antecedent made a parallell be drawne to the second consequent of the makers, unto one of the assigned infinitely continued: then the multiplied proportions shall be,

The Antecedent, the Consequent; the Antecedent, the Consequent of the second of the makers; every way the reason or rate is of Equallity.

The Antecedent; the Consequent of the first of the makers; the Parallell; the Antecedent of the second of the makers, by the 32. e. Therefore by multiplication of proportions, the reason of the Parallell, unto the Consequent of the second of the makers, that is, by the fabricke or construction, and the 32. e. the reason of the Antecedent of the Product, unto the Consequent, is made of the reason, &c. after the manner above written.

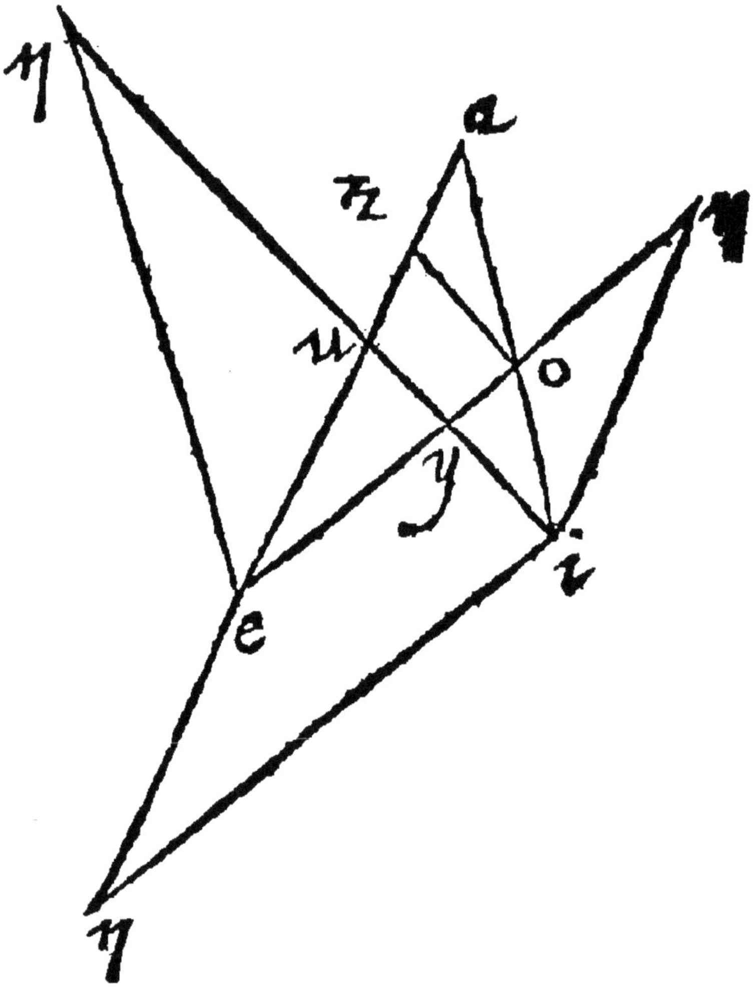

For examples sake, let the first speciall example be demonstrated. I say therefore, that the reason of *ia*, unto *ao*, is made of the reason of *iu*, unto *uy*, multiplied by the reason of *ye*, unto *eo*. For from the beginning of the Antecedent of the product, to wit, from the point *i*, let a line be drawne parallell to the right line *ey*, which shall meete with *ae*, continued or drawne out infinitely in *n*. Therefore, by the 32. e, as *ia*, is to *ao*: so is the parallell drawne to *eo*, the Consequent of the second of the makers. Therefore now the multiplied proportions are thus *iu, uy, in, ey*, by the 32. e: *ye, eo, ey, eo*. Therefore as the product of *iu*, by *ye*, is unto the product of *uy*, by *eo*: So *in*, is to *eo*, that is, *ia*, to *ao*.

So let the second of *Ptolemy* to be taught, which in our Table aforegoing is the fifth. I say therefore that the reason of *io*, unto *oa*; is made of the reason of *iy*, unto *yu*, and the reason of *ue*, unto *ea*. For now againe, from the beginning of the Antecedent of the Product *i*, let a line be drawne parallell unto *ea*, the Consequent of the second of the Makers, which shall meete with *eo*, drawne out at length, in *n*: therefore, by the 32. e. as *io*, is to *ao*; so is *en*, unto *ea*. Therefore now again the multiplied proportions are thus:

ue,	*ea*,	*ue*,	*ea*.
iy,	*yu*,	*en*,	*ue*;

by the 32. e. Therefore, by multiplication of proportions, the reason of *en*, unto *ea*, that is, of *io*, unto *oa*, is made of the reason of *iy*, unto *yu*, by the reason of *ue*, unto *ea*.

It shall not be amisse to teach the same in the examples of *Theon*. Let us take therefore the reason of the Reflex, unto the Segment; And of the segments betweene themselves; to wit, the 4. and 6. examples of our foresaid draught: I say therefore, that the reason of *oe*, unto *ey*, is made of the reason *oa*, unto *ai*, by the reason of *iu*, unto *uy*. For from the end *o*, to wit, from the beginning of the Antecedent of the product, let the right line *no*, be drawne parallell to *uy*. It shall be by the 32. e. as *oe*, is to *ey*: so the parallell *no*, shall be to *uy*: but the reason of *no*, unto *uy*, is made of the reason of *oa*, unto *ai*, and of *iu*, unto *uy*: for the multiplied proportions are,

iu,	*uy,*	*iu,*	*uy.*
oa,	*ai,*	*on,*	*iu.*

by the 32. e.

Againe, I say, that the reason of *ey*, unto *yo*, is compounded of the reason of *eu*, unto *ua*, and of *ai*, unto *io*.

Theon here draweth a parallell from *o*, unto *ui*. By the generall fabricke it may be drawne out of *e*, unto *ui*.

It shall be therefore as *ey*, is unto *yo*, so *en*, shall be unto *oi*. Now the proportions multiplied are,

ai,	*io,*	*ai,*	*io.*
eu,	*ua,*	*en,*	*ai.*

by the 32. e.

Therefore the reason of *en*, unto *io* , that is of *ey*, unto *yo*, shall be made of the foresaid reasons.

Of the segments of divers right lines, the *Arabians* have much under the name of *The rule of sixe quantities.* And the *Theoremes* of *Althindus*, concerning this matter, are in many mens hands. And *Regiomontanus* in his *Algorithmus*: and *Maurolycus* upon the 1 p iij. of *Menelaus*, doe make mention of them; but they containe nothing, which may not, by any man skillfull in Arithmeticke, be performed by the multiplication of proportions. For all those wayes of theirs are no more but speciall examples of that kinde of multiplication.

Of Geometry, *the sixt Booke, of a Triangle.*

1. Like plaines have a double reason of their homologall sides, and one proportionall meane, out of 20 p vj. and xj. and 18. p viij.

Or thus; Like plaines have the proportion of their correspondent proportionall sides doubled, & one meane proportionall: Hitherto wee have spoken of plaine lines and their affections: Plaine figures and their kindes doe follow in the next place. And first, there is premised a common corollary drawne out of the 24. e. iiij. because in plaines there are but two dimensions.

2. A plaine surface is either rectilineall or obliquelineall, [or rightlined, or crookedlined. H.]

Straightnesse, and crookednesse, was the difference of lines at the 4. e. ij. From thence is it here repeated and attributed to a surface, which is geometrically made of lines. That made of right lines, is rectileniall: that which is made of crooked lines, is Obliquilineall.

3. A rectilineall surface, is that which is comprehended of right lines.

A plaine rightlined surface is that which is on all sides inclosed and comprehended with right lines. And yet they are not alwayes right betweene themselves, but such lines as doe lie equally betweene their owne bounds, and without comparison are all and every one of them right lines.

4. A rightilineall doth make all his angles equall to right angles; the inner ones generally to paires from two forward: the outter always to foure.

Or thus: A right lined plaine maketh his angles equall unto right angles: Namely the inward angles generally, are equall unto the even numbers from two forward, but the outward angles are equall but to 4. right angles. *H.*

The first kinde I meane of rectilineals, that is a triangle doth make all his inner angles equall to two right angles, that is, to a binary, the first even number of right angles: the second, that is a quadrangle, to the second even number, that is, to a quaternary or foure: The third, that is, a Pentangle, of quinqueangle to the third, that is a senary of right angles, or 6. and so farre forth as thou seest in this Arithmeticall progression of even numbers,

2.	4.	6.	8.	10.	12.
3.	4.	5.	6.	7.	8.

Notwithstanding the outter angles, every side continued and drawne out, are alwayes equall to a quaternary of right angles, that is to foure. The former part being granted (for that is not yet demonstrated) the latter is from thence concluded: For of the inner angles, that of the outter, is easily proved. For the three angles of a triangle are equall to two right angles. The foure of a quadrangle to foure: of a quinquangle, to sixe: of a sexe angle, to eight: Of septangle, to tenne, and so forth, form a binarie by even numbers: Whereupon, by the 14. e. V. a perpetuall quaternary of the outer angles is concluded.

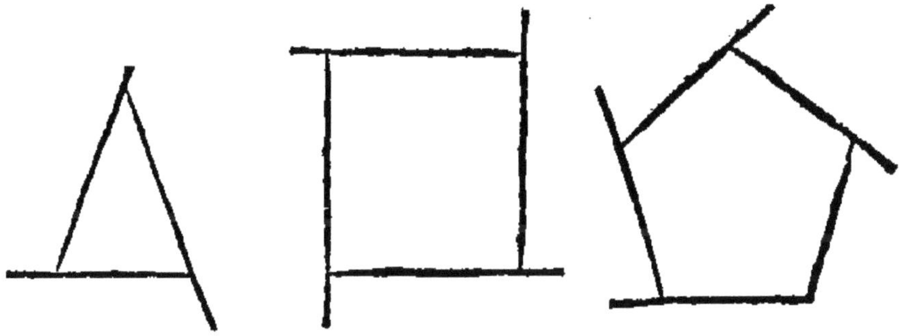

5. A rectilineall is either a Triangle or a Triangulate.

As before of a line was made a lineate: so here in like manner of a triangle is made a triangulate.

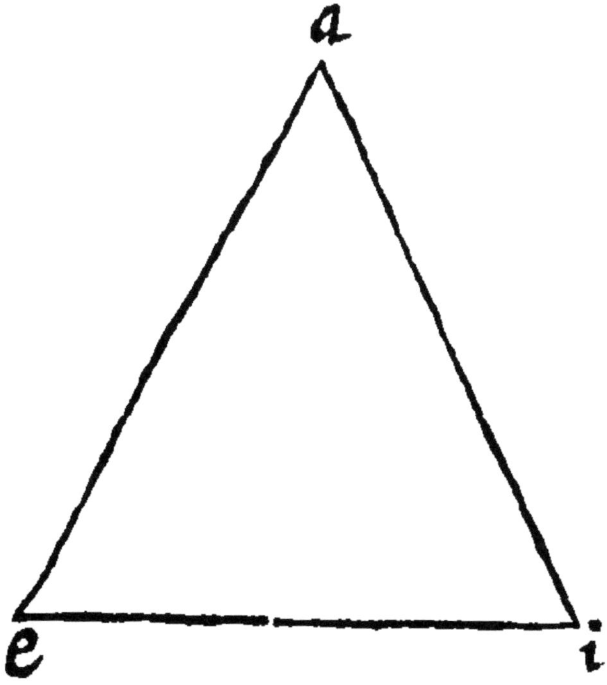

6. A triangle is a rectilineall figure comprehended of three rightlines. 21. d j.

As here *aei.* A triangular figure is of *Euclide* defined from the three sides; whereupon also it might be called *Trilaterum*, that is three sided, of the cause: rather than *Trianglum*, three cornered, of the effect; especially seeing that three angles, and three sides are not reciprocall or to be converted. For a triangle may have foure sides, as is *Acidoides*, or *Cuspidatum*, the barbed forme, which *Zonodorus* called *Cœlogonion*, or *Cavangulum*, an hollow cornered figure. It may also have both five, and sixe sides, as here thou seest. The name therefore of *Trilaterum* would more fully and fitly expresse the thing named: But use hath received and entertained the name of a triangle for a trilater: And therefore let it be still retained, but in that same sense:

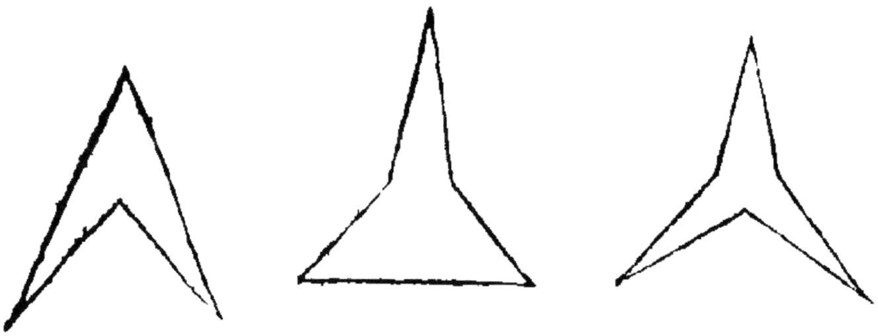

7. A triangle is the prime figure of rectilineals.

A triangle or threesided figure is the prime or most simple figure of all rectilineals. For amongst rectilineall figures there is none of two sides: For two right lines cannot inclose a figure. What is meant by a prime figure, was taught at the 11. e. iiij.

And

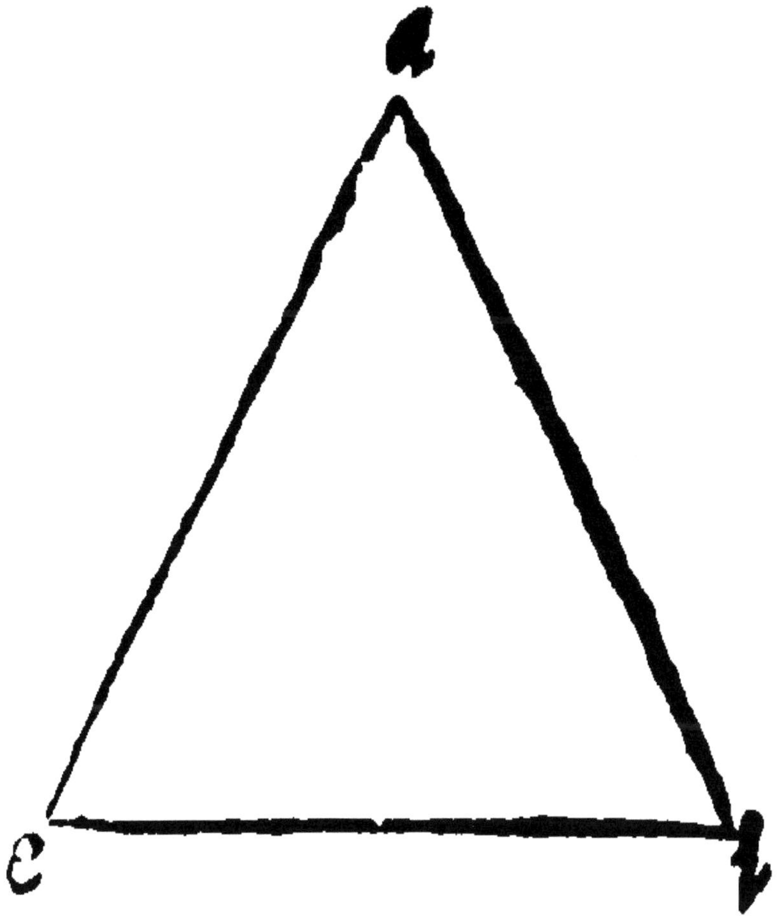

8. If an infinite right line doe cut the angle of a triangle, it doth also cut the base of the same: Vitell. 29. t j.

9. Any two sides of a triangle are greater than the other.

Thus much of the difinition of a triangle; the reason or rate in the sides and angles of a triangle doth follow. The reason of the sides is first.

Let the triangle be *aei*; I say, the side *ai*, is shorter, than the two sides *ae*, and *ei*, because by the 6. e ij, a right line is betweene the same bounds the shortest.

Therefore

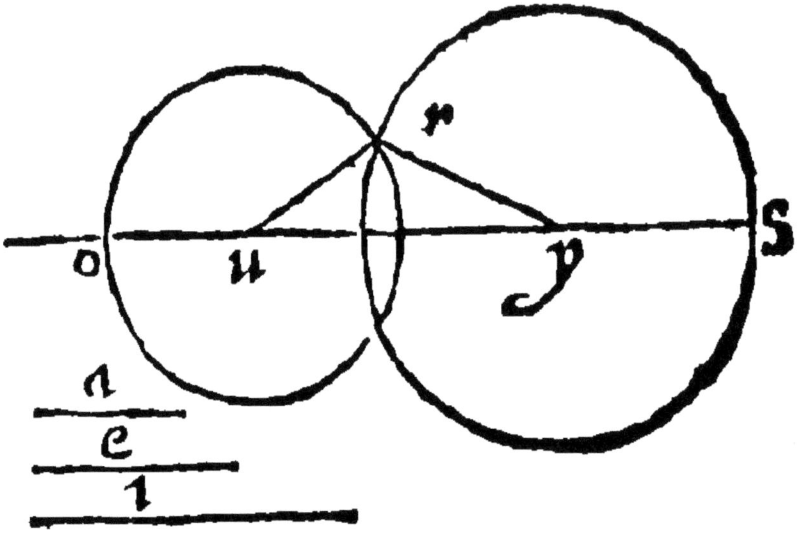

10. If of three right lines given, any two of them be greater than the other, and peripheries described upon the ends of the one, at the distances of the other two, shall meete, the rayes from that meeting unto the said ends, shall make a triangle of the lines given.

Let it be desired that a triangle be made of these three lines, *aei*, given, any two of them being greater than the other: First let there be drawne an infinite right; From this let there be cut off continually three portions, to wit, *ou*, *uy*, and *ys*, equall to *ae*, and *i*, the three lines given. Then upon the ends *y*, and *u*, at the distances *ou*, and *ys*; let two peripheries meet in the point *r*. The rayes from that meeting unto the said ends, *u*, and *y*, shall make the triangle *ury*: for those rayes shall be equall to the right lines given, by the 10. e v.

And

11. If two equall peripheries, from the ends of a right line given, and at his distance, doe meete, lines drawne from the meeting, unto the said ends, shall make an equilater triangle upon the line given. 1 p. j.

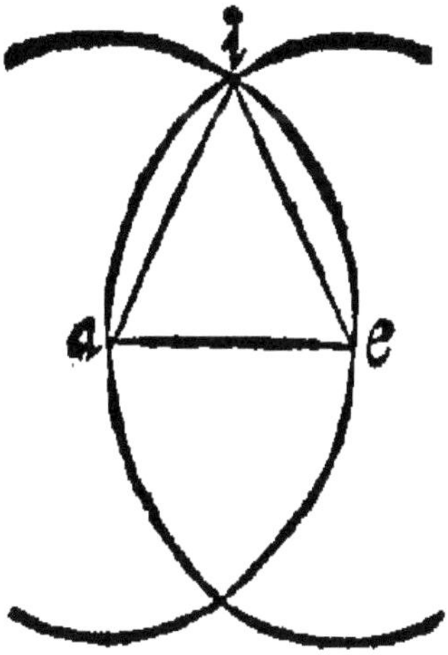

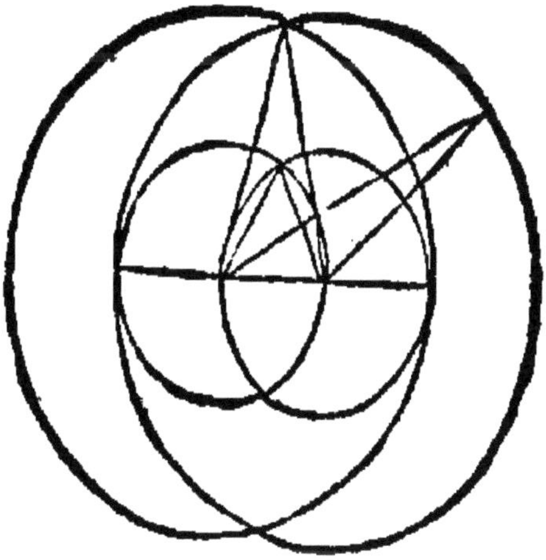

As here upon *ae*, there is made the equilater triangle, *aei*; And in like manner may be framed the construction of an equicrurall triangle, by a common ray, unequall unto the line given; and of a scalen or various triangle, by three diverse raies; all which are set out here in this one figure. But these specialls are contained in the generall probleme: neither doe they declare or manifest unto us any new point of Geometry.

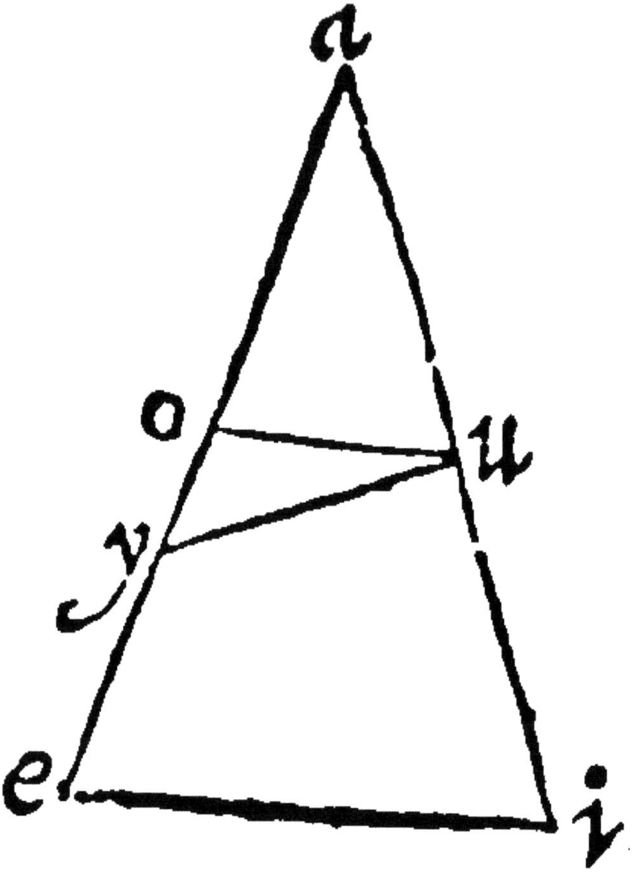

12. If a right line in a triangle be parallell to the base, it doth cut the shankes proportionally: And contrariwise. 2 p vj.

Such therefore was the reason or rate of the sides in one triangle; the proportion of the sides followeth.

As here in the triangle *aei*, let *ou*, be parallell to the base; and let a third parallel be understood to be in the toppe *a*; therefore, by the 28. e. v. the intersegments are proportionall.

The converse is forced out of the antecedent: because otherwise the whole should be lesse than the part. For if *ou*, be not parallell to the base *ei*, then *yu*, is: Here by the grant, and by the antecedent, seeing *ao*, *oe*, *ay*, *ye*, are proportionall: and the first *ao*, is lesser than *ay*, the third: *oe*, the second must be lesser than *ye*, the fourth, that is the whole then the part.

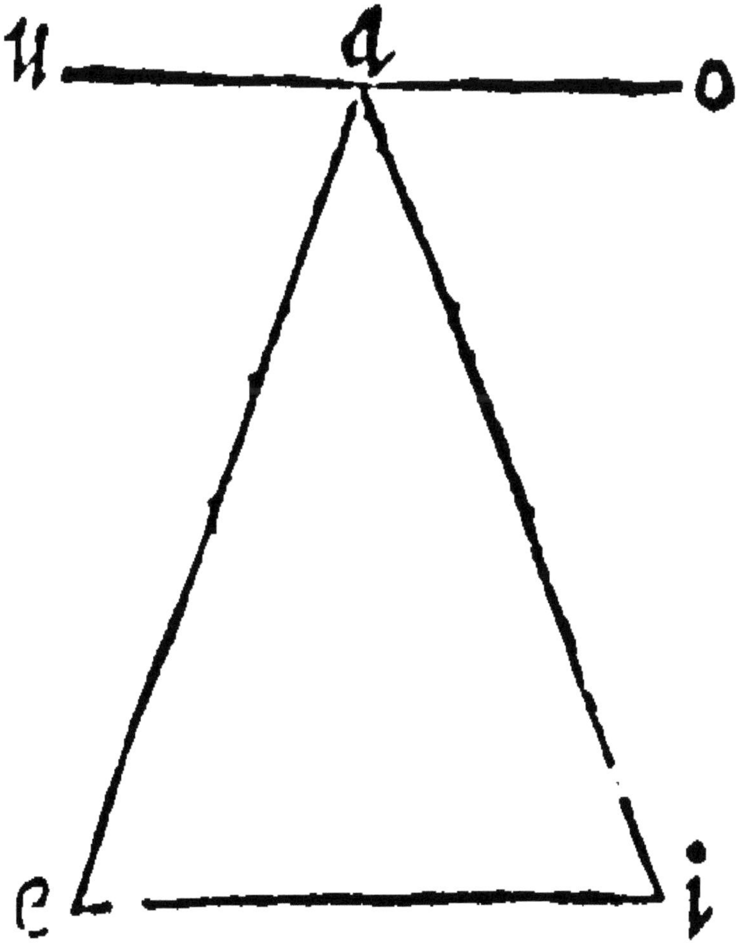

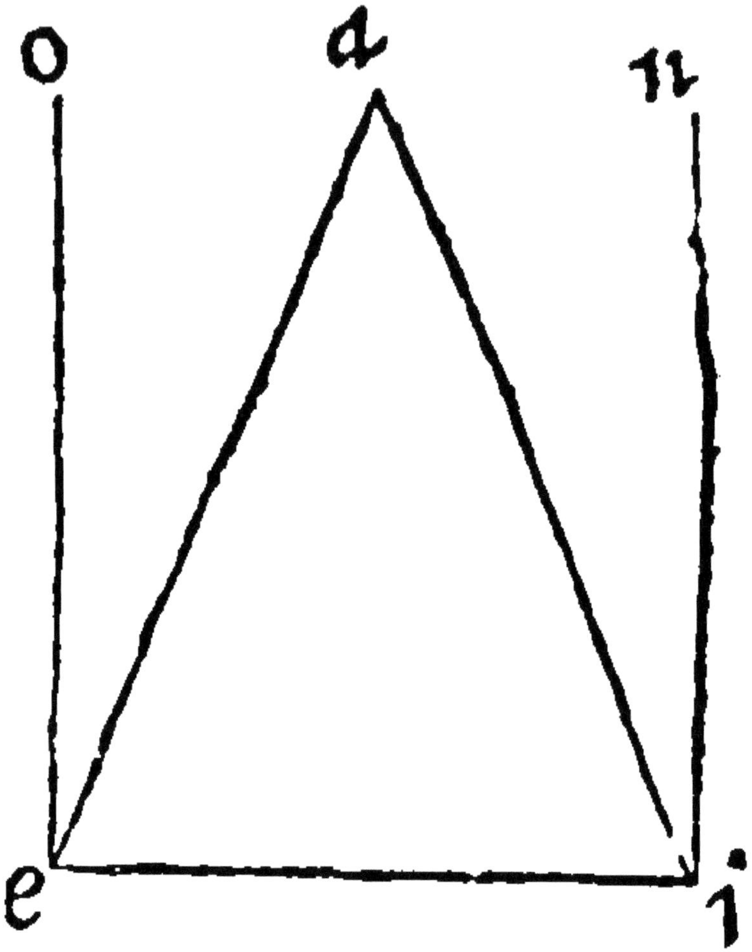

13. The three angles of a triangle, are equall to two right angles. 32. p j.

Hitherto therefore is declared the comparison in the sides of a triangle. Now is declared the reason or rate in the angles, which joyntly taken are equall to two right angles.

The truth of this proposition, saith *Proclus*, according to common notions, appeareth by two perpendiculars erected upon the ends of the base: for looke how much by the leaning of the inclination, is taken from two right angles at the base, so much is assumed or taken in at the toppe, and so by that requitall the equality of two right angles is made; as in the triangle *aei*, let, by the 24. e v, *ou*, be parallell against *ie*. Here three particular angles, *iao, iae, eau*, are equall to two right lines; by the 14. e v. But the inner angles are equall to the same three: For first, *eai*, is equall to it selfe: Then the other two are equall to their alterne angles, by the 24. e v.

Therefore

14. Any two angles of a triangle are lesse than two right angles.

For if three angles be equall to two right angles, then are two lesser than two right angles.

And

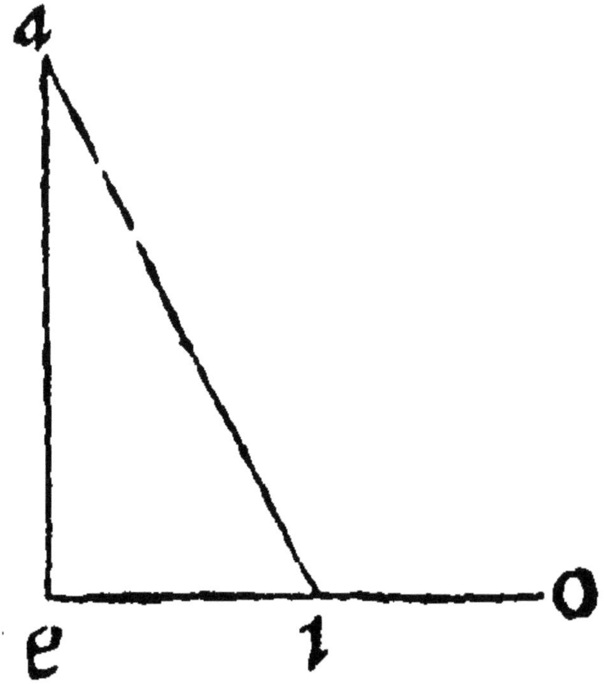

15. The one side of any triangle being continued or drawne out, the outter angle shall be equal to the two inner opposite angles.

This is the rate of the inner angles in one and the same triangle: The rate of the outter with the inner opposite angles doth followe. As in the triangle *aei*, let the side *ei*, be continued or drawne out unto *o*; the two angles on each side *aio* and *aie*, are by the 14 e v. equall to two right angles: and the three inner angles, are by the 13. e. equall also to two right angles; take away *aie*, the common angle, and the outter angle *aio*, shall be left equall to the other two inner and opposite angles.

Therefore

16. The said outter angle is greater than either of the inner opposite angles. 16. p j.

This is a consectary following necessarily upon the next former consectary.

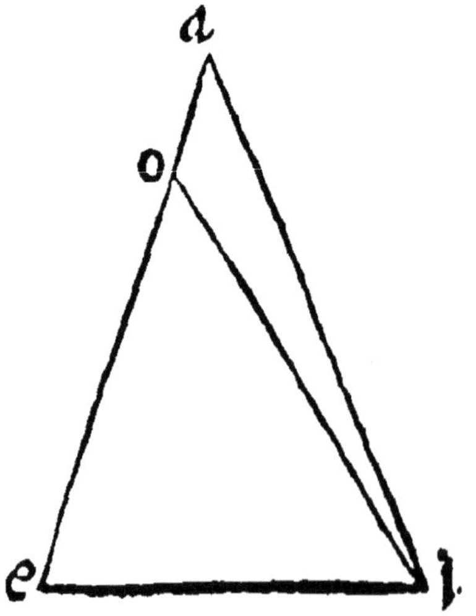

17. If a triangle be equicrurall, the angles at the base are equall: and contrariwise, 5. and 6. p. j.

The antecedent is apparent by the 7. e iij. The converse is apparent by an impossibilitie, which otherwise must needs follow. For if any one shanke be greater than the other, as *ae*: Then by the 7. e v. let *oe*, be cut off equall to it: and let *oi*, be drawne: then by 7. e iij. the base *oi*, must be equal to the base *ae*; but the base *oi*, is lesser than *ae*. For by the 9. e, *ia*; and *ao*, (to which *ae*, is equall, seeing that *oe*, is supposed to be equall to the same *ai*: and *ao*, is common to both) are greater than the said *oi*; therefore the same, *oi*, must be equall to the same *ae*, and lesser than the same, which is impossible. This was first found out by *Thales Milesius*.

Therefore

18. If the equall shankes of a triangle be continued or drawne out, the angles under the base shall be equall betweene themselves.

For the angles *aei*, and *ieo*: Item *aie*, and *eiu*, are equall to two right angles, by the 14. e v. Therefore they are equall betweene themselves: wherefore if you shall take away the inner angles, equall betweene themselves, you shall leave the outter equall one to another.

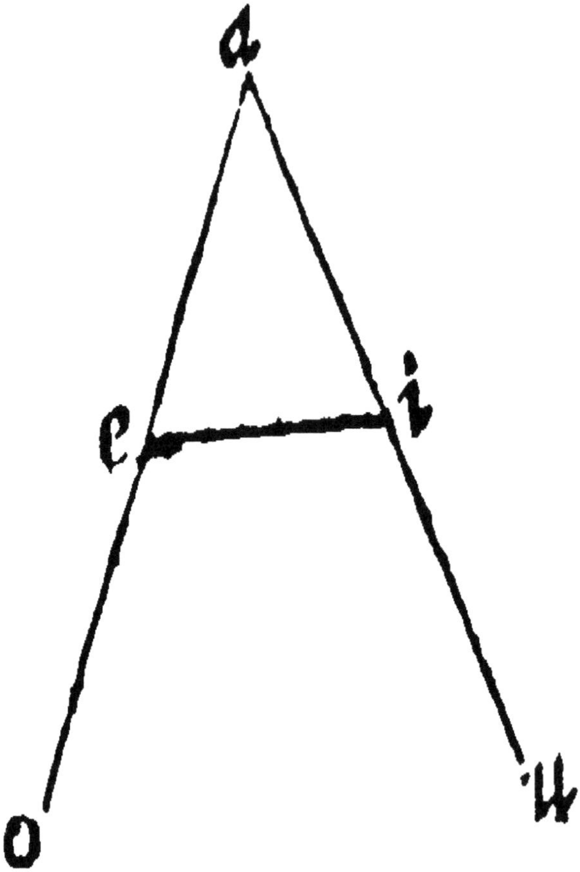

And

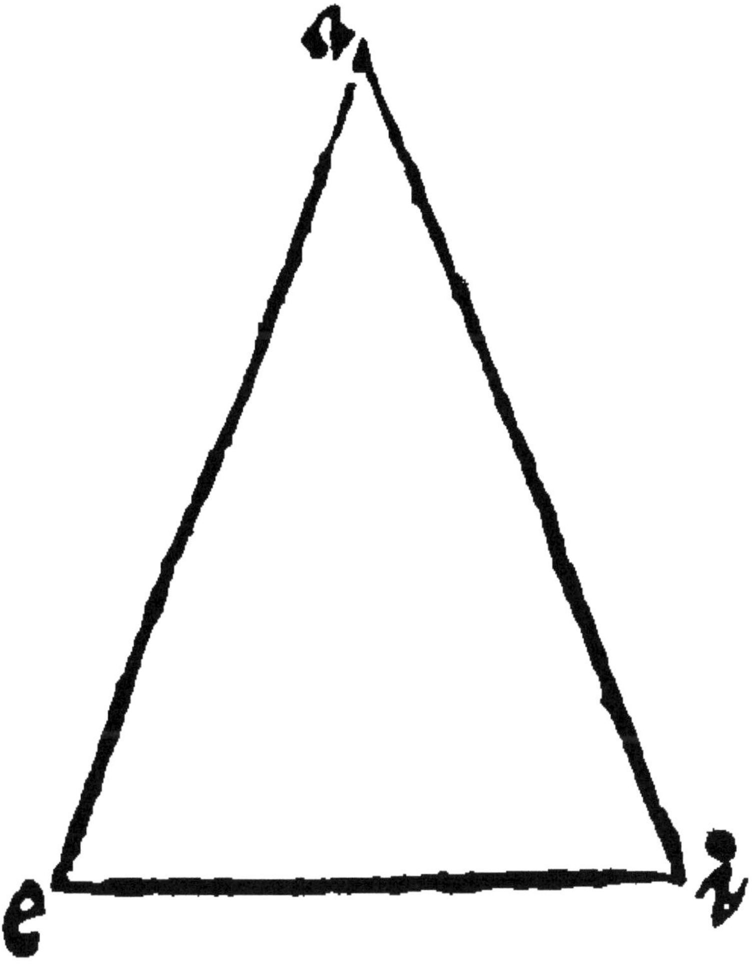

19. If a triangle be an equilater, it is also an equiangle: And contrariwise.

It is a consectary out of the condition of an equicrurall triangle of two, both shankes and angles, as in the example *aei*, shall be demonstrated.

And

20. The angle of an equilater triangle doth countervaile two third parts of a right angle. Regio. 23. p j.

For seeing that 3. angles are equall to 2. 1. must needs be equall to 2/3.

And

21. Sixe equilater triangles doe fill a place.

As here. For 2/3. of a right angle sixe lines added together doe make 12/3. that is foure right angles; but foure right angles doe fill a place by the 27. e. iiij.

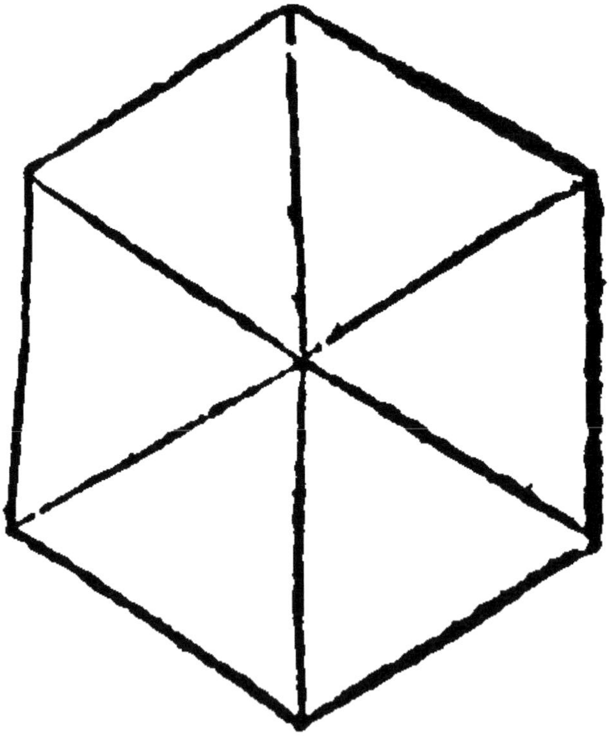

22. The greatest side of a triangle subtendeth the greatest angle; and the greatest angle is subtended of the greatest side. 19. and 18. p j.

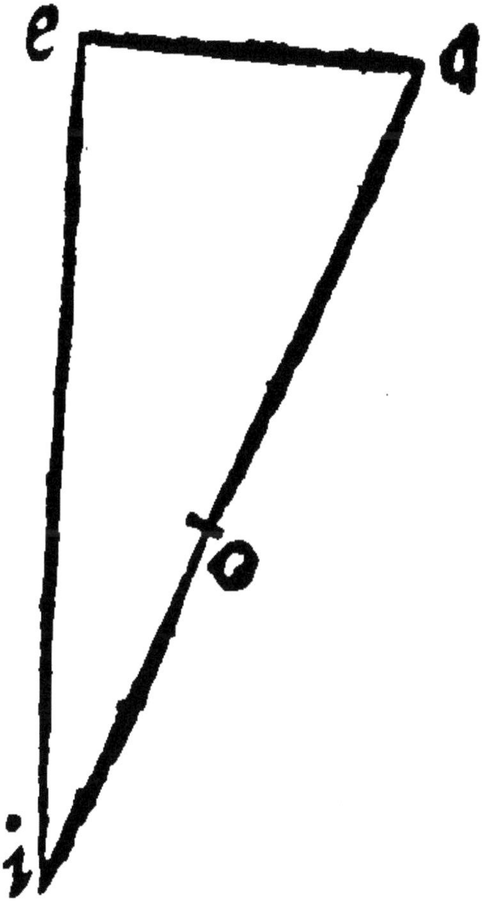

Subtendere, to draw or straine out something under another; and in this place it signifieth nothing else but to make a line or such like, the base of an angle, arch, or such like. And *subtendi*, is to become or made the base of an angle, arch, of a circle, or such like: As here, let *ai*, be a greater side than *ae*, I say the angle at *e*, shall be greater than that at *i*. For let there be cut off from *ai*, a portion equall to *ae*,; and let that be *io*: then the angle *aei*, equicrurall to the angle *oie*, shall be greater in base, by the grant. Therefore the angle shall be greater, by the 9 e iij.

The converse is manifest by the same figure: As let the angle *aei*, be greater than the angle *aie*. Therefore by the same, 9 e iij. it is greater in base. For what is there spoken of angles in generall, are here assumed specially of the angles in a triangle.

23. If a right line in a triangle, doe cut the angle in two equall parts, it shall cut the base according to the reason of the shankes; and contrariwise. 3. p vj.

The mingled proportion of the sides and angles doth now remaine to be handled in the last place.

Let the triangle be *aei*; and let the angle *eai*, be cut into two equall parts, by the right line *ao*: I say, as *ea*, is unto *ai*, so *eo*, is unto *oi*. For at the angle *i*, let the parallell *iu*, by the 24. e v. be erected against *ao*; and continue or draw out *ea*, infinitly; and it shall by the 20. e v. cut the same *iu*, in some place or other. Let it therefore cut it in *u*. Here, by the 28. e v. as *ea*, is to *au*, so is *eo*, to *oi*. But *au*, is equall to *ai*, by the 17. e. For the angle *uia*, is equall to the alterne angle *oai*, by the 21. e v. And by the grant it is equall to *oae*, his equall: And by the 21. e v. it is equall to the inner angle *aui*; and by that which is concluded it is equall to *uia*, his equall. Therefore by the 17. e, *au*, and *ai*, are equall. Therefore as *ea*, is unto *ai*, so is *eo*, unto *oi*.

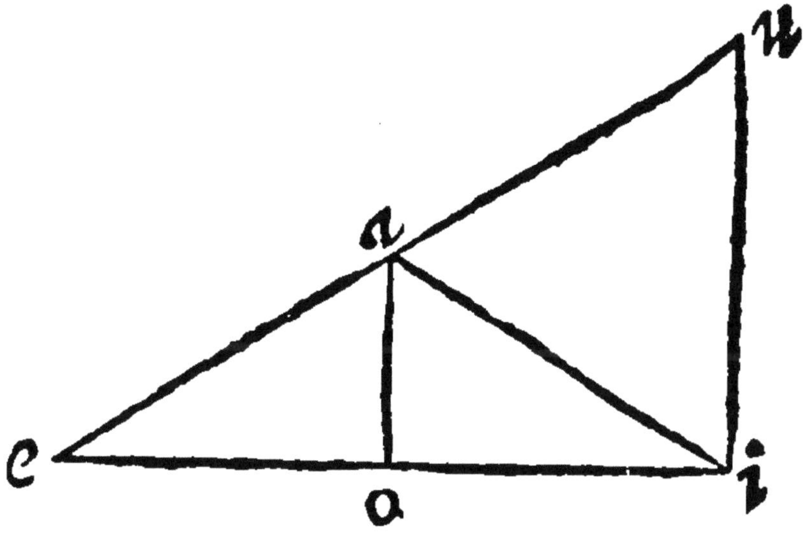

The Converse likewise is demonstrated in the same figure. For as *ea*, is to *ai*; so is *eo*, to *oi*: And so is *ea*, to *au*, by the 12 e: therefore *ai*, and *au*, are equall, Item the angles *eao*, and *oai*, are equall to the angles at *u*, and *i*, by the 21. e v. which are equall betweene themselves by the 17. e.

Of Geometry, the seventh Booke, Of the comparison of Triangles.

1. Equilater triangles are equiangles. 8. p. j.

Thus forre of the Geometry, or affections and reason of one triangle; the comparison of two triangles one with another doth follow. And first of their rate or reason, out of their sides and angles: Whereupon triangles betweene themselves are said to be equilaters and equiangles. First out of the equality of the sides, is drawne also the equalitie of the angles.

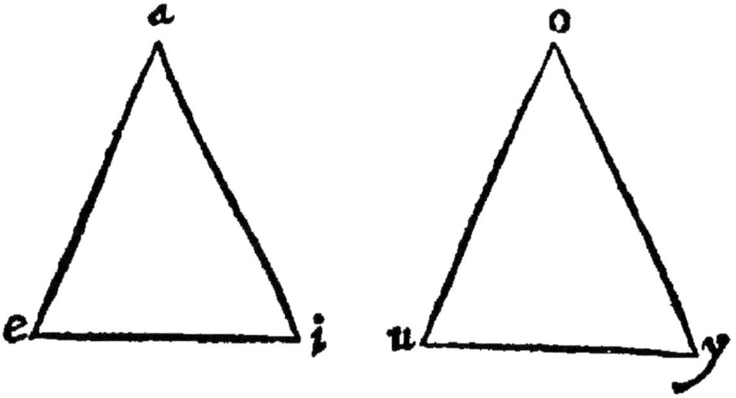

Triangles therefore are here jointly called equilaters, whose sides are severally equall, the first to the first, the second, to the second, the third to the third; although every severall triangle be inequilaterall. Therefore the equality of the sides doth argue the equality of the angles, by the 7. e iij. As here.

2. If two triangles be equall in angles, either the two equicrurals, or two of equall either shanke, or base of two angles, they are equilaters, 4. and 26. p j.

Or thus; If two triangles be equall in their angles, either in two angles contained under equall feet, or in two angles, whose side or base of both is equall, those angles are equilater. *H.*

This element hath three parts, or it doth conclude two triangles to be equilaters three wayes. 1. The first part is apparent thus: Let the two triangles be *aei*, and *ouy*; because the equall angles at *a*, and *o*, are equicrurall, therefore they are equall in base, by the 7. e iij.

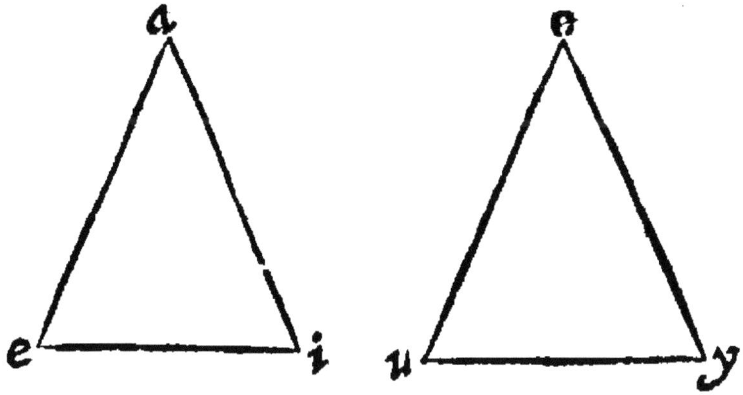

2 The second thus: Let the said two triangles *aei*, and *ouy*, be equall in two angles a peece, at *e*, and *i*, and at *u*, and *y*. And let them be equall in the shanke *ei*, to *uy*. I say, they are equilaters. For if the side *ae*, (for examples sake) be greater than the side *ou*, let *es*, be cut off equall unto it; and draw the right line *is*. Here by the antecedent, the triangles *sei*, and *ouy*, shall be equiangles, and the angles *sie*, shall be equall to the angle *oyu*, to which also the whole angle *aie*, is equall, by the grant. Therefore the whole and the part are equall, which is impossible. Wherefore the side *ae*, is not unequall but equall to the side *ou*: And by the antecedent or former part, the triangles *aei*, and *ouy*, being equicrurall, are equall, at the angle of the shanks: Therefore also they are equall in their bases *ai*, and *oy*.

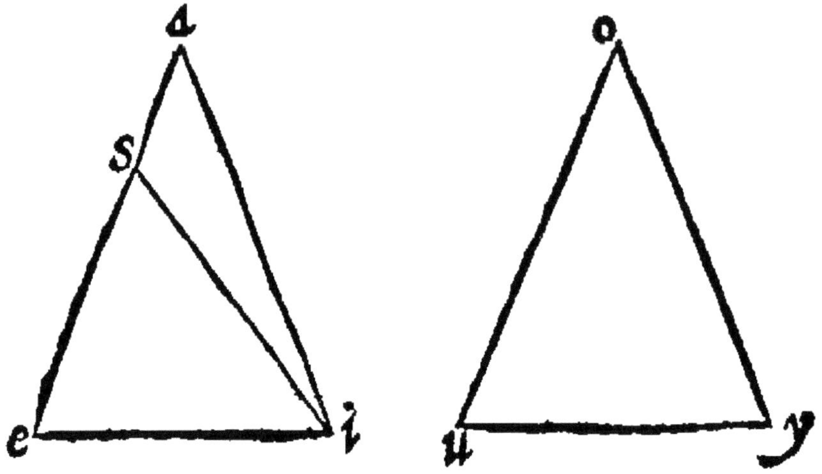

3 The third part is thus forced: In the triangles *aei*, and *ouy*, let the angles at *e*, and *i*, and *u*, and *y*, be equall, as afore: And *ae*, the base of the angle at *i*, be equall to *ou*, the base of angle at *y*: I say that the two triangles given are equilaters. For if the side *ei*, be greater than the side *uy*, let *es*, be cut off equall to it, and draw the right line *as*. Therefore by the antecedent, the two triangles, *aes*, and *ouy*, equall in the angle of their equall shankes are equiangle: And the angle *ase*, is equall to the angle *oyu*, which is equall by the grant unto the angle *aie*. Therefore *ase*, is equall to *aie*, the outter to the inner, contrary to the 15. e. vj. Therefore the base *ei*, is not unequall to the base *uy*, but equall. And therefore as above was said, the two triangles *aei*, and *ouy*, equall in the angle of their equall shankes, are equilaters.

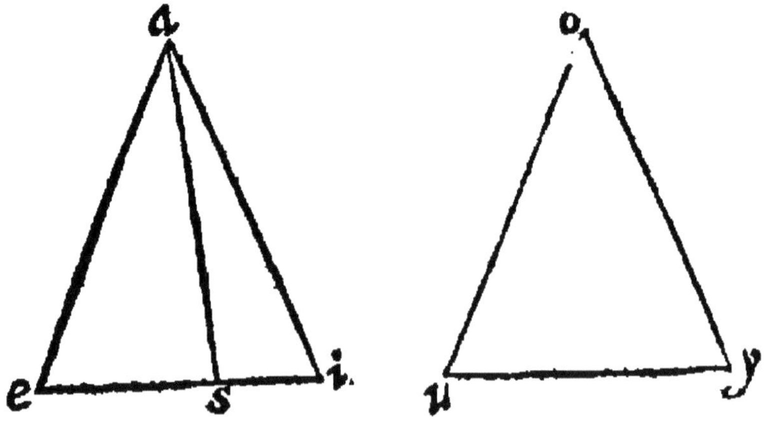

3. Triangles are equall in their three angles.

The reason is, because the three angles in any triangle are equall to two right angles, by the 13. e vj. As here, the greatest triangle, all his corners joyntly taken, is equall to the least.

And yet notwithstanding it is not therefore to be thought to be equiangle to it: For Triangles are then equiangles, when the severall angles of the one, are equall to the severall angles of the other: Not when all joyntly are equall to all.

Therefore

4. If two angles of two triangles given be equall, the other also are equall.

All the three angles, are equall betweene themselves, by the 3 e. Therefore if from equall you take away equall, those which shall remaine shall be equall.

5. If a right triangle equicrurall to a triangle be greater in base, it is greater in angle: And contrariwise. 25. and 24. p j.

Thus farre of the reason or rate of equality, in the sides and angles of triangles: The reason of inequality, taken out of the common and generall inequality of angles, doth follow. The first is manifest, by the 9. e iij. as here thou seest in *aei* and *ouy*.

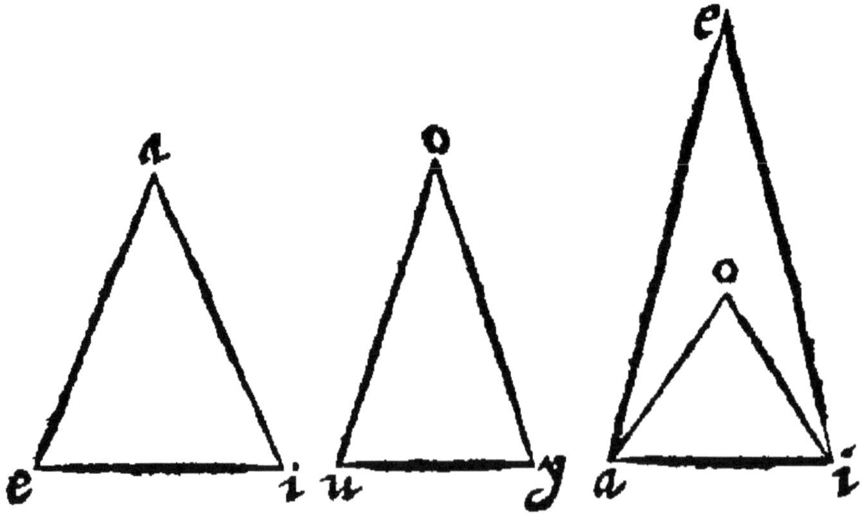

6. If a triangle placed upon the same base, with another triangle, be lesser in the inner shankes, it is greater in the angle of the shankes.

This is a consectary drawne also out of the 10 e iij. As here in the triangle *aei*, and *aoi*, within it and upon the same base. Or thus: If a triangle placed upon the same base with another triangle, be lesse then the other triangle, in regard of his feet, (those feete being conteined within the feete of the other triangle) in regard of the angle conteined under those feete, it is greater: *H*.

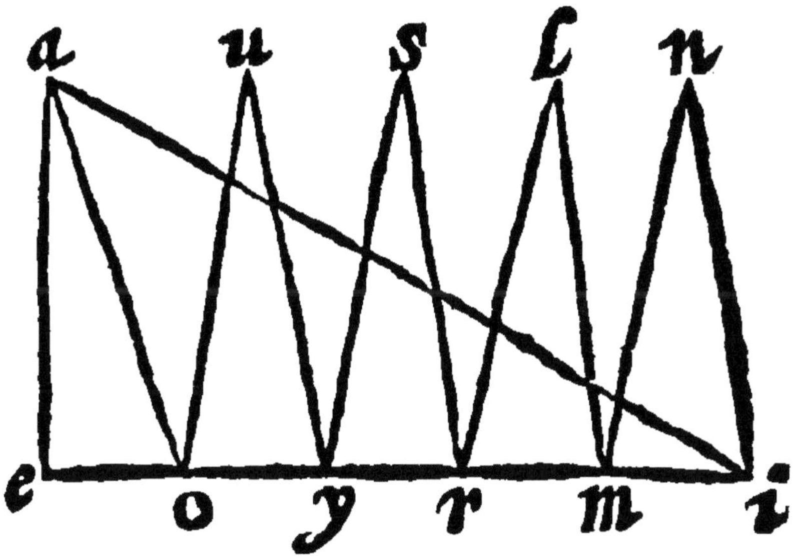

7. Triangles of equall heighth, are one to another as their bases are one to another.

Thus farre of the Reason or rate of triangles: The proportion of triangles doth follow; And first of a right line with the bases. It is a consectary out of the 16 e iiij.

Therefore

8. Upon an equall base, they are equall.

This was a generall consectary at the 16. e iiij: From whence *Archimedes* concluded, If a triangle of equall heighth with many other triangles, have his base equall to the bases of them all, it is equall to them all: as here thou seest *aei* to be equall to the triangles *aeo, uoy, syr, lrm, nmi*. Here hence also thou mayst conclude, that *Equilater* triangles are equall: Because they are of equall heighth, and upon the same base.

And

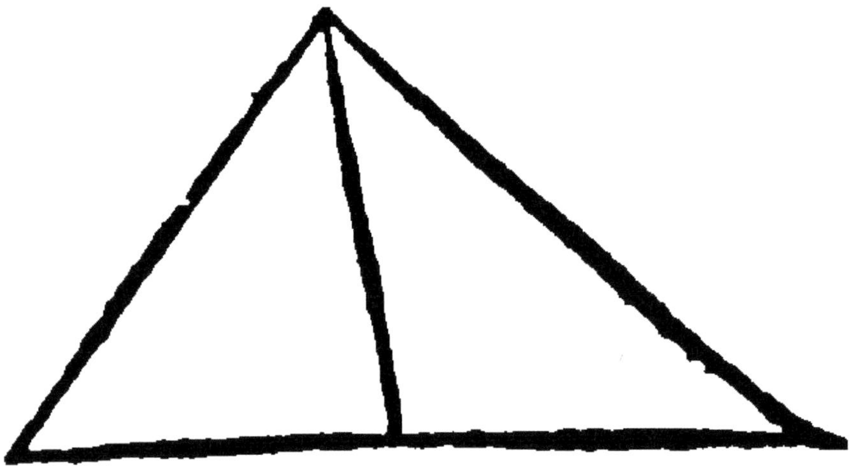

9. If a right line drawne from the toppe of a triangle, doe cut the base into two equall parts, it doth also cut the triangle into two equall parts: and it is the diameter of the triangle.

As here thou seest: For the bisegments, or two equall portions thus cut are two triangles of equall heighth that is to say, they have one toppe common to both, within the same parallels) and upon equall bases: Therefore they are equall: And that right line shall be the diameter of the triangle, by the 5 e iiij, because it passeth by the center.

10. If a right line be drawne from the toppe of a triangle, unto a point given in the base (so it be not in the middest of it) and a parallell be drawne from the middest of the base unto the side, a right line drawne from the toppe of the sayd parallell unto the sayd point, shall cut the triangle into two equall parts.

Let the triangle given be *aei*: And let *ao*, cut the base *ei*, in *o* unequally: And let *uy* be parallell from *u*, the middest of the base, unto the sayd *ao*. I say that *yo* shall divide the triangle into two equall portions. For let *au* be knit together with a right line: That line, by the 9 e, shall divide the triangle into two equall parts. Now the two triangles *ayu*, and *you*, are equall by the 8 e; because they are of equall height, and upon the same base.

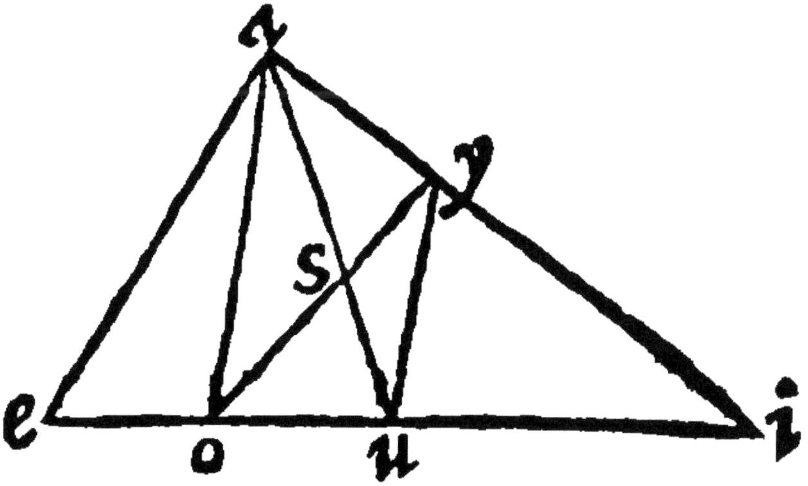

Take away *ysu*, the common triangle; And you shall leave *asy*, and *osu*, equall betweene themselves: The common right lined figure *ysui*, let it be added to both the sayd equall triangles: And then *oyi*, shall be equall to *aui*, the halfe part; And therefore *aeoy*, the other right lined figure, shall be the halfe of the triangle given.

11. If equiangled triangles be reciprocall in the shankes of the equall angle, they are equall: And contrariwise. 15. p. vj. Or thus, as the learned Mr. Brigges hath conceived it: If two triangles, having one angle, are reciprocall, &c.

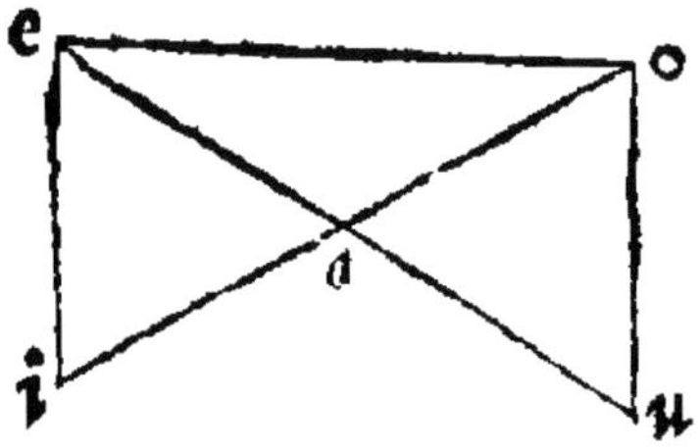

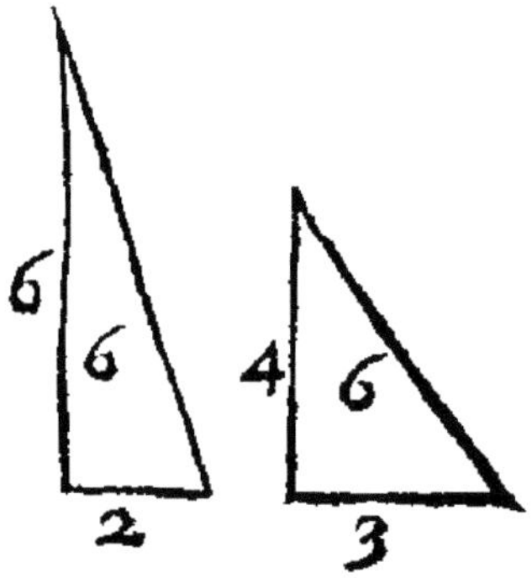

Direct proportion in triangles, is such as hath in the former beene taught: Reciprocall proportion followeth. It is a consectary drawne out of the 18 e iiij; which is manifest, as oft as the equall angle is a right angle: For then those shankes, [comprehending the equall angles,] are the heights and the bases; As here thou seest in the severed triangles. Notwithstanding in obliquangle triangles, although the shankes are not the heights, the cause of the truth hereof is the same. Yet if any man shall desire a demonstration of it, it is thus: Let therefore the diagramme or figure bee in the triangles *aei*, and *aou*: And the angles *oau*, and *eai*, let them be equall: And as *ua* is to *ae*, so let *ia* be unto *ao*: I say that the triangles *aou*, and *eai*, are equall. For *eo* being knit together with a right line, *uao* is unto *oae*, as *ua* is unto *ae*, by the 7 e: And *ia*, unto *ao*, by the grant, is as *eai* is unto *eao*. Therefore *uao*, and *eai*, are unto *eao* proportionall: And therefore they are equall one to another.

The converse, is concluded by the same sorites, but by saying all backward. For *ua* unto *ae* is, as *uao* is unto *oae*, by the 7 e: And as *eai*, by the grant: Because they are equall: And as *ia* is unto *ao*, by the same, Wherefore *ua* is unto *ae*, as *ia* is unto *ao*.

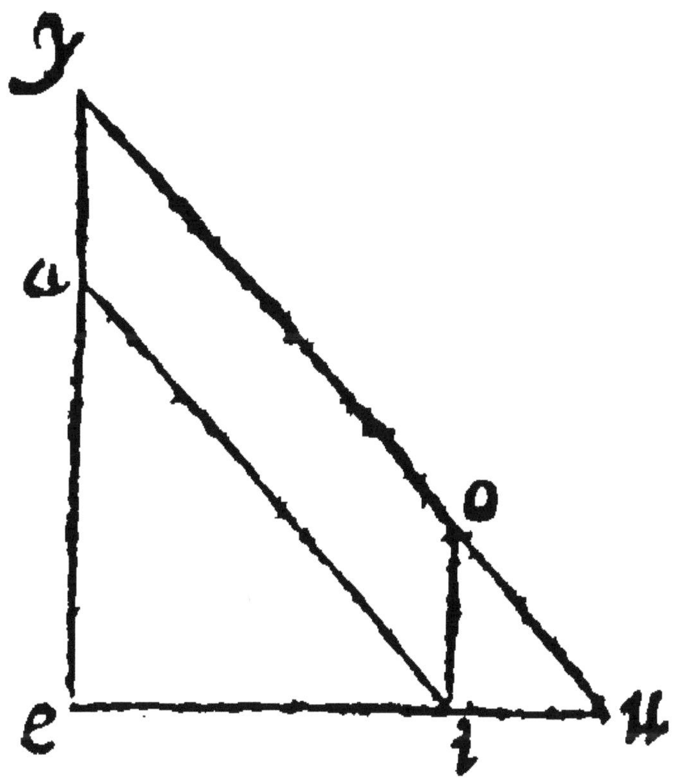

12. If two triangles be equiangles, they are proportionall in shankes: And contrariwise: 4 and 5. p. vj.

The comparison both of the rate and proportion of triangles hath in the former beene taught: Their similitude remaineth for the last place. Which similitude of theirs consisteth indeed of the reason, or rate of their angles and proportion of the shankes. Therefore for just cause was the reason of the angles set first: Because from thence not onely their reason, but also their latter proportion is gathered. Let *aei* and *iou*, be two triangles equiangled: And let them be set upon the same line *eiu*, meeting or touching one another in the common point *i*. Then, seeing that the angles at *e* and *i*, are granted to bee equall, the lines *oi*, and *ae*, are parallel, by the 21 e v. Therefore by the 22 e v *uo* and *ea*, being continued, shall meete. Item, The right lines *ai*, and *yu*, by the 21 e v, are parallel, because the angle *aie* is equall to *oui*, the inner opposite to it. Therefore seeing that *ai* is parallell to the base *yu*, by the 21 e v, *ea* shall be to *ay*, that is, by the 26 e v, to *io*, as *ei* is to *iu*: And alternly, or crosse wayes, *ea* shall be to *ei*, as *io* is to *iu*. This is the first proportion. Item, seeing that *io* is parallell to the base *ye*; *yo*, that is, by the 26 e v, *ai* shall bee unto *ou*, as *ei*, is unto *iu*: And crosse wise, as *ai* is unto *ie*, so is *ou* unto *ui*. This is the second proposition. Lastly, equiordinately: *ae* is to *ai*, as *oi* is to *ou*: wherefore if triangles be equiangled, they are proportional in shankes.

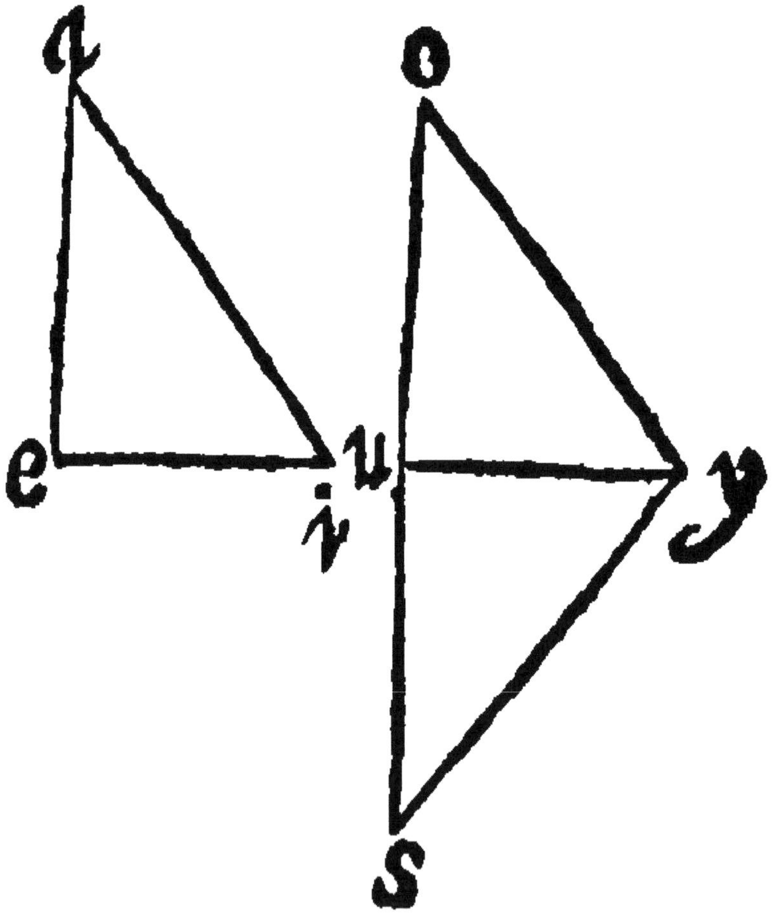

This converse is thus demonstrated. Let there be two triangles *aei*, and *ouy*, proportionall in shankes: And as *ae* is to *ei*; so let *ou*, be to *uy*: And as *ai* is to *ie*; so let *oy* bee to *yu*. Then at the points *u* and *y*, let angles be made by the 11 e iij. equall to the angles at *e* and *i*, and let the triangle *uys*, be made: for the other angles at *a* and *s*, shall be equall by the 4 e. And the triangle *yus*, shall be equiangled to the assigned *aei*. And by the antecedent, it shall be proportionall to it in shankes. Thus are two triangles *ouy*, by the grant; and *uys*, by the construction, proportionall in shanks to the same triangle *aei*: And as *ae*, is to *ei*, so is *ou*, to *uy*; so is *su*, to *uy*. Therefore seeing *ou* and *su*, are proportionall to the same *yu*, they are equall; Item, as *ai* is to *ie*: so is *oy* unto *yu*: so also is *sy* unto *yu*. Therefore *oy* and *sy*, seeing they are proportionall to the same *yu*, are equall. (*yu* is the common side.) The triangle therefore *ouy*, is equilater unto the triangle *syu*. And by the 1 e, it is to it equiangle: And therefore it is equiangled to the triangle *aei*, which was to be prooved. This was generally before taught at the 20 e iiij, of homologall sides subtending equall angles.

Therefore,

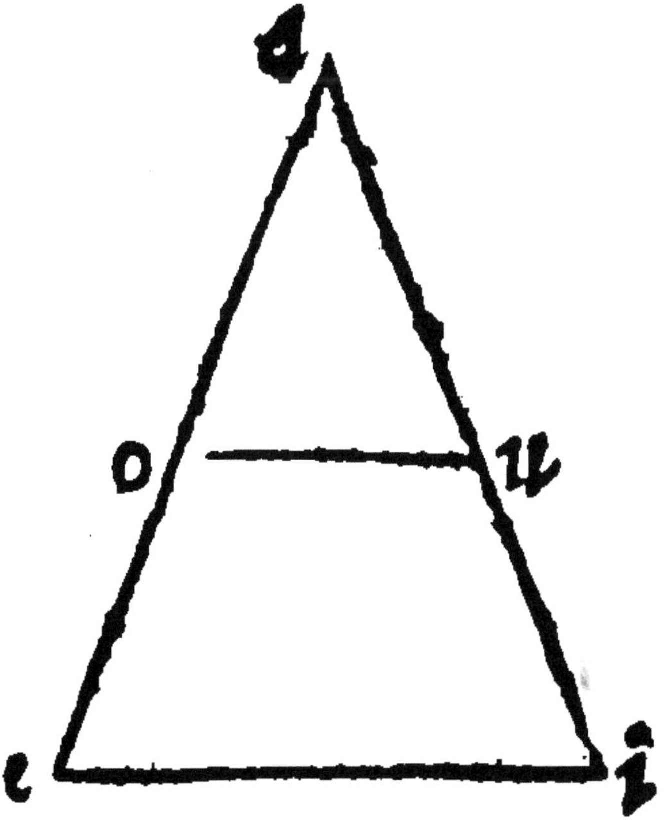

13. If a right line in a triangle be parallell to the base, it doth cut off from it a triangle equiangle to the whole, but lesse in base.

As in the triangle *aei*, the right line *ou*, doth cut off the triangle *aou*, equiangle, by the 21 e v, to the whole *aei*; But the base *ou*, is lesse than the base *ei*, as appeareth by the 21 e, and by the alternation of the sides.

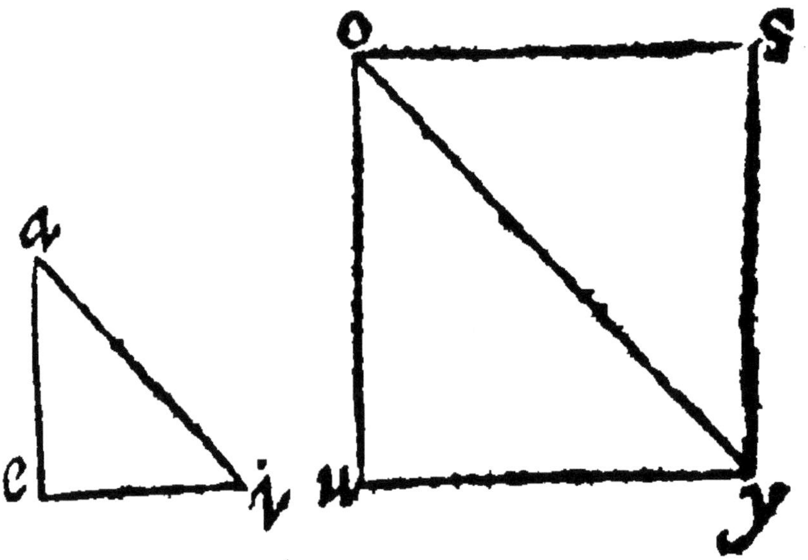

14. If two trangles be proportionall in the shankes of the equall angle, they are equiangles: 6 p vj.

Let therefore the triangles given be *aei*, and *ouy*, equall in their angles *a* and *o*: And in their shankes let *ea*, be unto *ai*, as *ou* is to *oy*: And by the 11 e iij, let the angles *soy*, and *oys*, be equall to the angles *eai*, and *eia*: The other at *s* and *e*, shall be equall, by the 4 e. Here thou seest that the triangle *aei*, is equiangle unto *oys*. Now, by the 12 e. as *ea* is to *ai*: so is *so* to *oy*: and therefore, by the grant, so is *uo* to *oy*. Therefore seeing that *uo*, and *os*, are proportionall to *oy*, they are both equall. Lastly, if the common shanke *oy* bee added to both the shankes *ou*, and *oy*, are equall to the shankes *so* and *oy*. [But by the construction the angles *oys* and *aie* are equall. And, by the 4 e, the other at *s* and *e* are equall. Therefore the first triangle *aei*, is made equiangled to the third. Now seeing the second triangle *uoy* is to the third *soy*, equall in the shanks of the equall angle, it is to the same equilater, and by the 1 e, equiangled: *Shon.*] Wherefore the second triangle *ouy* shall likewise be equiangled to *osy*, the third: And therefore if two triangles proportionall in shankes be equall in the angle of their shankes, they are equiangles.

15. If triangles proportionall in shankes, and alternly parallell, doe make an angle betweene them, their bases are but one right line continued. 32 p. vj.

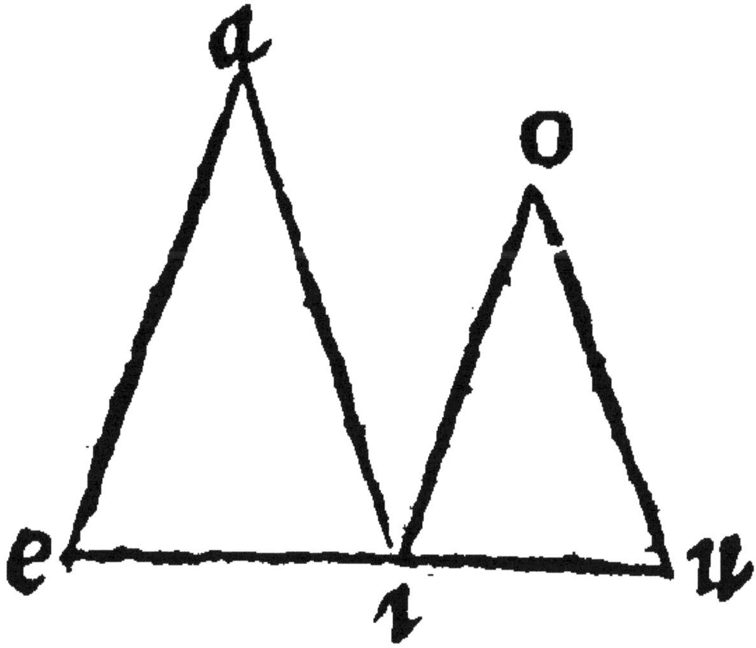

Or thus: If being proportionall in their feet, and alternately parallels, they make an angle in the midst betweene them, they have their bases continued in a right line: *H.*

The cause is out of the 14 e v. For they shall make on each side, with the falling line *ai*, two angles equall to two right angles.

Let the triangles *aei* and *oiu*, be proportionall in shanks: As *ae* is to *ai*, so let *io* be to *ou*: And let *ea* bee parallel to *io*: And *ai* to *ou*: Item, let them make the angle *aio*, betweene them, to wit, betweene their middle shankes *ai*, and *oi*, I say their bases *ei*, and *iu*, are but one right line continued. For seeing that by the grant *ae*, and *oi*, are parallels: Item *ai* and *uo*, the right line *ai* and *oi*, shall make, by the 21 e v, the angles at *a*, and *o*, equall to the alterne angle *aio*: And therefore they are equall betweene themselves: And then, by the 14 e, the triangles given are equiangles: Therefore the angle *oui*, is equall to the angle *aie*: Wherfore the three angles *oiu*, *oia*, and *aie*, by the 3 e, are equall to the three angles of the triangle *eai*, which are equall by the 13 e vj. Unto two right angles: And therefore they themselves also are equall to two right angles. Wherefore, by the 14 e v, *ei*, and *iu*, are one right line continued.

16. If two triangles have one angle equall, another proportionall in shankes, the third homogeneall, they are equiangles. 7. p. vj.

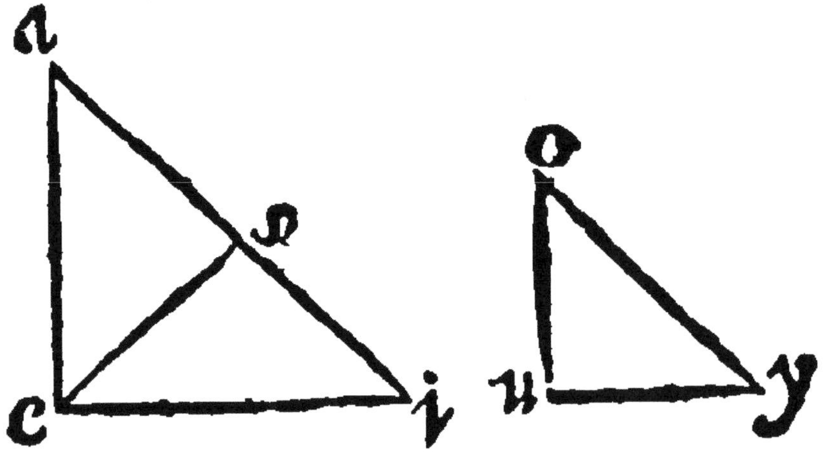

Let *aei*, and *ouy*, the triangles given be equall in their angles *a*, and *o*: and proportionall in the shankes of the angles *e*, and *u*: and their other angles, at *i*, and *y*, homogeneall, that is, let them be both, either acute, or obtuse, or right angles. But first let them be acute, I say, the other at *e*, & *u*, are equall. Otherwise let *aes*, by the 11 e iij. be made equall to the same *ouy*; Then have you them by the 4 e, equiangles; and the angles *ase*, shall be equall to the angle *oyu*; and both are acute angles: and by the 12. e, *aes*, and *ouy*, are proportionall in sides: and as *ae*, is to *es*; so shall *ou*, be to *uy*, that is, by the grant, so shall *ae*, be to *ei*. Therefore because the same *ea*, hath unto two, to wit, *es*, and *ei*, the same reason, the said *es*, and *ei*, are equall one to another: And therefore, by the 17. e. vj. the angles at the base in *s* and *i*, are equall. Therefore both of them are acute angles: And in like manner *ase*, is an acute angle, contrary to the 14. e v. The same will fall out altogether like to both the other, being either obtuse or right angles. The last part of a right angle is manifest by the 4 e of this Booke.

Of Geometry the eight Booke, of the diverse kindes of Triangles.

1. A triangle is either right angled, or obliquangled.

The division of a triangle, taken from the angles, out of their common differences, I meane, doth now follow. But here first a speciall division, and that of great moment, as hereafter shall be in quadrangles and prismes.

A right *2. A right angled triangle is that which hath one right angle: An obliquangled is that which hath none. 27. d j.*

angled triangle in Geometry is of speciall use and force; and of the best Mathematicians it is called *Magister matheseos*, the master of the Mathematickes.

Therefore

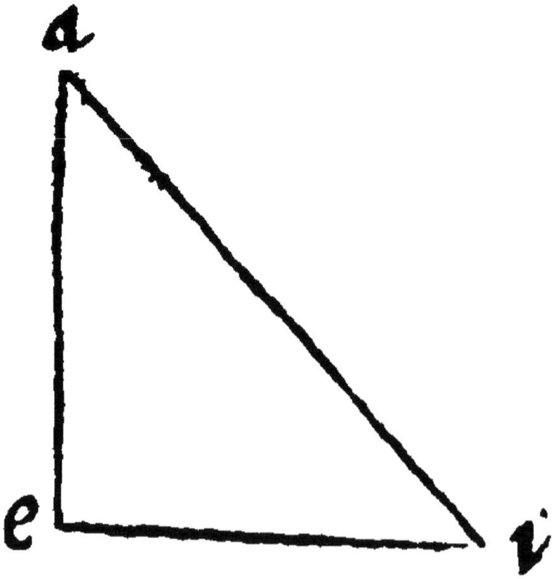

3. If two perpendicular lines be knit together, they shall make a right angled triangle.

As here in *aei*. This construction and manner of making of a right angled triangle, is drawne out of the definition of a right angle. For right lines perpendicular are the makers of a right angle, as is manifest by the 13. e iij.

4. If the angle of a triangle at the base, be a right angle, a perpendicular from the toppe shall be the other shanke: [and contrariwise Schon.]

As is manifest in the same example.

5. If a right angled triangle be equicrurall, each of the angles at the base is the halfe of a right angle: And contrariwise.

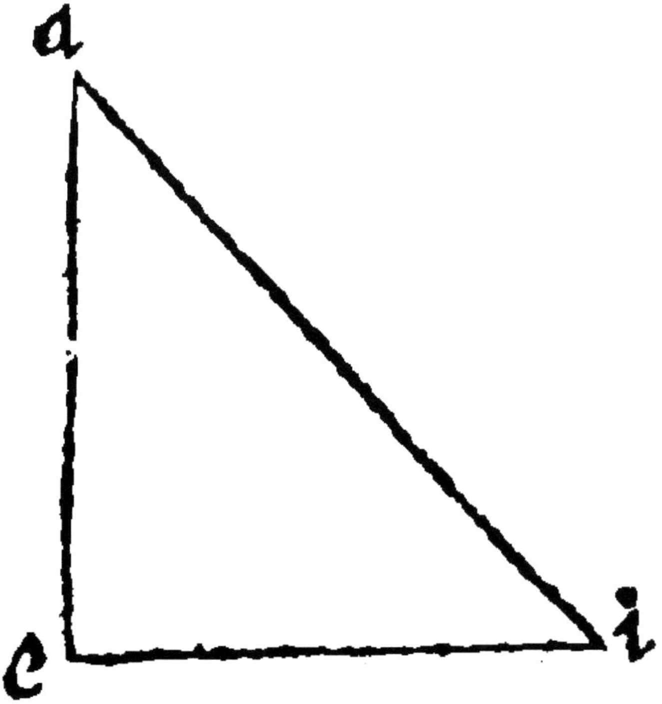

As in the triangle *aei*: For they are both equall to one right angle, by the 13. e. vj. And betweene themselves, by the 17. e. vj.

Therefore

6. If one angle of a triangle be equall to the other two, it is a right angle [And contrariwise Schon.]

Because it is equall to the halfe of two right angles, by the 13. e vj.

And

7. If a right line from the toppe of a triangle cutting the base into two equall parts be equall to the bisegment, or halfe of the base, the angle at the toppe is a right angle: [And contrariwise Schon.]

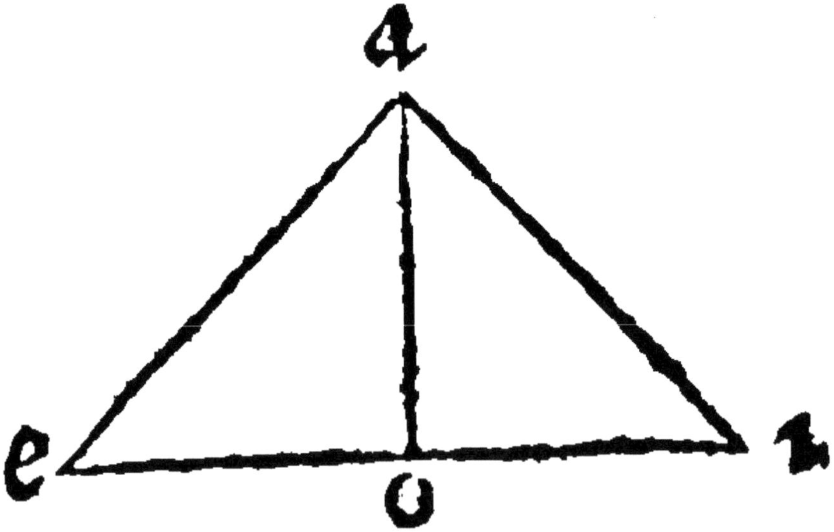

As in the triangle *aei*, the right line *ao*, cutting the base *ei*, in *o*, into two equall parts, is equall to *eo*, or *oi*, the halfe of the base maketh two equicrural triangles; and the severall angles at the top equall to the angles at the ends, *viz. e*, and *i*, by the 17. e. vj. Therefore the angle at the toppe is equall to the other two: wherefore by the 6 e, it is a right angle.

8. A perpendicular in a triangle from the right angle to the base, doth cut it into two triangles, like unto the whole and betweene themselves, 8. p vj. [And contrariwise Schon.]

As in the triangle *aei*, the perpendicular *ao*, doth cut the triangles *aoe*, and *aoi*, like unto the whole *aei*, because they are equiangles to it; seeing that the right angle on each side is one, and another common in *i*, and *e*: Therefore the other is equall to the remainder, by 4. e vij. Wherefore the particular triangles are equiangles to the whole: As proportionall in the shankes of the equall angles, by the 12. e vij. But that they are like betweene themselves it is manifest by the 22. e iiij.

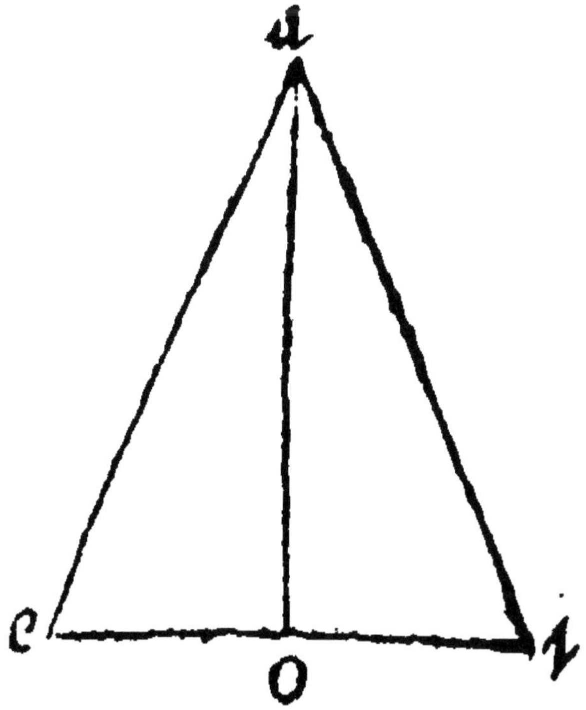

Therefore

9. The perpendicular is the meane proportionall betweene the segments or portions of the base.

As in the said example, as *io*, is to *oa*: so is *oa*, to *oe*, because the shankes of equall angles are proportionall, by the 8. e. From hence was *Platoes* Mesographus invented.

And

10. Either of the shankes is proportionall betweene the base, and the segment of the base next adjoyning.

For as *ei*, is unto *ia*, in the whole triangle, so is *ai*, to *io*, in the greater. For so they are homologall sides, which doe subtend equall angles, by the 23. e. iiij. Item, as *ie*, is to *ea*; in the whole triangle, so is *ae*, to *eo*, in the lesser triangle.

Either of the shankes is proportionall betweene the summe, and the difference of the base and the other shanke. And contrariwise. If one side be proportionall betweene the summe and the difference of the others, the triangle given is a rectangle. M. *H. Brigges*.

This is a consectary arising likewise out of the 4 e. of very great use.

In the triangle *ead*, the shanke *ad*, 12. is the meane proportionall betweene *bd*, 18. (the summe of the base *ae*, 13. and the shanke *ed*, 5.) and 8. the difference of the said base and shanke: For if thou shalt draw the right lines *ba*, and *ac*, the angle *bac*, shall be by the 6. e, a rectangle; (because it is equall to the angles at *b*, and *c*, seeing that the triangles *bea*, and *eac*, are equicrurall.) And by the 9 e, *bd*, *da*, and *dc*, are continually proportionall.

If a quadrate of a number, given for the first shanke, be divided of another, the halfe of the difference of the divisour, and quotient shall be the other shanke, and the halfe of the summe shall be the base. Or thus, *The side of di-*

vided number doubled, and the difference of the divisour and quotient, shall be the two shankes, and the summe of them shall be the base.

Let the number given for the first shanke be 4. And let 8. divide 16. the quadrate of 4. by 2. The halfe of 8 - 2, that is 3. shall be the other shanke: And the halfe of 8 + 2, that is 5. shall be the base.

Therefore

If any one number shall divide the quadrate of another, the side of the divided, and the halfe of the difference of the divisour and the quotient, shall be the two shankes of a rectangled triangle, and the halfe of the summe of them shall be the base thereof.

Let the two numbers given be 4. and 6. The square of 6. let it be 36. and the quotient of 36. by 4. be 9: And the side is 6. for the one shanke. Now 9 - 4. that is, 5. is the difference of the divisour and quotient, whose halfe 2.½, is the other shanke. And 9 + 4. that is 13. is the summe the said devisour and quotient, whose halfe 6.½, is the base.

Againe let 4. and 8. be given. The quadrate of 8. is 64. And the quotient of 64 is 16. and the side of 64. is 8. for the one shanke. The halfe 16 - 4. that is 6. is the other shanke. And the halfe of 16 + 4. that is 10, is the base.

11. If the base of a triangle doe subtend a right-angle, the rectilineall fitted to it, shall be equall to the like rectilinealls in like manner fitted to the shankes thereof: And contrariwise, out of the 31. p. vj.

Or thus: If the base of a triangle doe subtend a right angle, the right lined figure made upon the base, is equall to the right lined figures like, and in like manner situate upon the feete: *H.*

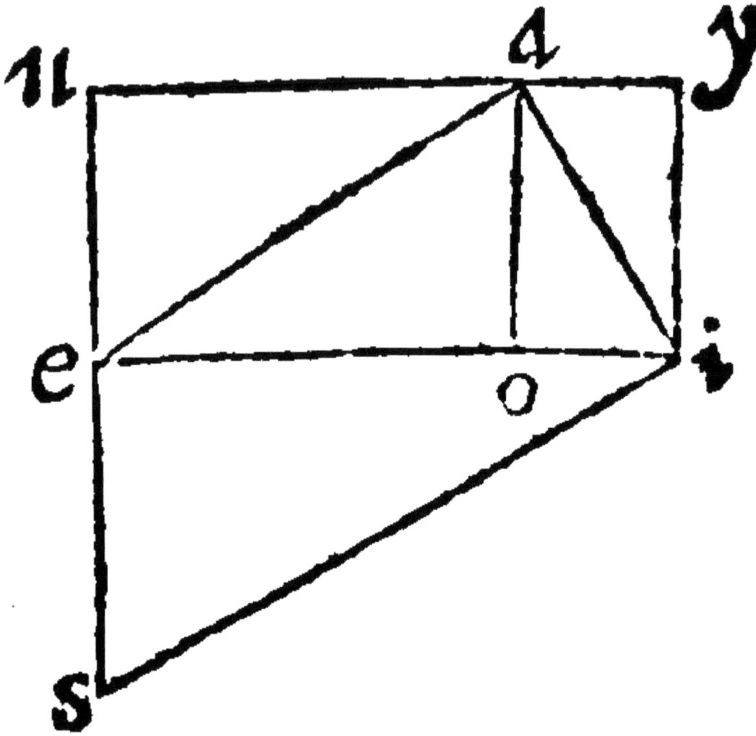

Let the right angled triangle be *aei*: and let there be also the triangles *eau*, and *aiy*, and to them upon the base of the said right angle, by the 23 e iiij. let the triangle *ies*, be made like, and in like manner situate. I say, that *eis*, is equall joyntly to *eau*, and *aiy*. Let *ao*, a perpendicular fall from the right angle *a*, to the base *ei*: This by the *ioe*, doth yeeld us twise three proportionals, to wit, *ie*, *ea*, *eo*: Item, *ei*, *ia*, *io*: Therefore, by the 25. e. iiij, as *ie*, is to *eo*: so is the triangle *ies*, to the triangle *eau*; And as *ei*, is to *oi*, so is the triangle *eis*, to the triangle *aiy*: But *ei*, is equall to *eo*, and *oi*, the whole, to wit, to his parts. Wherefore by the second composition in Arithmeticke (9. c. ij.) the triangle *eis*, is equall to the triangles *eau*, and *iay*.

The Converse is thus proved: Let the triangle be *aei*: And let the perpendicular *eo*, be erected upon *ae*, equall to *ei*: And draw a right line from *o* to *a*: Here by the former, the rectilinealls situate at *oe*, and *ea*, that is by the construction, at *ae*, and *ie*, are equall to the rightilineall at *ao*, made alike and situate alike: And by the graunt they are equall, to the rectilineall at *ai*, made alike and situated alike. Therefore seeing the like rectilineals at *ao*, and *ai*, are equall; they have by the 20 e iiij, their homologall sides equall: And the two triangles are equiliters: And by the 1 e vij, equiangles. But *aeo*, is a right angle, by the construction: And *aei*, is proved to be equall to the same *aeo*: Therefore, by the 13 e v. *aei*, also is a right angle.

12. An obliquangled triangle is either Obtusangled or Acutangled.

The division of an obliquangled triangle is taken from the speciall differences of an oblique angle. For at the 15 e iij, we were taught that an oblique angle was either obtuse or acute: Therefore an obliquangled triangle is an obtuseangle, and an Acutangle.

13. An obtusangle is that triangle which hath one blunt corner. 28.d i.

There can be but one right angle in a triangle, by the 2 e. Therefore also in it there can be but one blunt angle.

Therefore

14. If the obtuse or blunt angle be at the base of the triangle given, a perpendicular drawne from the toppe of the triangle, shall fall without the figure: And contrarywise.

As here in *aei*, the perpendicular *io*, falleth without: This is manifest by the 4 e.

And

15. If one angle of a triangle be greater than both the other two, it is an obtuse angle: And contrariwise.

This is plaine by the 6 e.

And

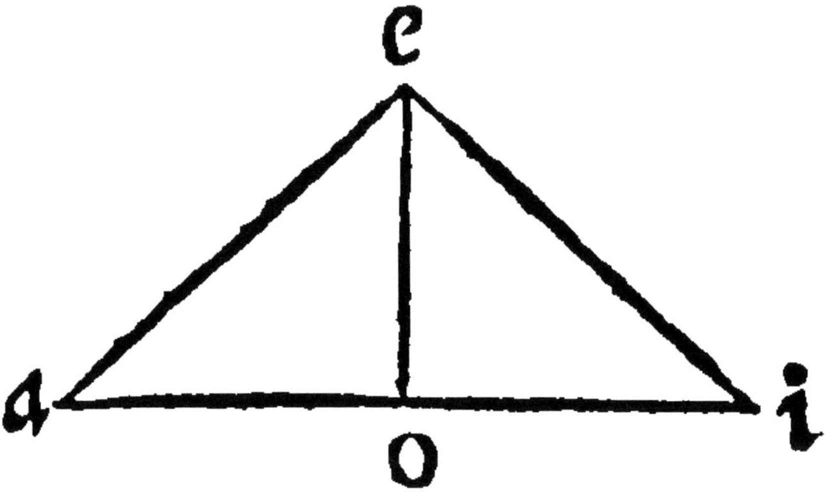

16. If a right line drawne from the toppe of the triangle cutting the base into two equall parts, be lesse than one of those halfes, the angle at the toppe is a blunt-angle. And contrariwise.

As in *aei*, the perpendicular *eo*, cutting the base *ai* into two equall parts *ao*, and *oi*: And the said *eo* is lesse than either *ao*, or *oi*: Therefore the angle *aei*, is a blunt angle

17. An acutangled triangle is that which hath all the angles acute. 29 d j.

Therefore

18. A perpendicular drawne from the top falleth within the figure: And contrariwise.

As in *aei*, the perpendicular *ao* falleth within as is plaine by the 4 e.

And

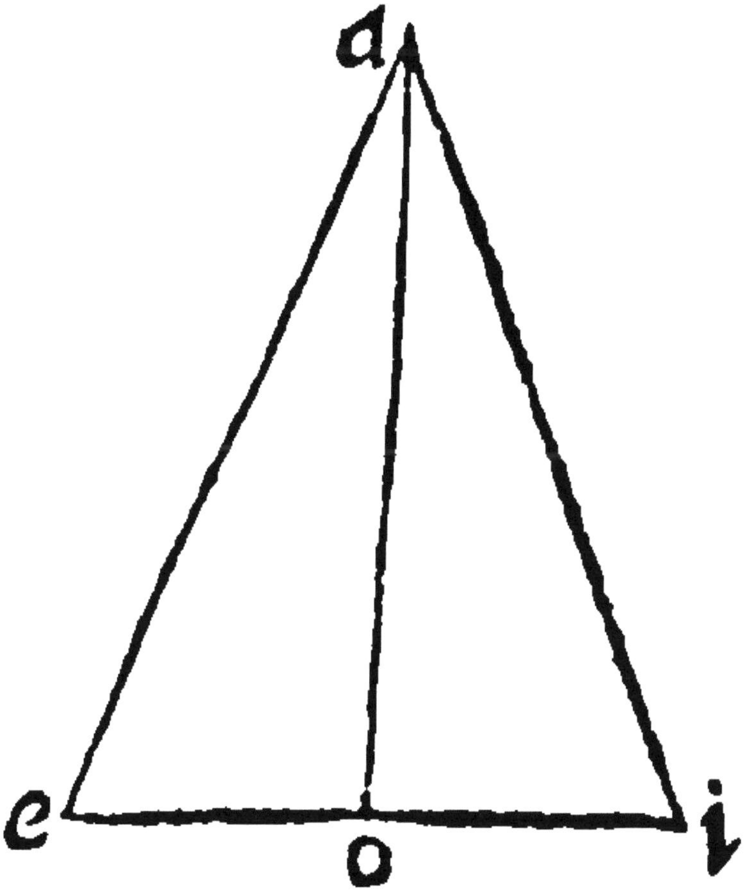

19. If any one angle of triangle be lesse then the other two, it is acute: And contrariwise.

As is manifest by the 6 e.

And

20. If a right line drawne from the toppe of the triangle; cutting the base into two equall parts, be greater than either of those portions, the angle at the toppe is an acute angle: And contrariwise.

As in *aei*, let *ao* cutting the base *ei* into two equall parts, be greater than any one of those parts, the angle at the toppe is an acuteangle, as appeareth by the 7 e.

The ninth Booke, of P. Ramus *Geometry, which intreateth of the measuring of right lines by like right-angled* triangles.

The Geometry of like right-angled triangles, amongst many other uses that it hath, it doth especially afford us the geodæsy or measuring of right lines: And that mastery, which before (at the 2 e viij) attributed the right angled triangles, shall here be found to be a true mastery indeed. For it shall containe the geodesy of right lines; and afterward the geodesy of plaines and solides, by the measuring of their sides, which are right lines.

1. For the measuring of right lines; we will use the Iacobs staffe, which is a squire of unequall shankes.

Radius, commonly called *Baculus Iacob, Iacobs* staffe, as if it had been long since invented and practised by that holy Patriarke, is a very auncient instrument, and of all other Geometricall instruments, commonly used, the best and fittest for this use. *Archimedes* in his book of the Number of the sand, seemeth to mention some such thing: And *Hipparchus*, with an instrument not much unlike this, boldly attempted an haynous matter in the sight of God, as *Pliny* thinketh, namely to deliver unto posterity the number of the starres, and to assigne or fixe them in their true places by the *Norma*, the squire or *Iacobs* staffe. And indeed true it is that the Radius is not onely used for the measuring of the earth and land: But especially for the defining or limiting of the starres in their places and order: And for the describing and setting out of all the regions and waies of the heavenly city. Yea and *Virgill* the famous Poet, in his 3 *Ecloge, Ecquis fuit alter, Descripsit radio totum, qui gentibus orbem?* and againe afterward in the 6 of his *Eneiades*, hath noted both these uses. *Cœliquè meatus. Describent radio & surgentia sidera dicent.* Long after this the *Iewes* and *Arabians*, as *Rabbi Levi*; But in these latter daies, the *Germaines* especially, as *Regiomontanus*; *Werner, Schoner*, and *Appian* have grac'd it: But above all other the learned *Gemma Phrisius* in a severall worke of that argument onely, hath illustrated and taught the use of it plainely and fully.

The *Iacobs* staffe therefore according to his owne, and those Geometricall parts, shall here be described (The astronomicall distribution wee reserve to his time and place.) And that done, the use of it shall be shewed in the measuring of lines.

This instrument, at the discretion of the measurer may be greater or lesser. For the quantity of the same can no otherwayes be determined.

2. The shankes of the staffe are the Index and the Transome.

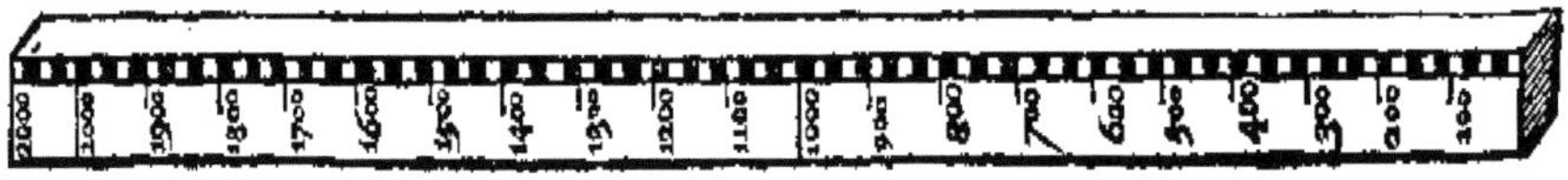

The principall parts of this instrument are two, the *Index*, or *Staffe*, which is the greater or longer part: and the *Transversarium*, or Transome, and is the lesser and shorter.

3. The Index is the double and one tenth part of the transome.

Or thus: The Index is to the transversary double and 1/10 part thereof. *H.* As here thou seest.

> *4. The Transome is that which rideth upon the Index, and is to be slid higher or lower at pleasure.*

Or, The transversary is to be moved upon the Index, sometimes higher, sometimes lower: *H.* This proportion in defining and making of the shankes of the instrument is perpetually to be observed: as if the transome be 10. parts, the Index must be 21. If that be 189. this shall be 90. or if it be 2000. this shall be 4200. Neither doth it skill what the numbers be, so this be their proportion. More than this, That the greater the numbers be, that is the lesser that the divisions be, the better will it be in the use. And because the Index must beare, and the transome is to be borne; let the index be thicker, and the transome the thinner.

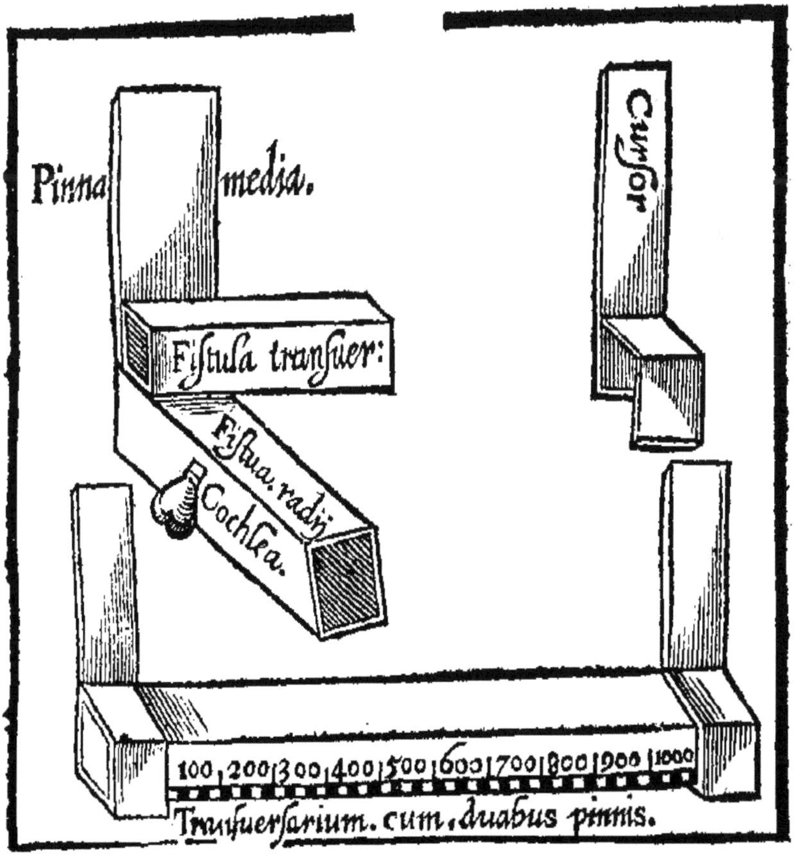

But of what matter each part of the staffe be made, whether of brasse or wood it skilleth not, so it be firme, and will not cast or warpe. Notwithstanding, the transome will more conveniently be moved up and downe by brasen pipes, both by it selfe, and upon the Index higher or lower right angle wise, so touching one another, that the alterne mouth of the one may touch the side of the other. The thrid pipe is to be moved or slid up and downe, from one end of the transome to the other; and therefore it may be called the *Cursor*. The fourth and fifth pipes, fixed and immoveable, are set upon the ends of the transome, are unto the third and second of equall height with finnes, to restraine when neede is, the opticke line, and as it were, with certaine points to define it in the transome.

The three first pipes may, as occasion shall require, be fastened or staied with brasen scrues. With these pipes therefore the transome may be made as great, as need shall require, as here thou seest.

The fabricke or manner of making the instrument hath hitherto beene taught, the use thereof followeth: unto which in generall is required: First, a just distance. For the sight is not infinite. Secondly, that one eye be closed: For the optick faculty conveighed from both the eyes into one, doth aime more certainely; and the instrument is more fitly applied and set to the cheeke bone, then to any other place. For here the eye is as it were the center of the circle, into which the transome is inscribed. Thirdly, the hands must be steady; for if they shake, the proportion of the Geodesy must needes be troubled and uncertaine. Lastly, the place of the station is from the midst of the foote.

5. If the sight doe passe from the beginning of one shanke, it passeth by the end of the other: And the one shanke is perpendicular unto the magnitude to be measured, the other parallell.

These common and generall things are premised. That the sight is from the beginning of the Index by the end of the transome; Or contrariwise, From the beginning of the transome, unto the end of the Index. And that the Index is right, that is, perpendicular to the line to be measured, the transome parallell. Or contrariwise. Now the perpendicularity of the Index, in measurings of lengthts, may be tried by a plummet of lead appendent; But in heights and breadths, the eye must be trusted; although a little varying of the plummet can make no sensible errour. By the end of the transome, understand that which is made by the line visuall, whether it be the outmost finne, or the Cursour in any other place whatsoever.

6. Length and Altitude have a threefold measure; The first and second kinde of measure require but one distance, and that by granting a dimension of one of them, for the third proportionall: The third two distances, and such onely is the dimension of Latitude.

Geodesy of right lines is two fold; of one distance, or of two. Geodesy of one distance is when the measurer for the finding of the desired dimension doth not change his place of standing. Geodesy of two distances is when the measurer by reason of some impediment lying in the way betweene him and the magnitude to be measured, is constrained to change his place, and make a double standing.

Here observe, That length and heighth, may be joyntly measured both with one, and with a double station: But breadth may not be measured otherwise than with two.

7. If the sight be from the beginning of the Index right or plumbe unto the length, and unto the farther end of the same, as the segment of the Index is, unto the segment of the transome, so is the heighth of the measurer unto the length.

Let therefore the segment of the Index, from the toppe, I meane, unto the transome be 6. parts. The segment of the transome, to wit, from the Index unto the opticke line be 18. The Index, which here is the heighth of the measurer, 4. foote: The length, by the rule of three, shall be 12. foote. The figure is thus, for as *ae*, is to *ei*, so is *ao*, unto *ou*, by the 12. e vij. For they are like triangles. For *aei*, and *aou*, are right angles: And that which is at *a* is common to them both: Wherefore the remainder is equall to the remainder, by the 4. e vij.

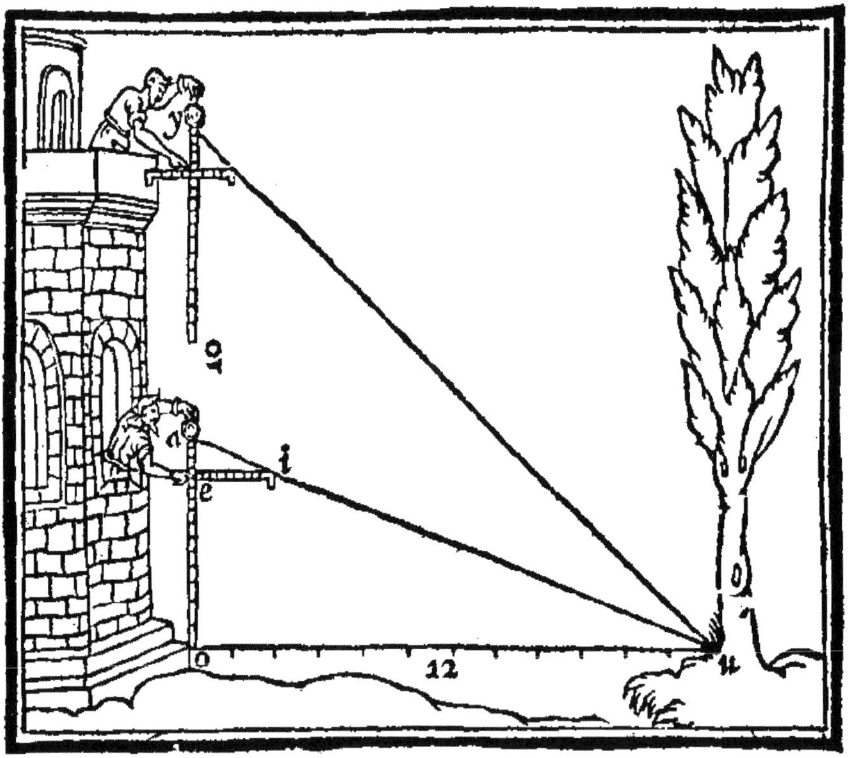

The same manner of measuring shall be used from an higher place; as out of *y*, the segment of the Index is 5. parts; the segment of the transome 6: and then the height be 10 foote: the same Length shall be found to bee 12 foote.

Neither is it any matter at all, whether the length in a plaine or levell underneath: Or in an ascent or descent of a mountaine, as in the figure under written.

Thus mayest thou measure the breadths of Rivers, Valleys, and Ditches. For the Length is alwayes after this manner, so that one may measure the distance of shippes on the Sea, as also *Thales Milesius*, in *Proclus* at the 26 p j, did measure them. An example thou hast here.

Hereafter in the measuring of Longitude and Altitude, sight is unto the toppe of the heighth. Which here I doe now forewarne thee of, least afterward it should in vaine be reitered often.

The second manner of measuring a Length is thus:

8. If the sight be from the beginning of the index parallell to the length to be measured, as the segment of the transome is, unto the segment of the index, so shall the heighth given be to the length.

As if the segment of the Transome be 120 parts: the height given 400 foote: The segment of the Index 210 parts: The length, by the golden rule shall be 700 foote. The figure is thus. And the demonstration is like unto the former; or indeed more easier. For the triangles are equiangles, as afore. Therefore as *ou* is to *ua*: so is *ei* to *ia*.

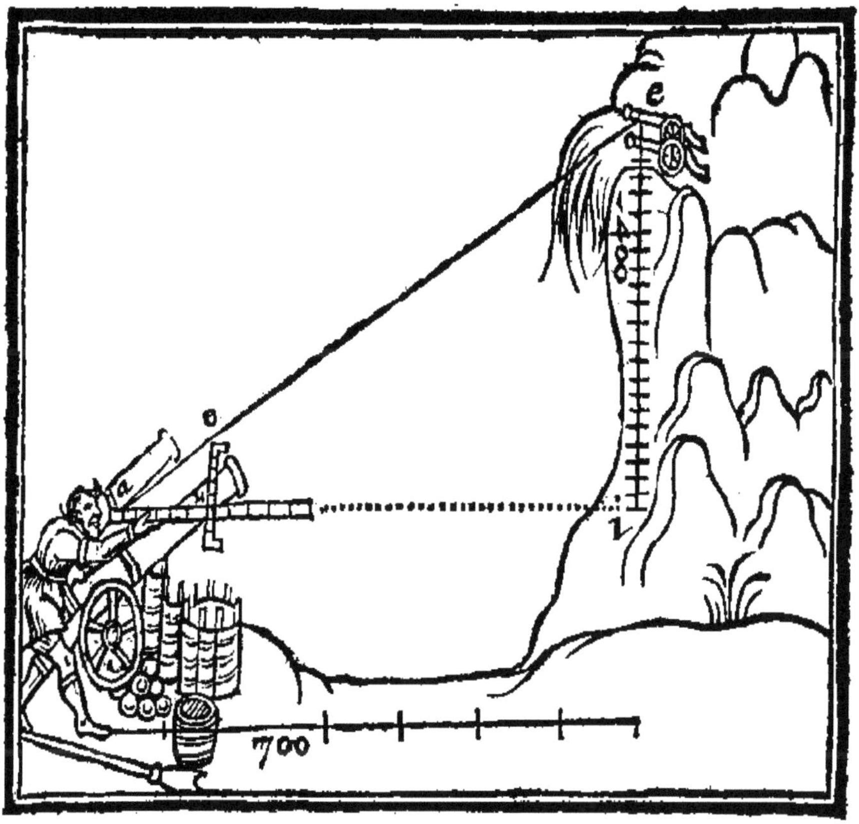

This is the first and second kinde of measuring of a Longitude, by one single distance or station: The third which is by a double distance doth now follow. Here the transome, if there be roome enough for the measurer to goe farre enough backe, must be put lower, in the second distance.

> *9. If the sight be from the beginning of the transverie parallell to the length to be measured, as in the index the difference of the greater segment is unto the lesser; so is the difference of the second station unto the length.*

This kinde of Geodæsy is somewhat more subtile than the former were. The figure is thus; in which let the first ayming be from *a*, the beginning of the transome, and out of *ai* the length sought by *o*, the end of the Index, unto *e*, the toppe of the heighth: And let the segment of the Index be *ou*: The second ayming let it be from *y*, the beginning of the transome, out of a greater distance by *s*, the end of the Index, unto *e*, the same note of the heighth: And let the segment of the Index be *sr*.

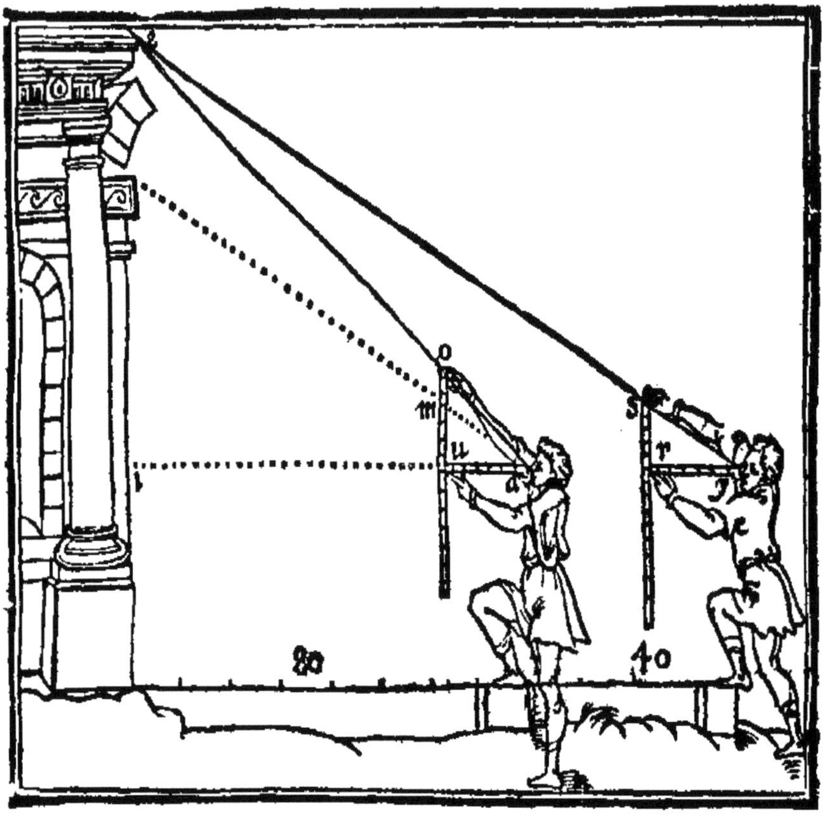

Here the measuring performed, is the taking of the difference betweene *ou* and *sr*. The rest are faigned onely for demonstrations sake. Therefore in the first station let *aml*, be from the beginning of the transome, be parallell to *ye*. Here first *mu*, is equall to *sr*. For the triangles *mua*, and *sry*, are equall in their shankes *ua*, and *ry*, by the grant (Because the transome standeth still in his owne place:) And the angles at *mua*, *uam*, are equall to the angles: And all right angles are equall, by the 14 e iij. These are the outter and inner opposite one to another: And such are equall by the 1 e v. Therefore they are equilaters, by the 2 e vij; And *om*, is the difference of the segments of the Index. Then as *om* is to *mu*, so is *el*, to *li*; as the equation of three degrees doth shew. For, by the 12 e vij, as *om* is to *ma*: so is *el* to *la*: And as *ma* is to *mu*; so is *la*, to *li*. Therefore by right, as *om*, is to *mu*: so is *el*, to *li*: And by the 12 e vj, so is *ya*, to *ai*: As if the difference of the first segment be 36 parts: The second segment be 72 parts: The difference of the second station 40 foote. The length sought shall be 80 foote. And here indeed is no heighth definitely given, that may make any bound of the principall proportion. Notwithstanding the Heighth, although it be of an unknowne measure, is the bound of the length sought: And therefore it is an helpe and meanes to argue the question. Because it is conceived to stand plumbe upon the outmost end of the length.

Therefore that third kinde of measuring of length is oftentimes necessary, when by neither of the former wayes the length may possibly be taken, by reason of some impediment in the way, to wit of a wall, or tree, or house, or mountaine, whereby the end of the length may not be seene, which was the first way: Nor an height next adjoyning to the end of the length is given, which is the second way.

Hitherto we have spoken of the threefold measure of longitude, the first and second out of an heighth given the third cut of a double distance: The measuring of heighth followeth next, and that is also threefold. Now heighth is a perpendicular line falling from the toppe of the magnitude, unto the ground or plaine whereon the measurer doth stand, after which manner Altitude or heighth was defined at the 9 e iiij. The first geodesy or manner of measuring of heighths is thus.

10. If the sight be from the beginning of the transome perpendicular unto the height to be measured, as the segment of the transome, is unto the segment of the Index, so shall the length given be to the height.

Let the segment of the transome be 60 parts: the segment of the Index 36: the Length given 120 foote: the height sought shall be, by the golden rule, 72 foote.

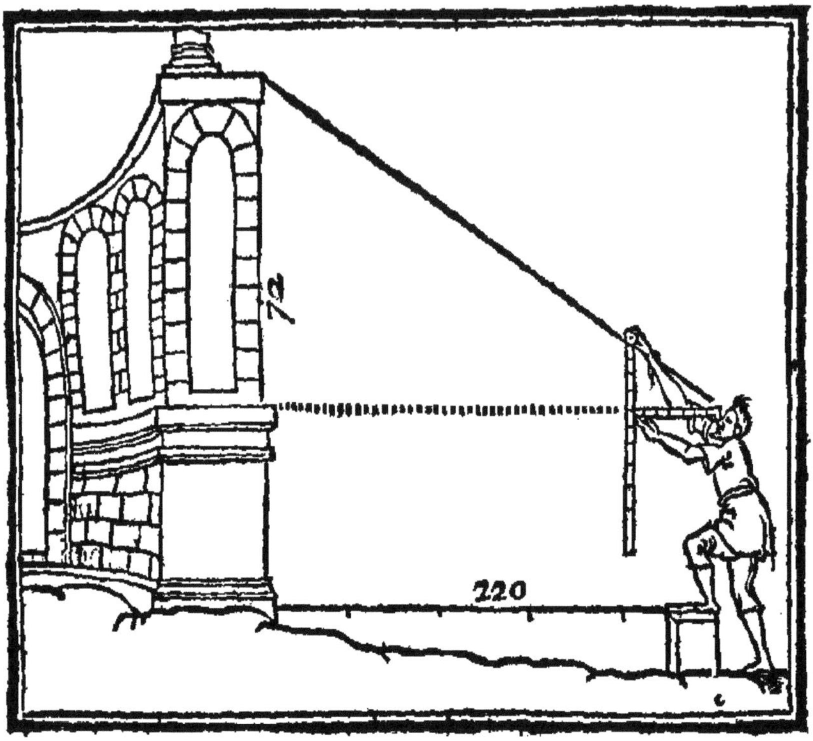

The Figure is thus: And the demonstration is by the 12 e vij, as afore: but here is to be added the height of the measurer; which if it be 4 foot, the whole height shall be 76 foote.

Therefore in an eversed altitude

11. If the sight be from the beginning of the Index parallell to the height, as the segment of the transome is, unto the segment of the index, so shall the length given be, unto the height sought.

Eversa altitudo, An eversed altitude (Reversed, *H*:) is that which we call depth, which indeed is nothing else, in the Geometers sense, but heighth turned topsie turvie, as we say, or with the heeles upward. For out of the heighth concluded by subducting that which is above ground, the heighth or depth of a Well shall remaine.

Let the segment of the transome *ae*, be 5 parts: the segment of the Index *ei*, be 13: the diameter of the Well (which now standeth for the length:) be 10 foote, which at toppe is supposed to be equall to that at bottome: the opposite height, by the 12 e vij, and the golden rule shall be 26 foote: From whence you must take the segment of the Index reaching over the mouth of the Well: And the true height (or depth) shall remaine; as if that segment of 13 parts be as much as 2 foote, the height sought shall be 24 foote. The second manner of measuring of heights followeth.

12. If the sight be from the beginning of the Index perpendicular to the heighth to be measured, as the segment of the Index is unto the segment of the Transome, so shall the length given be to the heighth.

As if the segment of the Index be 60 parts: and the segment also of the transome be 60: And the Length given be 250 foote: By the Rule of three, the height also shall be 250 foote: as thou seest in the example underneath: For as *ae* is to *ei*; so is *aeo* to *ou*, by the 12 e vij. But here unto the height found, you must adde the height of the measurer: Which if it be 4 foot, the whole height shall be 254 foote.

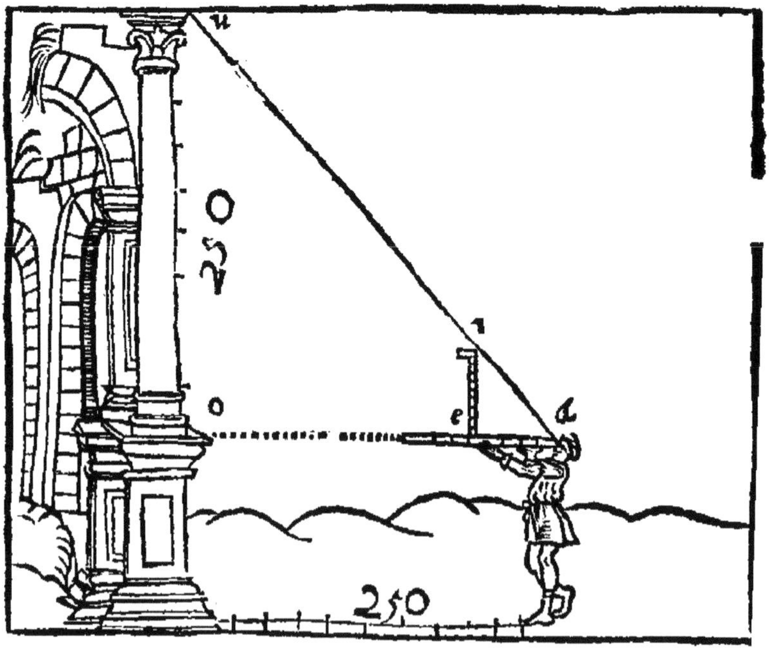

Therefore

> *13. If the sight be from the beginning of the Index (perpendicular to the magnitude to be measured) by the names of the transome, unto the ends of some known part of the height, as the distance of the Names is, unto the rest of the transome above them, so shall the known part be unto the part sought.*

Or thus: If the sight passe from the beginning of the Index being right, by the vanes of the transversary, to the tearmes of some parts; as the distance of the vanes is unto the rest of the transversary above the index, so is the part knowne unto the remainder: *H.*

This is a consectary of a knowne part of an height, from whence the rest may be knowne, as in the figure.

As *ou* is unto *uy*, so is *ei* to *is*. For as *ou*, is unto *ua*: so is *ei* unto *ia*, by the 12 e vij. And as *ua*, is to *uy*, so is *ia* unto *is*; and by right, as *ou*, is to *uy*, so is *ei*, to *is*. Here thou hast three bounds of the proportion. Let therefore *ou*, be 20 parts: *uy* 30: And *ei*, the knowne part, let it be 15 foote: Therefore thou shalt conclude *is*, the rest to be 22½.

The first and second kinde of measuring of heights is thus: The third followeth.

14 If the sight be from the beginning of the Index perpendicular to the heighth, as in the Index the difference of the segment, is unto the difference of the distance or station; so is the segment of the transome unto the heighth.

Hitherto you must recall that subtilty, which was used in the third manner of measuring of lengths.

Let the first aime be taken from *a*, the beginning of the Index perpendicular unto the height to be measured: And from an unknowne length *ai*, by *o*, the end of the transome, unto *e*, the toppe of the height *ei*: And let the segment of the Index be *ua*. The second ayme, let it be taken from *y*, the beginning of the same Index; and out of a greater distance, by *s*, the end of the transome, unto the same toppe *e*. And the segment of the Index let it be *ry*.

Here, as afore, the measuring is performed and done, by the taking of the difference of the said *yr*, above *au*: Now the demonstration is concluded, as in the former was taught. Let the parallell *lsm*, be erected against *aoe*.

Here first the triangles *oua*, & *srl*, are equilaters, by the 2 e vij.; (seeing that the angles at *a*, and *l*, the externall and internall, are equall in bases *ou*, and *sr*, for the segment in each distance is the same still:) Therefore *ua*, is equall to *rl*. Now the rest is concluded by a sorites of foure degrees: As *yr*, is unto *yi*: so by the 12. e vij. is *sr*, that is, *ou*, unto *ei*: And as *ou*, is unto *ei*, so is *au*, that is, *lr*, unto *ai*. Therefore the remainder *yl*, unto the remainder *ya*; shall be as *yr*, is unto the whole *yi*, and therefore from the first unto the last, as *sr*, is to *ei*.

Therefore let the difference of the Index be 23. parts: The difference of the distance 30. foote: The segment of the transome 44. parts: The height shall be 57.9/23. or foote.

Therefore

15 Out of the Geodesy of heights, the difference of two heights is manifest.

Or thus: By the measure of one altitude, we may know the difference of two altitudes: *H*.

For when thou hast taken or found both of them, by some one of the former wayes, take the lesser out of the greater; and the remaine shall be the heighth desired. From hence therefore by one of the towers of unequall heighth, you may measure the heighth of the other. First out of the lesser, let the length be taken by the first way: Because the height of the lesser, wherein thou art, is easie to be taken, either by a plumbe-line, let fall from the toppe to the bottom, or by some one of the former waies. Then measure the heighth, which is above the lesser: And adde that to the lesser, and thou shalt have the whole heighth, by the first or second way. The figure is thus, and the demonstration is out of the 12. e vij. For as *ae*, is to *ei*, so is *ao*, to *ou*. Contrariwise out of an higher Tower, one may measure a lesser.

16 If the sight be first from the toppe, then againe from the base or middle place of the greater, by the vanes of the transome unto the toppe of the lesser heighth; as the said parts of the yards are unto the part of the first yard; so the heighth betweene the stations shall be unto his excesse above the heighth desired.

Let the unequall heights be these, *as*, the lesser, and *uy*, the greater: And out of the assigned greater *uy*, let the lesser, *as*, be sought. And let the sight be first from *u*, the toppe of the greater, unto *a*, the toppe of the lesser, making at the shankes of the staffe the triangle *urm*. Then againe let the same sight be from the base, or from the lower end of *uy*, the heighth given, unto *a*, the same toppe of the lesser, making by the shankes of the staffe the triangle *yln*, so that the segments of the yard be, the upper one, I meane, *ur*, the neather one *ul*: I say the whole of *ur*, and *nl*, is unto *ur*: so is the *uy*, greater heighth assigned, unto *as*, the lesser sought.

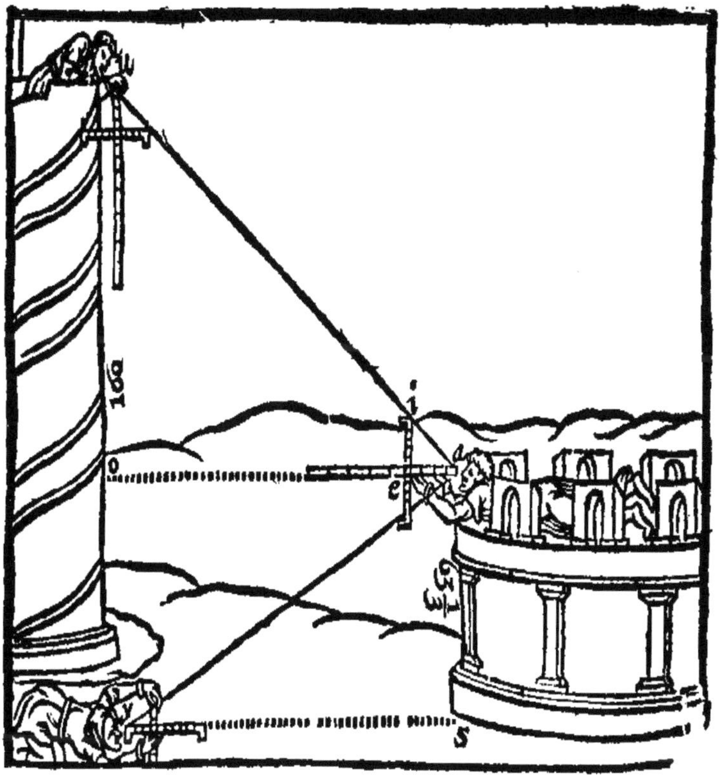

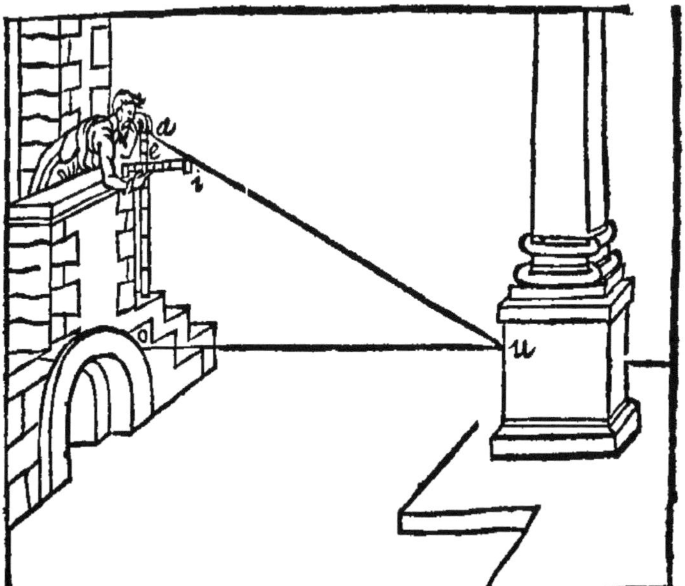

The Demonstration, by drawing of *ao*, a perpendicular unto *uy*, is a proportion out of two triangles of equall heighth. For the forth of the totall equally heighted triangles *uao*, and *yas*, although they be reciprocall in situation, they have their bases *uo*, and *as*, as if their were *oy*. Then they have the same with the whole triangles; as also the subducted triangles *urm*, and *ynl*, of equal heighth; to wit whose common heighth is the segment of the transome remained still in the same place, there *rm*, here *yl*. And therefore the bases of these, namely, the segments of the yards *ur*, and *nl*, have the same rate with *uo*, unto *oy*. As therefore *uo*, is unto *oy*: so is, *ur*, unto *nl*. And backward, as *nl*, is to *ur*; so is, *yo*, unto *ou*, as here thou seest:

nl,———*ur*: *yo*,———*ou*.

Therefore furthermore by composition of the Antecedent with the Consequent unto the Consequent, by the 5 c 9 ij. Arith. As *nl*, and *ur*, are unto *ur*: so are *yo*, and *ou*, unto *ou*, that is *yu*, unto *ou*, on this manner.

nl,———	*ur*,	*yo*,———	*ou*,
nr,		*ou*,	
...	*ur*,	*yu*,	*ou*,

there is given *nl*, and *ur*, for the first proportionall: *ur*, for the second: and *yu*, for the third: Therefore there is also given *ou*, for the fourth: Which *ou*, subducted out of *uy*, there remaineth *oy*, that is, *as*, the lesser altitude sought.

For let the parts of the yard be 12. and 6. and the summe of them 18. Now as 18. is unto 12. so is the whole altitude *uy*, 190. foote, unto the excesse 126⅔ foote. The remainder therefore 63⅓ foote, shall be *as*, the lesser heighth sought.

But thou maist more fitly dispose and order this proportion thus: As *ur*, is unto *nl*: so is *uo* unto *oy*. Therefore by Arithmeticall composition, as *ur*, and *nl*, are unto *nl*: so *uo*, and *oy*, that is, the whole *uy*, is unto *oy*, that is, unto *as*. For here a subduction of the proportion, after the composition is no way necessary, by the crosse rule of societia, thus:

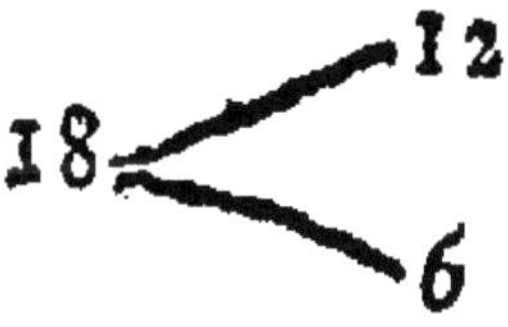

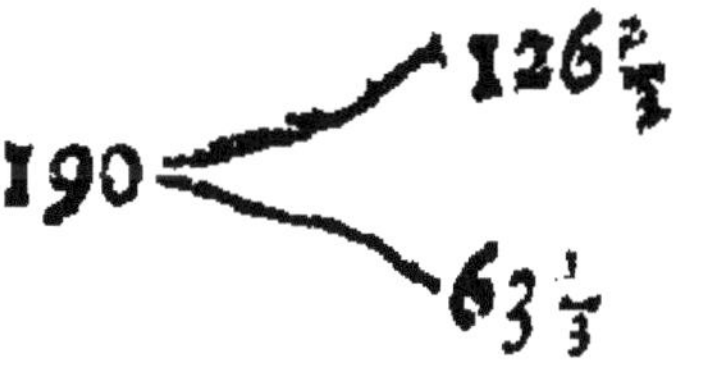

The second station might have beene in *o*, the end of the perpendicular from *a*. But by taking the ayme out of the toppe of the lesser altitude, the demonstration shall be yet againe more easie and short, by the two triangles at the yard *aei*, and *aef*, resembling the two whole triangles *aou*, and *aoy*, in like situation, the parts of the shanke cut, are on each side the segments of the transome.

One may againe also out of the toppe of a Turret measure the distance of two turrets one from another: For it is the first manner of measuring of longitudes, neither doth it here differ any whit from it, more than the yard is hang'd without the heighth given. The figure is thus: And the Demonstration is by the 12. e vij. For as *ae*, the segment of the yard, is unto *ei* the segment of the transome: so is the assigned altitude *ao*, unto the length *ou*.

The geodesy or measuring of altitude is thus, where either the length, or some part of the length is given, as in the first and second way: Or where the distance is double, as in the third.

17 If the sight be from the beginning of the yard being right or perpendicular, by the vanes of the transome, unto the ends of the breadth; as in the yard the difference of the segment is unto the differēce of the distance, so is the distance of the vanes unto the breadth.

The measuring of breadth, that is of a thwart or crosse line, remaineth. The Figure and Demonstration is thus: The first ayming, let it be *aei*, by *o*, and *u*, the vanes of the transome *ou*. The second, let it be *yei*, by *s*, and *r*, the vanes of the transome *sr*. Then by the point *s*, let the parallell *lsm*, be drawne against *aoe*. Here first, the triangles *oua*, and *sil*, are equilaters, by the 2 e vij. Because the angles at *n* and *j*, are right angles: And *uao*, and *jls*, the outter and inner, are equall in their bases *ou*, and *sj*, by the grant: Because here the segment of the transome remaineth the same: Therefore *ua*, is equall to *jl*. These grounds thus laid, the demonstration of the third altitude here taken place. For as *yl*, is unto *ya*: so is *sj*, unto *er*: And, because parts are proportion-all unto their multiplicants, so is *sr*, unto *ei*: for the rest doe agree.

The same shall be the geodesy or manner of measuring, if thou wouldest from some higher place, measure the breadth that is beneath thee, as in the last example. But from the distance of two places, that is, from latitude or breadth, as of Trees, Mountaines, Cities, Geographers and Chorographers do gaine great advantages and helpes.

Wherefore the geodesy or measuring of right lines is thus in length, heighth, and breadth, from whence the Painter, the Architect, and Cosmographer, may view and gather of many famous place the windowes, the statues or imagery, pyramides, signes, and lastly, the length and heighth, either by a single or double: the breadth by a double dimension onely, that is, they may thus behold and take of all places the nature and symmetry; as in the example next following thou mayst make triall when thou pleasest.

The tenth Booke of Geometry, of a Triangulate and Parallelogramme.

And thus much of the geodesy of right lines, by the meanes of rectangled triangles: It followeth now of the triangulate.

1. A triangulate is a rectilineall figure compounded of triangles.

As before (for the dichotomies sake) of a line was made a Lineate, to signifie the *genus* of surface and a Body: so now is for the same cause of a triangle made a Triangulate, to declare and expresse the *genus* of a Quadrilater and Multilater, and indeed more justly, then before in a Lineate. For triangles doe compound and make the triangulate, but lines doe not make the lineate.

Therefore

2. The sides of a triangulate are two more than are the triangles of which it is made.

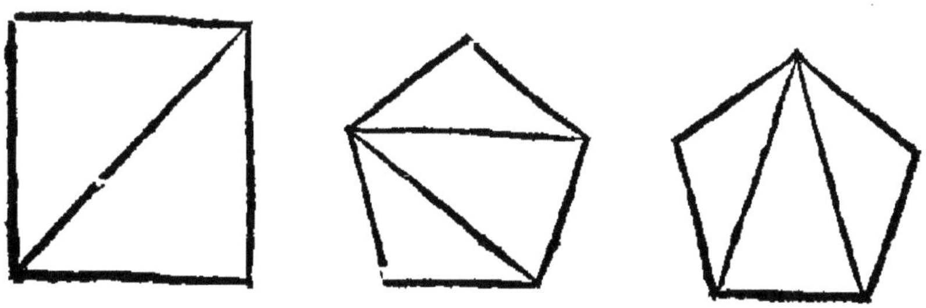

As the sides of a Quadrangle are 4. Therefore the triangles which doe make the same foure-sided figure are but 2. The sides of a Quinquangle are 5, Therefore the triangles are 3, and so forth of the rest, as here thou seest. And that indeed is the least: For even a triangle it selfe, may be cut into as many triangles as one please.

That both the inner and outter are equall to right angles, in every kinde of right line figure, it was manifest at the 4 e vj. The inner in a Quadrangle, are equall to 4. In a Quinquangle, to 6: In an Hexangle, to 8; and so forth.

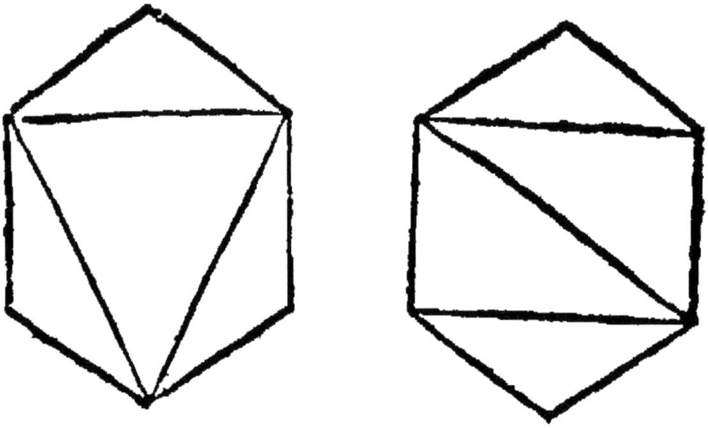

But the outter, in every right-lined figure, are equall to 4 right angles: as here may be demonstrated, by the 14 e v and 13 e vj.

And

3. Homgeneall Triangulates are cut into an equall number of triangles, è 20 p vj.

For if they be Quadrangles, they be cut into two triangles: If Quinquangles, into 3. If Hexangles, into 4, and so forth.

4. Like triangulates are cut into triangles alike one to another and homologall to the whole è 20 p vj.

Or thus: Like Triangulates are divided into triangles like one unto another, and in porportion correspondent unto the whole: *H.*

As in these two quinquangles. First the particular triangles are like betweene themselves. For the shankes of *aeu* and *ysm*, equall angles are proportionall, by the grant. Therefore the triangles themselves are equiangles, by 14 e vij. And therefore alike, by the 12 e vij, and so forth of the rest.

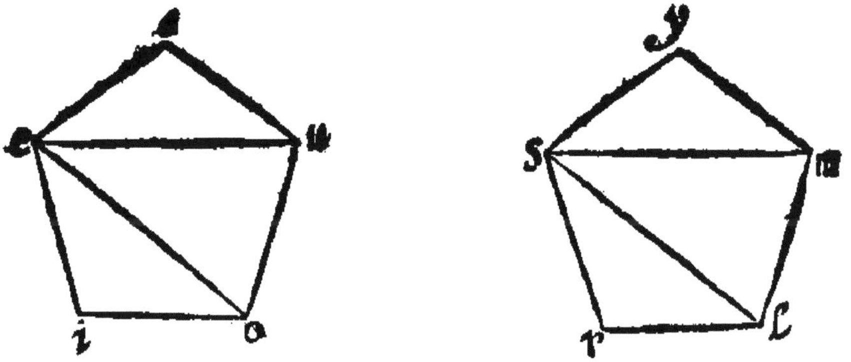

The middle triangles, the equall angles being substracted shall have their other angles equall: And therefore they also shall be equiangles and alike, by the same.

Secondarily, the triangles *aeu*. and *ysm*: *eio* and *srl*; *eou*, and *slm*, to wit, alike betweene themselves, are by the 1 e vj, in a double reason of their homologall sides *eu, sm, eo, sl,* which reason is the same, by meanes of the common sides. Therefore three triangles are in the same reason: And therefore they are proportionall: And, by the third composition, as one of the antecedents is, unto one of the consequents; so is the whole quinquangle to the whole.

5. A triangulate is a Quadrangle or a Multangle.

The parts of this partition are in Euclide, and yet without any shew of a division. And here also, as before, the species or severall kinds have their denomination their angles, although it had beene better and truer to have beene taken from their sides; as to have beene called a Quadrilater, or a Multilater. But in words use must bee followed as a master.

6. A Quadrangle is that which is comprehended of foure right lines. 22 d j.

As here thou seest. But a Quadrangle may also bee a sphearicall, and a conicall, and a cylindraceall, and that those differences are common, we doe foretell at the 3 e v. And a Quadrangle may be a plaine, which is not a quadrilater, as here.

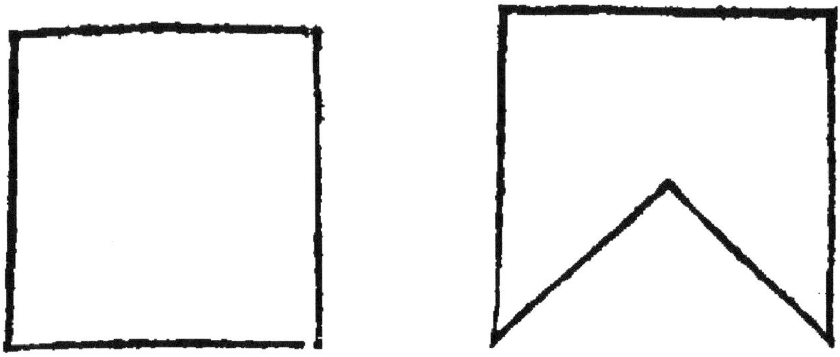

7. *A quadrangle is a Parallelogramme, or a Trapezium.*

This division also in his parts is in the Elements of Euclide, but without any forme or shew of a division. But the difference of the parts shall more fitly be distinguished thus: Because in generall there are many common parallels.

8. A Parallelogramme is a quadrangle whose opposite sides are parallell.

As in the example, the side *ae*, is parallell to the side *io*: And the side *ei*, is parallell to opposite side *ao*.

Therefore

9. If right lines on one and the same side, doe joyntly bound equall and parallall lines, they shall make a parallelogramme.

The reason is, because they shall be equall and parallell betweene themselves, by the 26. e v.

And

10 A parallelogramme is equall both in his opposite sides, and angles, and segments cut by the diameter.

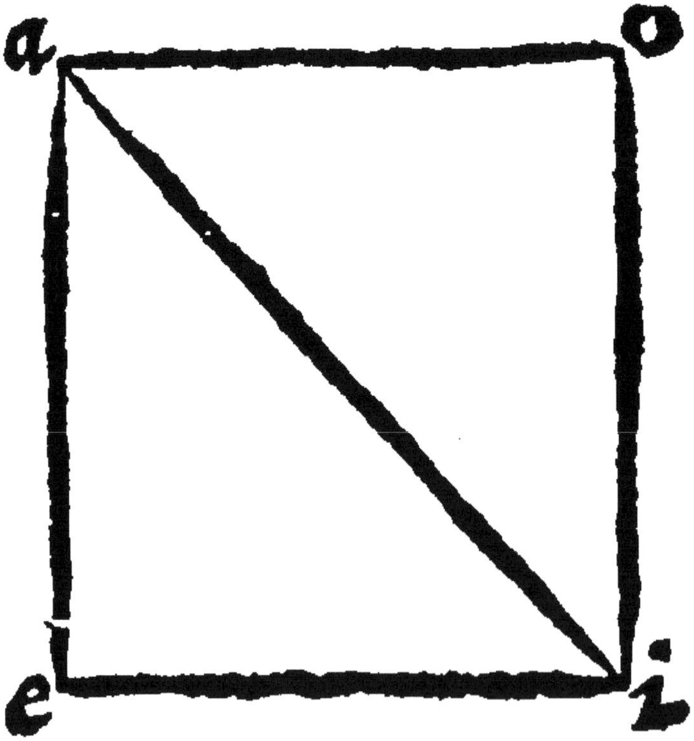

Or thus: The opposite, both sides, and angles, and segments cut by the diameter are equall. Three things are here concluded: The first is, that the opposite sides are equall: This manifest by the 26 e v. Because two right lines doe jointly bound equall parallells.

The second, that the opposite angles are equall, the Diagonall *ai*, doth shew. For it maketh the triangles *aei*, and *ioa*, equilaters: And therefore also equiangles: And seeing that the particular angles at *a*, and *i*, are equall, the whole is equall to the whole. This part is the 34. p j;

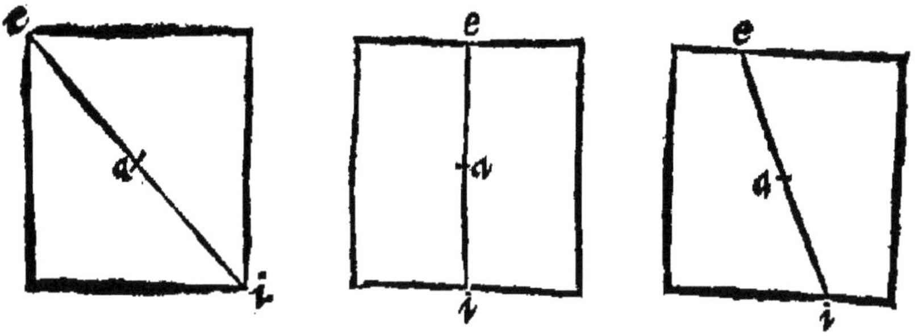

The third: The segments cut by the diameter are alwayes equall, whether they be triangles, or any manner of quadrangles, as in the figures. For the Diameter doth cut into two equall parts, the parallelogramme by the Angles, or by the opposite sides, or by the alternall equall segments of the sides.

And

11. The Diameter of a parallelogramme is cut into two by equall raies.

As in the three figures *aei*, next before: This a parallelogramme hath common with a circle, as was manifest at the 28. e iiij.

And

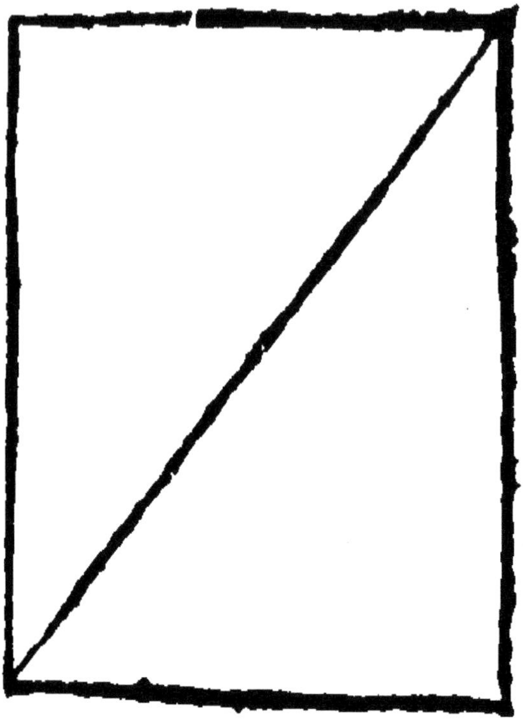

12 A parallelogramme is the double of a triangle of a trinangle of equall base and heighth, 41. p j.

The comparison first in rate of inequality of a parallelogramme with a triangle, doth follow: As here thou seest in this diagramme. For a parallelogramme is cut into two equall triangles, by the antecedent. Therefore it is the double of the halfe.

And

13 A parallelogramme is equall to a triangle of equall heighth and double base unto it: è 42. p j.

As to *aei*, the triangle, the parallelogramme *aoiu*, is equall: because halfe of the parallelogramme is equall to the triangle: Therefore the halfes being equall, whole also shall be equall.

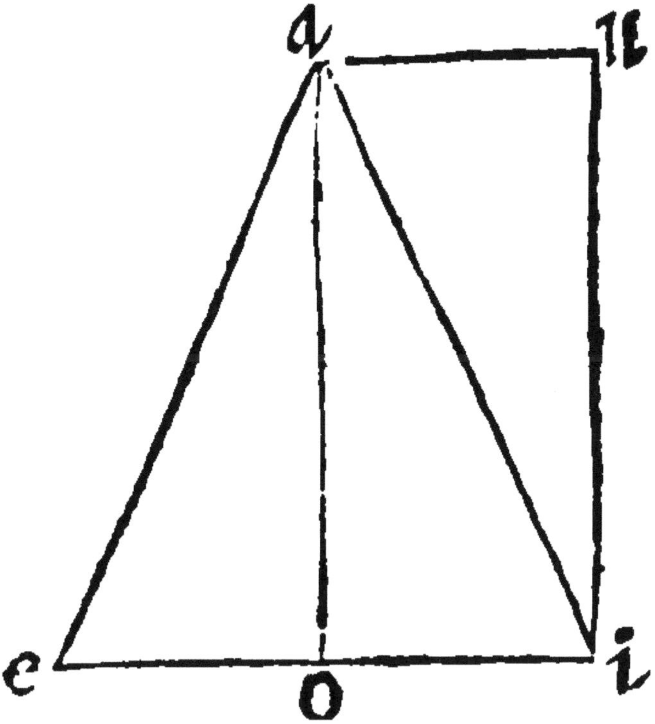

From whence one may

14 To a triangle given, in a rectilineall angle given, make an equall parallelogramme.

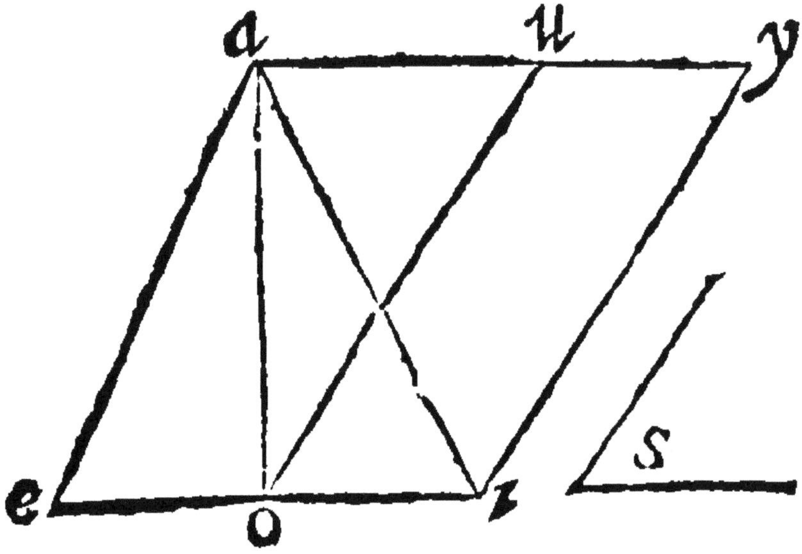

As here to the triangle, *aei*, given in *s*, the right lined angle given, you may equall the parallelogramme *ouyi*.

15 A parallelogramme doth consist both of two diagonals, and complements, and gnomons.

For these three parts of a parallelogramme are much used in Geometricall workes and businesses, and therefore they are to be defined.

16 The Diagonall is a particular parallelogramme having both an angle and diagonall diameter common with the whole parallelogramme.

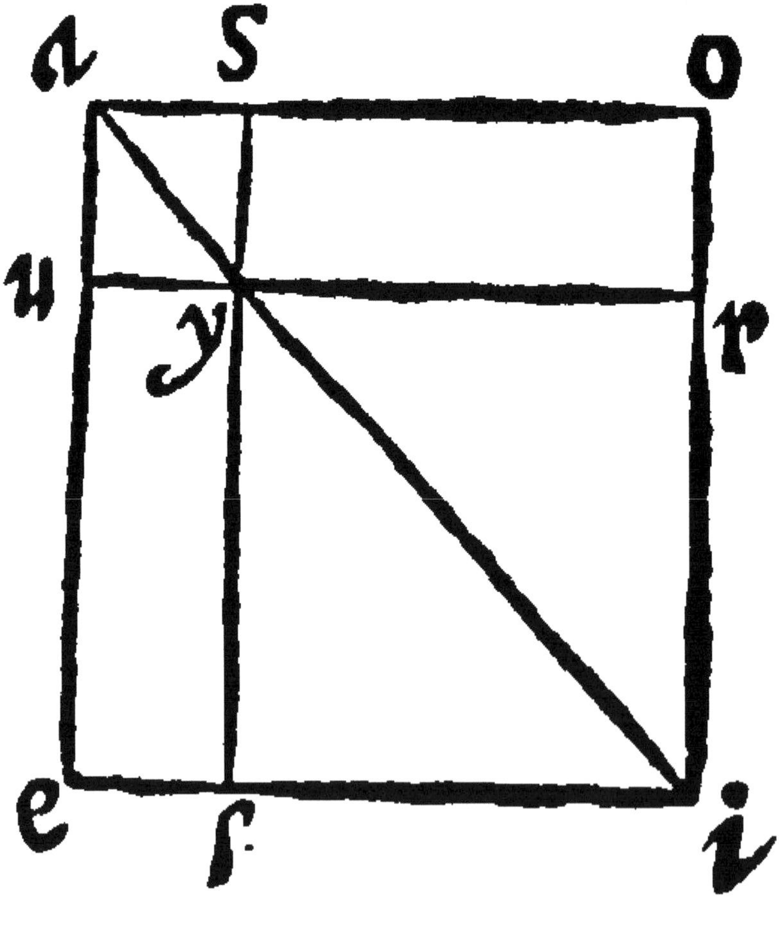

First the Diagonall is defined: As in the whole parallelogramme *aeio*, the diagonals are *auys*, and *ylir*; Because they are parts of the whole, having both the same common angles at *a*, and *i*: and diagonall diameter *ai*, with the whole parallelogramme: Not that the whole diagonie is common to both: But because the particular diagonies are the parts of the whole diagony. Therefore the diagonalls are two.

17 The Diagonall is like, and alike situate to the whole parallelogramme: è 24. p vj.

There is not any, either rate or proportion of the diagonall propounded, onely similitude is attributed to it, as in the same figure, the Diagonall *auys*, is like unto the whole parallelogramme *aeio*. For first it is equianglar to it. For the angle at *a*, is common to them both: And that is equall to that which is at *y*, (by the 10. e x:) And therefore also it is equall to that at *i* by the 10. e x. Then the angles *auy*, and *asy*, are equall, by the 21. e v. to the opposite inner angles at *e*, and *o*. Therefore it is equiangular unto it.

Againe, it is proportionall to it in the shankes of the equall angles. For the triangles *auy*, and *aei*, are alike, by the 12 e vij, because *uy* is parallell to the base. Therefore as *au* is *uy*; so is *ai* to *ei*: Then as *uy* is to *ya*; so is *ei* to *ia*. Againe by the 21 e v, because *sy* is parallell to the base *io*, as *ay* is to *ys*: so is *ai*, to *io*: Therefore equiordinately, as *uy* is to *ys*: so is *ei* to *io*: Item as *sy* is to *ya*, so is *io* to *ia*: And as *ya* is to *as*: so is *ia* to *ao*. Therefore equiordinately, as *ys* is to *sa*: so is *io* to *oa*. Lastly as *sa* is unto *ay*; so is *oa* unto *ai*: And as *ay* is to *au*; so is *ai* unto *ae*. Therefore equiordinately, as *sa* is to *au*: so is *ao*, to *ae*. Wherefore the Diagonall *su* is proportionall in the shankes of equall angles to the parallelogramme *oe*.

The demonstration shall be the same of the Diagonall *rl*. The like situation is manifest, by the 21 e iiij. And from hence also is manifest, That the diagonall of a Quadrate, is a Quadrate: Of an Oblong, an Oblong: Of a Rhombe, a Rhombe: Of a Rhomboides, a Rhomboides: because it is like unto the whole, and a like situate.

Now the Diagonalls seeing they are like unto the whole and a like situate, they shall also be like betweene themselves and alike situate one to another, by the 21 and 22 e iiij.

Therefore

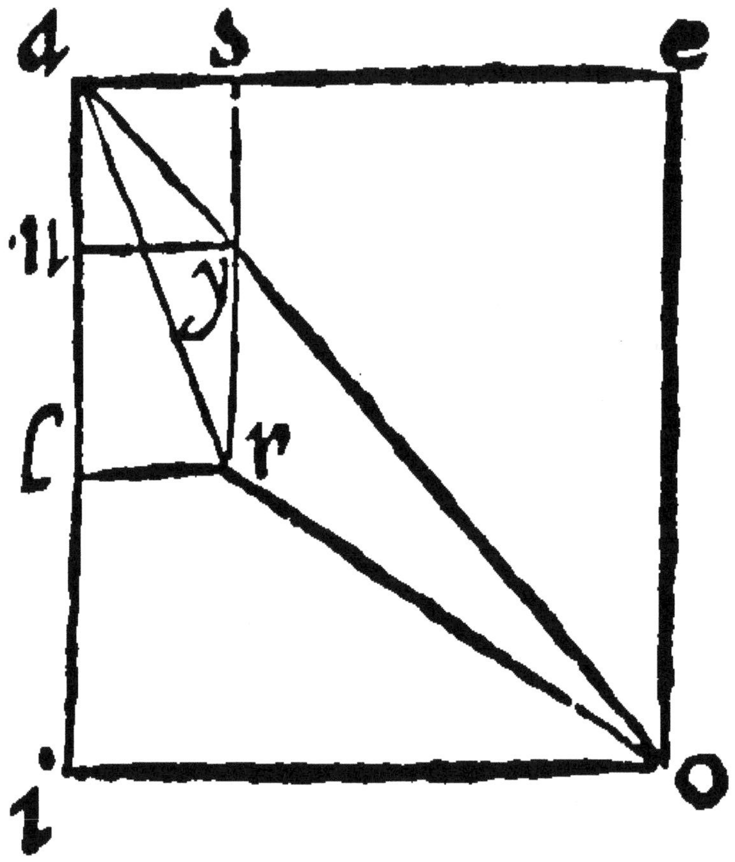

18. If the particular parallelogramme have one and the same angle with the whole, be like and alike situate unto it, it is the Diagonall. 26 p vj.

This might have beene drawn, as a consectary, out of the former: But it may also as it is by Euclide be forced, by an argument *ab impossibili*. For otherwise the whole should be equall to the part, which is impossible.

As for example, Let the particular parallelogramme *auys*, be coangular to the whole parallelogramme *aeio*; And let it have the same angle with it at *a*; like unto the whole and alike situate unto it; I say it is the Diagonall.

Otherwise, let the diverse Diagony be *aro*: And let *lr* be parallell against *ae*: Therefore *alrs*, shall bee the Diagonall, by the 6 e [16.] Now therefore it shall be, by 8 e [17 e,] as *ea* is to *ai*: so is *sa* unto *al*: Againe,by the grant, as *ea* is unto *ai*: so is *sa* to *au*: Therefore the same *sa* is proportionall to *al*, and to *au*: And *al* is equall to *au*, the part to the whole, which is impossible.

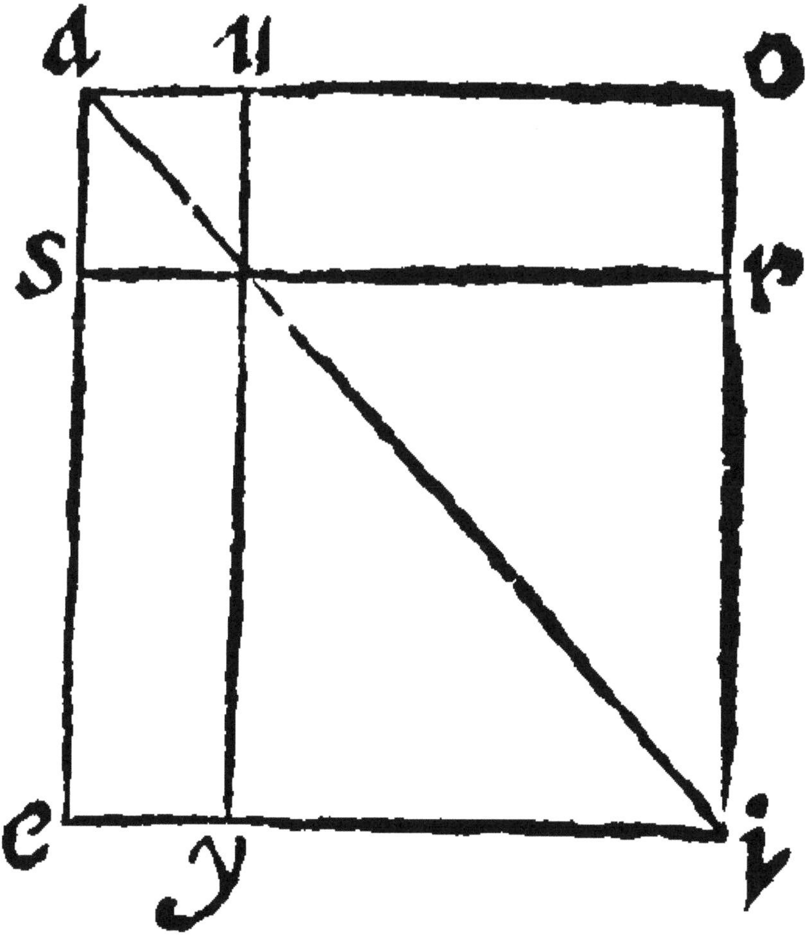

19. The Complement is a particular parallelogramme, comprehended of the conterminall sides of the diagonals.

Or thus: It is a particular parallelogramme conteined under the next adjoyning sides of the diagonals.

As in this figure, are *ur*, and *sy*: For each of them is comprehended of the continued sides of the two diagonals. And therefore are they called Complements, because they doe with the Diagonals *complere*, that is, fill or make up the whole parallelogramme. Neither in deed may the two diagonals be described, but withall the complements must needes be described.

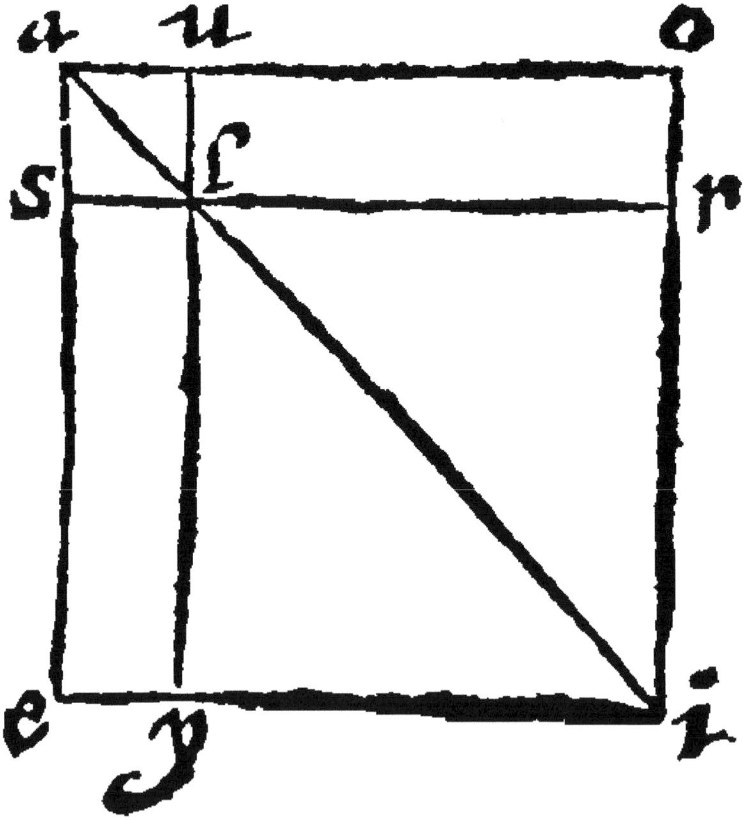

20. The complements are equall. 43 p j.

As in the same figure, are the sayd *ur*, and *sr*: For the triangles *aei*, and *aoi*, are equall, by the 12 e. Item, so are *asl*, and *aul*: Item, so are *lui*, and *lri*. Therefore if you shall on each side take away equall triangles from those which are equall, you shall leave the Complements equall betweene themselves.

Therefore

21. If one of the Complements be made equall to a triangle given, in a right-lined angle given, the other made upon a right line given shall be in like manner equall to the same triangle. 44 p j.

As if thou shouldest desire to have a parallelogramme upon a right line given, and in a right lined angle given, to be made equall to a triangle given, this proposition shall give satisfaction.

Let aei *be the Triangle given: The Angle be* o*: And the right line given be* iu*: And the Parallelogramme* ay *is equall to* aei*, triangle given in the angle assigned, by the 13 e. Then let the side* ay*, bee continued to* r*, equally to* iu*, the line given: And let* ru *be knit by a right line: And from* r *drawne out a diagony untill it doe meete with* as*, infinitely continued; which shall meete with it, by the 19 e v, in* l*. And the sides* yi*, and* ru*, let them be continued equally to* sl*. in* m *and* n*. And knit* ln *together with a right line.*

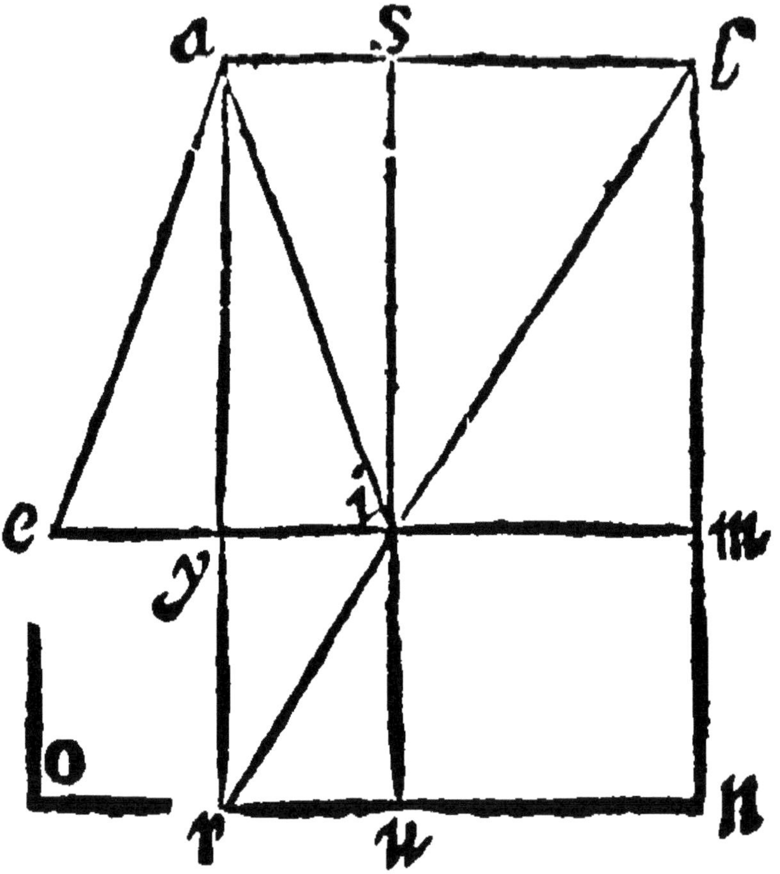

This complement *mu*, is equall to the complement *ys*, which is equall to the Triangle assigned, by the former, and that in a right lined angle given.

And

> *22 If parallelogrammes be continually made equall to all the triangles of an assigned triangulate, in a right lined angle given, the whole parallelogramme shall in like manner be equall to the whole triangulate. 45 p j.*

This is a corollary of the former, of the Reason or rate of a Parallelogramme with a Triangulate; and it needeth no farther demonstration; but a ready and steddy hand in describing and working of it.

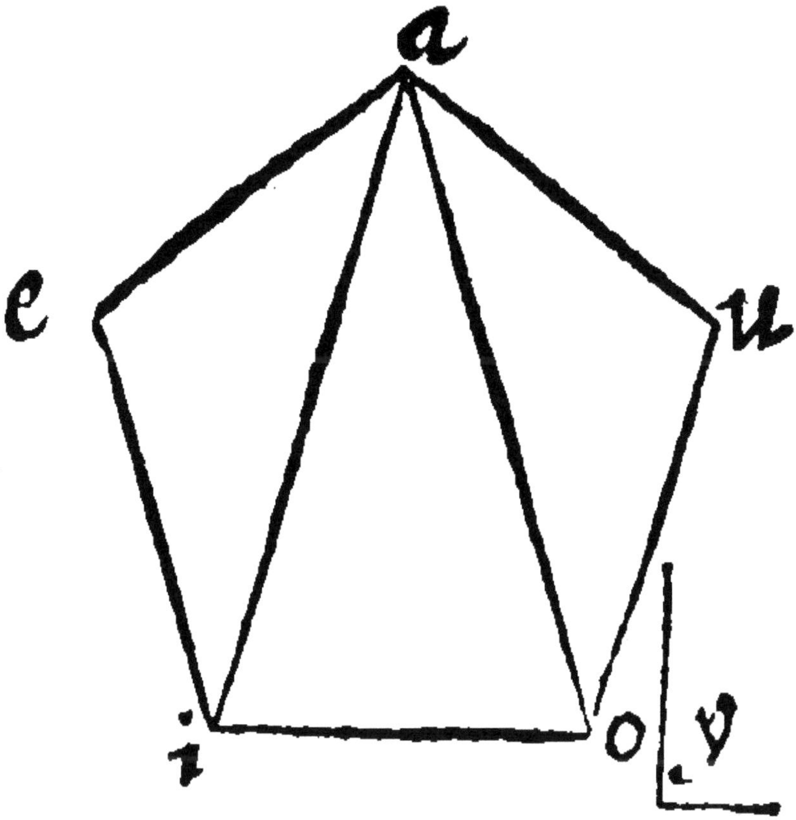

Take therefore an infinite right line; upon the continue the particular parallelogrammes, As if the Triangulate *aeiou*, were given to be brought into a parallelogramme: Let it be resolved into three triangles, *aei*, *aio*, and *aou*: And let the Angle be *y*: First in the assigned Angle, upon the Infinite right line, make by the former the Parallelogramme *ae*, in the angle assigned, equall to *aei*, the first triangle. Then the second triangle, thou shalt so make upon the said Infinite line, that one of the shankes may fall upon the side of the equall complement; The other be cast on forward, and so forth in more, if neede be.

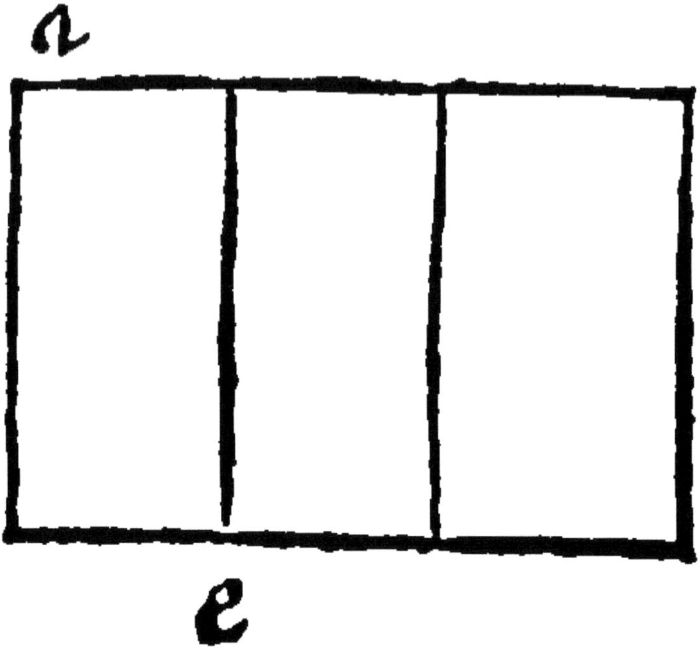

Here thou hast 3 complements continued, and continuing the Parallelogramme: But it is best in making and working of them, to put out the former, and one of the sides of the inferiour or latter Diagonall, least the confusion of lines doe hinder or trouble thee.

Therefore

23. A Parallelogramme is equall to his diagonals and complements.

For a Parallelogramme doth consist of two diagonals, and as many complements: Wherefore a Parallelogramme is equall to his parts: And againe the parts are equall to their whole.

24. The Gnomon is any one of the Diagonall with the two complements.

There is therefore in every Parallelogramme a double Gnomon; as in these two examples. Of all the space of a parallelogramme about his diameter, any parallelogramme with the two complements, let it be called the Gnomon. Therefore the gnomon is compounded, or made of both the kindes of diagonall and complements.

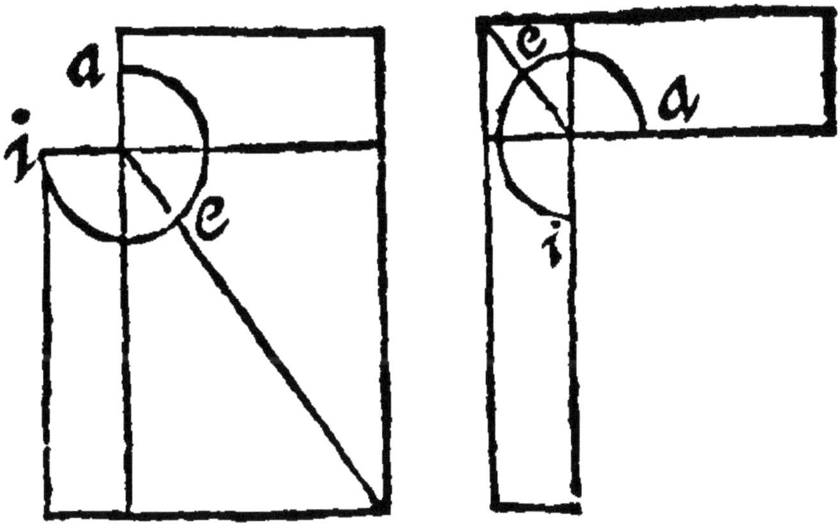

In the Elements of Geometry there is no other use, as it seemeth of the gnomons than that in one word three parts of a parallelogramme might be signified and called by three letters *aei*. Otherwise gnomon is a perpendicular.

25. Parallelogrames of equall height are one to another as their bases are. 1 p vj.

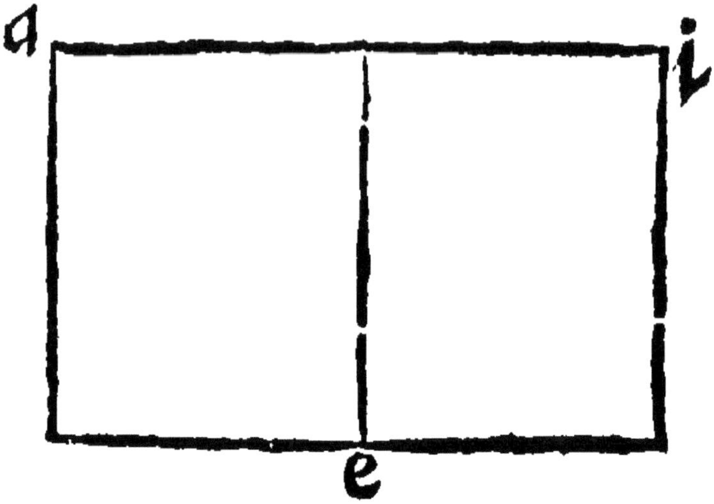

As is apparent, by the 16 e iiij. Because they be the double of Triangles, by the 12 e, of first figures: As *ae*, and *ei*.

Therefore

26 Parallelogrammes of equall height upon equall bases are equall. 35. 36 pj.

As is manifest in the same example.

27 If equiangle parallelogrammes be reciprocall in the shankes of the equall angle, they are equall: And contrariwise. 15 p vj.

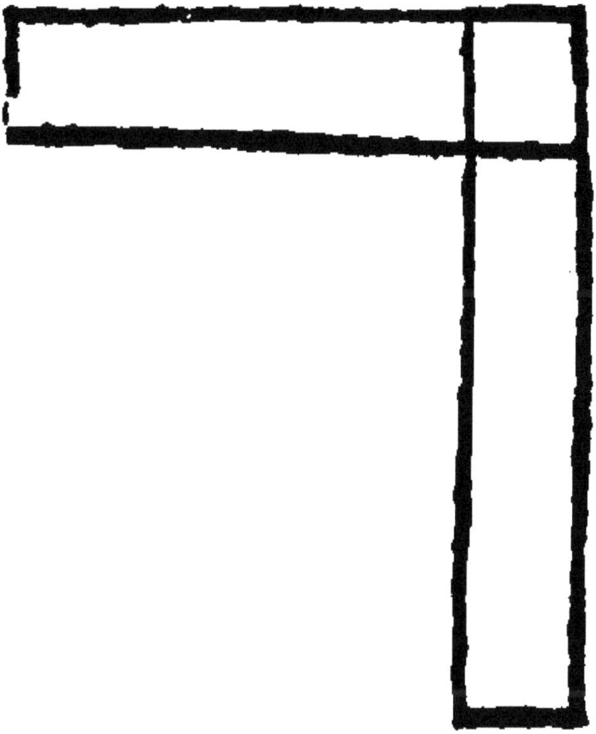

It is a consectary drawne out of the 11 e vij: As here thou seest: And yet indeed both that (as there was sayd) and this is rather a consectary of the 18 e iiij, which here also is more manifest.

Therefore

28 If foure right lines be proportionall, the parallelogramme made of the two middle ones, is equall to the equiangled parallelogramme made of the first and last: And contrariwise, e 16 p vj.

For they shall be equiangled parallelogrammes reciprocall in the shankes of the equall angle.

And

> *29 If three right lines be proportionall, the parallelogramme of the middle one is equall to the equiangled parallelogramme of the extremes: And contrariwise.*

It is a consectary drawne out of the former.

Of Geometry, *the eleventh Booke, of a Right angle.*

1. A Parallelogramme is a Right angle or an Obliquangle.

Hitherto we have spoken of certaine common and generall matters belonging unto parallelogrammes: specials doe follow in Rectangles and Obliquangles, which difference, as is aforesaid, is common to triangles and triangulates. But at this time we finde no fitter words whereby to distinguish the generals.

2. A Right angle is a parallelogramme that hath all his angles right angles.

As in *aeio*. And here hence you must understand by one right angle that all are right angles. For the right angle at *a*, is equall to the opposite angle at *i*, by the 10 e x.

And therefore they are both right angles, by the 14 e iij. The other angle at *e*, and *o*, by the 4 e vj, are equall to two right angles: And they are equall betweene themselves, by the 10 e x. Therefore all of them are right angles. Neither may it indeed possible be, that in a parallelogramme there should be one right angle, but by and by they must be all right angles.

Therefore

3 A rightangle is comprehended of two right lines comprehending the right angle 1. d ij.

Comprehension, in this place doth signifie a certaine kind of Geometricall multiplication. For as of two numbers multiplied betweene themselves there is made a number: so of two sides (*ductis*) driven together, a right angle is made: And yet every right angle is not rationall, as before was manifest, at the 12. e iiij, and shall after appeare at the 9 e.

And

4 Foure right angles doe fill a place.

Neither is it any matter at all whether the foure rectangles be equall, or unequall; equilaters, or unequilaters; homogeneals, or heterogenealls. For which way so ever they be turned, the angles shall be right angles: And therefore they shall fill a place.

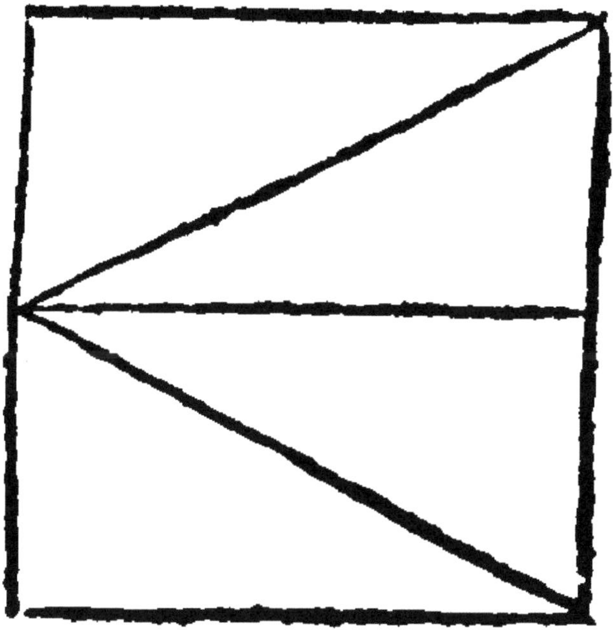

5 If the diameter doe cut the side of a right angle into two aquall parts, it doth cut it perpendicularly: And contrariwise.

As here appeareth by the 1 e vij. by drawing of the diagonies of the bisegments. The converse is manifest, by the 2 e vij. and 17. e vij.

Therefore

6 If an inscribed right line doe perpendicularly cut the side of the right angle into two equall parts, it is the diameter.

The reason is, because it doth cut the parallelogramme into two equall portions.

7 A right angle is equall to the rightangles made of one of his sides and the segments of the other.

As here the foure particular right angles are equall to the whole, which are made of *ae*, one of his sides, and of *ei*, *io*, *ou*, *uy*, the segments of the other.

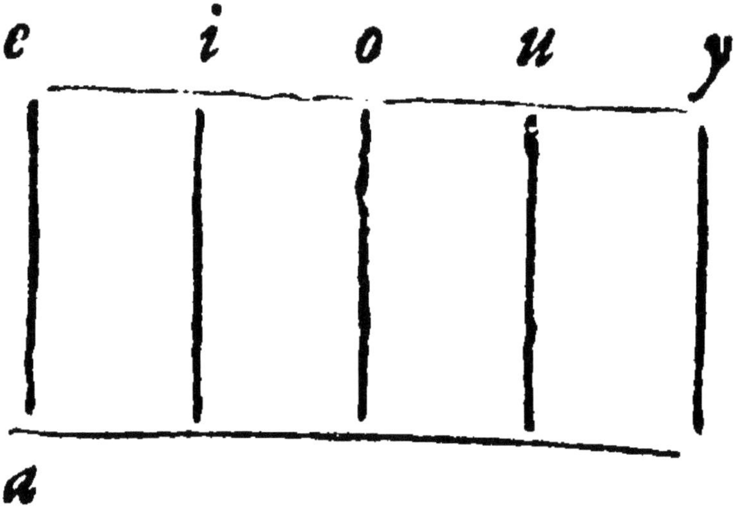

The Demonstration of this is from the rule of congruency: Because the whole agreeth to all his parts. But the same reason in numbers is more apparent by an induction of the parts: as foure times eight are 32. I breake or divide 8. into 5. and 3. Now foure times 5. are 20. And foure times 3. are 12. And 20. and 12. are 32. And 32. and 32. are equall. Therefore 20. and 12. are also equall to 32.

Lastly, every arithmeticall multiplication of the whole numbers doth make the same product, that the multiplication of the one of the whole numbers given, by the parts of the other shall make: yea, that the multiplication of the parts by the parts shall make. This proportion is cited by *Ptolomey* in the 9. Chapter of the 1 booke of his Almagest.

8 If foure right lines be proportionall, the rectangle of the two middle ones, is equall to the rectangle of the two extremes. 16. p vj.

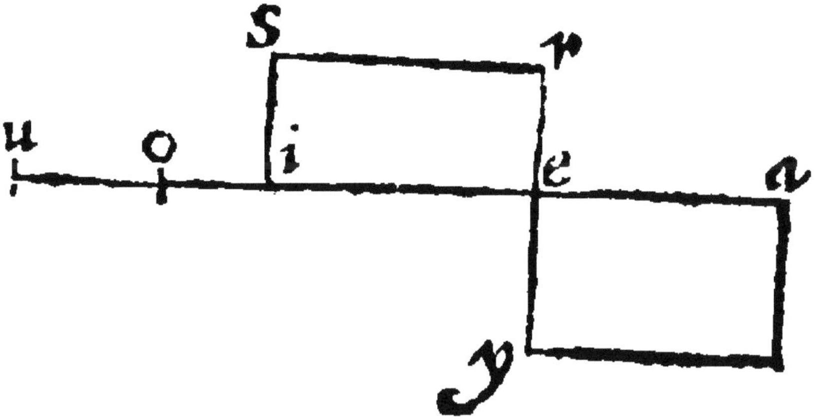

It is a speciall consectary out of the 28 e x. As here are foure right lines proportionall betweene themselves: And the rectangle of the extremes, or first and last let it be *ay*: Of the middle ones, let it be *se*.

9 The figurate of a rationall rectangle is called a rectinall plaine. 16. d vij.

A rationall figure was defined at the 12. e iiij. of which sort amongst all the rectilineals hitherto spoken of, we have not had one: The first is a Right angled parallelogramme; And yet not every one indifferently: But that onely whose base is rationall to the highest: And that reason of the base and heighth is expressable by a number, where also the Figurate is defined. A rectangle or irrational sides, such as were mentioned at the 9 e j. is irrationall. Therefore a rectangled rationall of rationall sides, is here understood: And the figurate thereof, is called, by the generall name, A *Plaine:* Because of all the kindes of *Plaines*, this kinde onely is rationall.

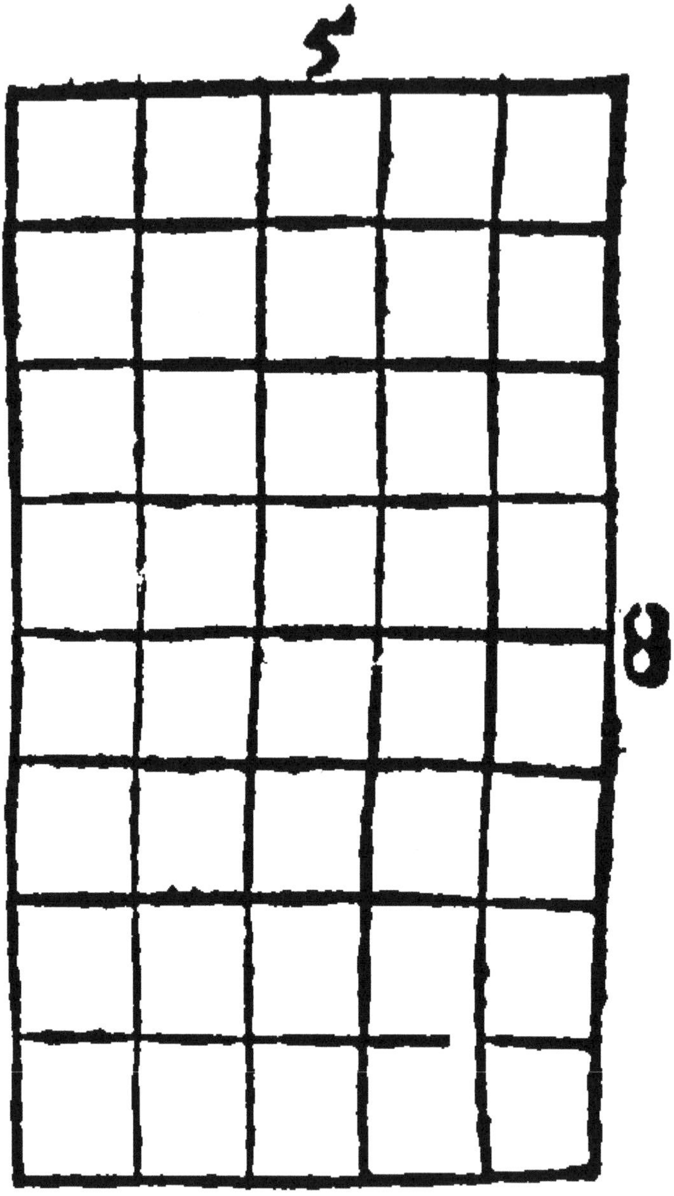
5
8

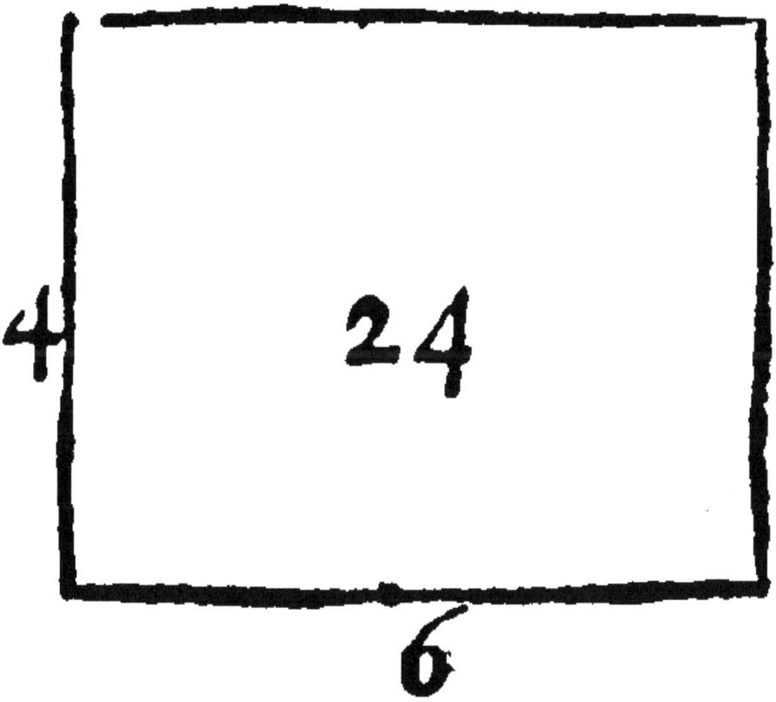

If therefore the Base of a Rectangle be 6. And the height 4. The plot or content shall be 24. And if it be certaine that the rectangles content be 24. And the base be 6. It shall also be certaine that the heighth is 4. The example is thus. And this multiplication, as appeared at the 13. e iiij. is geometricall: As if thou dost multiply 5. by 8. thou makest 40. for the Plaine: And the sides of this Plaine, are 5. and 8. it is all one as if thou hadst made a rectangled parallellogramme of 40. square foote content, whose base should be 5. foote, and the heigth 8. after this manner.

This manner of multiplication, say I, is Geometricall: Neither are there here, of lines made lines, as there of unities were made unities; but a magnitude one degree higher, to wit, a surface, is here made.

Here hence is the *Geodesy* or manner of measuring of a rectangled triangle made knowne unto us. For when thou shalt multiply the shankes of a right angle, the one by the other, thou dost make the whole rectangled parallelogramme, whose halfe is a triangle, by the 12. e x.

Of Geometry the twelfth Booke, Of a Quadrate.

1 A Rectangle is a Quadrate or an Oblong.

This division is made in the proper termes: but the thing it selfe and the subject difference is common out of the angles and sides.

2 A Quadrate is a rectangle equilater 30. d j.

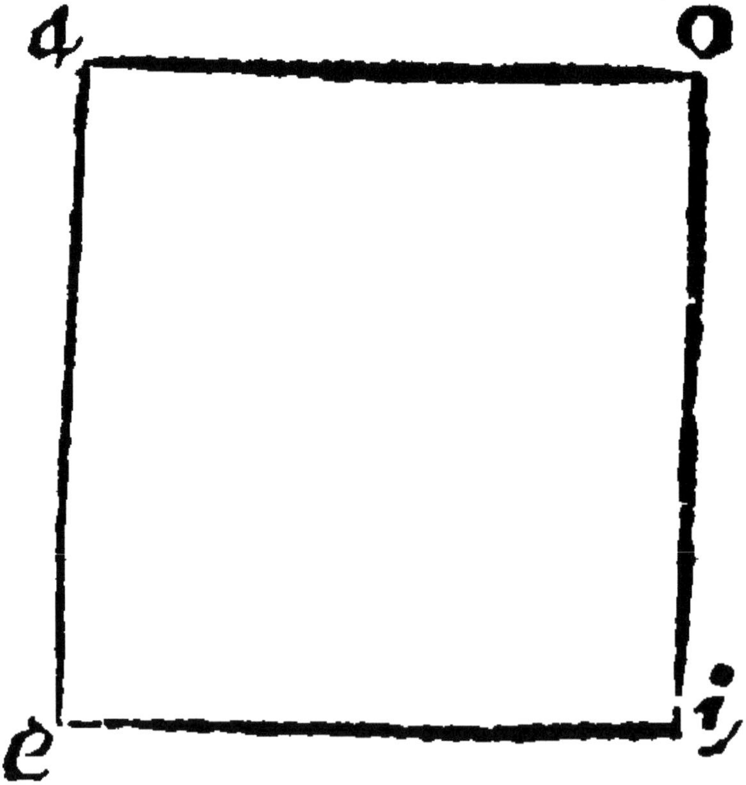

Quadratum, a Quadrate, or square, is a rectangled parallellogramme of equall sides: as here thou seest *aeio*, to be.

Plaines are with us, according to their diverse natures and qualities, measured with divers and sundry kindes of measures. Boord, Glasse, and Paving-stone are measured by the foote: Cloth, Wainscote, Painting, Paving, and such like, by the yard: Land, and Wood, by the Perch or Rodde.

Of Measures and sundry sorts thereof commonly used and mentioned in histories we have in the former spoken at large: Yet for the farther confirmation of some thing then spoken, and here againe now upon this particular occasion repeated, it shall not be amisse to heare what our Statutes speake of these three sorts here mentioned.

It is ordained, saith the Statute, That three Barley-cornes dry and round, doe make an *Ynch*: twelve ynches doe make a *Foote*: three foote doe make a *Yard*: Five yards and an halfe doe make a *Perch*: Fortie perches in length, and foure in breadth doe make an *Aker. 33. Edwardi 1. De Terris mensurandis.* Item, *De compositione Ulnarum & Perticarum.*

Moreover observe, that all those measures there spoken of were onely lengths: These here now last repeated, are such as the magnitudes by them measured are, in Planimetry, I meane, they are Plaines: In Stereometry they are solids, as hereafter we shall make manifest. Therefore in that which followeth, An *ynch* is not onely a length three barley-cornes long: but a plaine three barley-cornes long, and three broad. A *Foote* is not onely a length of 12. ynches: But a plaine also of 12. ynches square, or containing 144. square ynches: A *yard* is not onely the length of three foote: But it is also a plaine 3. foote square every way. A *Perch* is not onely a length of 5½. yards: But it is a plot of ground 5½. yards square every way.

A Quadrate therefore or square, seeing that it is equilater that is of equall sides: And equiangle by meanes of the equall right angles, of quadrangles that onely is ordinate.

Therefore

3 The sides of equall quadrates, are equall.

And

The sides of equall quadrates are equally compared: If therefore two or more quadrates be equall, it must needs follow that their sides are equall one to another.

And

4 The power of a right line is a quadrate.

Or thus: The possibility of a right line is a square H. A right line is said *posse quadratum*, to be in power a square; because being multiplied in it selfe, it doth make a square.

5 If two conterminall perpendicular equall right lines be closed with parallells, they shall make a quadrate. 46. p. j.

Or thus: If two equall perpendicular lines, ioyning one with another, be inclosed together by parallell lines they will make a square. *H*. As in *aeio*, let the perpendiculars *ae*, and *ei*, equall betweene themselves, be closed with two parallells, *ao*, against *ei*: And *oi*, against *ae*; they shall make the quadrate or square *aeio*. For it is a parallelogramme, by the grant: Because the opposite sides are parallell: And it is rectangled: because seeing the angle *aei*, of the perpendicular lines, is a right angle, they shall be all right angles by the 2 e xj. Then one side *ei*, is equall to all the rest. First to *ao*, that over against it, by the 8 e x. And then to *ea*, by the grant: And therefore to *oi*, to that over against it, by the 8 e x.

6 The plaine of a quadrate is an equilater plaine.

Or thus: The plaine number of a square, is a plaine number of equall sides, *H*.

A quadrate or square number, is that which is equally equall: Or that which is comprehended of two equall numbers, A quadrate of all plaines is especially rationall; and yet not alwayes: But that onely is rationall whose number is a quadrate. Therefore the quadrates of numbers not quadrates, are not rationalls.

Therefore

7 A quadrate is made of a number multiplied by it self.

Such quadrates are the first nine. 1, 4, 9, 16, 25, 36, 49, 64, 81, made of once one, twice two, thrise three, foure times foure, five times five, sixe times sixe, seven times seven, eight times eight, and nine times nine. And this is the summe of the making and invention of a quadrate number of multiplication of the side given by it selfe.

1 2 3 4 5 6 7 8 9. The sides given.
1 4 9 16 25 36 49 64 81. The quadrates found.

Hereafter diverse comparisons of a quadrate or square, with a rectangle, with a quadrate, and with a rectangle and a quadrate iointly. The comparison or rate of a quadrate with a rectangle is first.

And

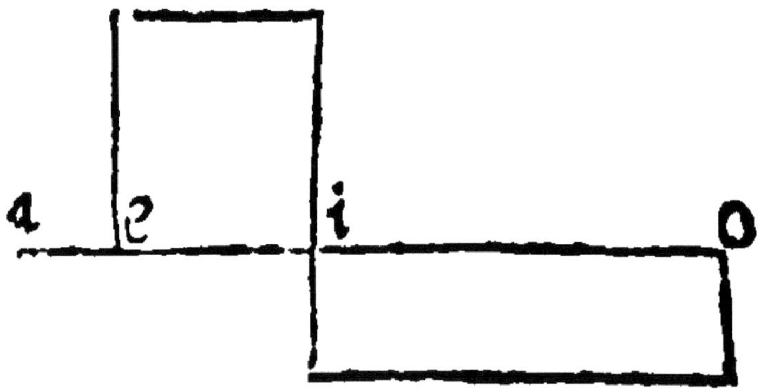

8 If three right lines be proportionall, the quadrate of the middle one, shall be equall to the rectangle of the extremes: And contrariwise: 17. p vj. and 20. p vij.

Or thus: If three lines be proportionall, the square made of the middle line is equall to the right angled parallelogramme made of the two outmost lines: *H*.

It is a corallary out of the 28. e x. As in *ae*, *ei*, *io*.

9 If the base of a triangle doe subtend a right angle, the powre of it is as much as of both the shankes: And contrariwise 47, 48. p j.

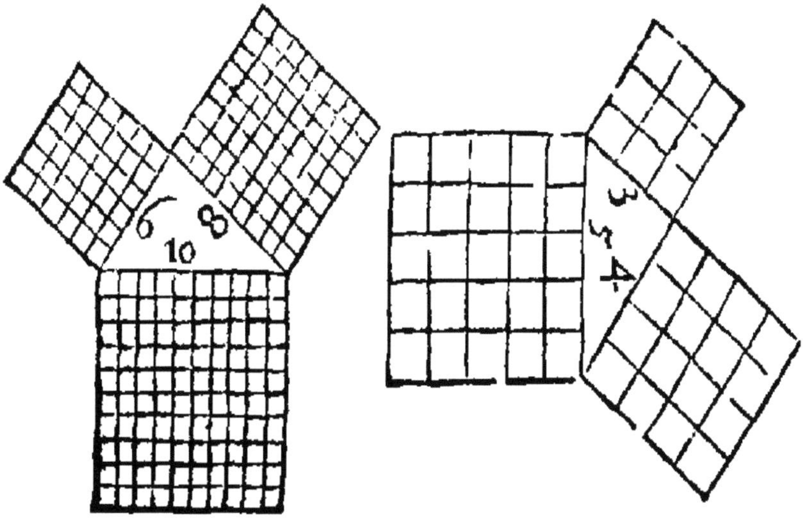

It is a consectary out of the 11. e viij. But it is sometime rationall, and to be expressed by a number: yet but in a scalene triangle onely. For the sides of an equicrurall right-angled triangle are irrationall; Whereas the sides of a scalene are sometime rationall; and that after two manners, the one of *Pythagoras*, the other of *Plato*, as *Proclus* teacheth, at the 47. p j. *Pythagora's* way is thus, by an odde number.

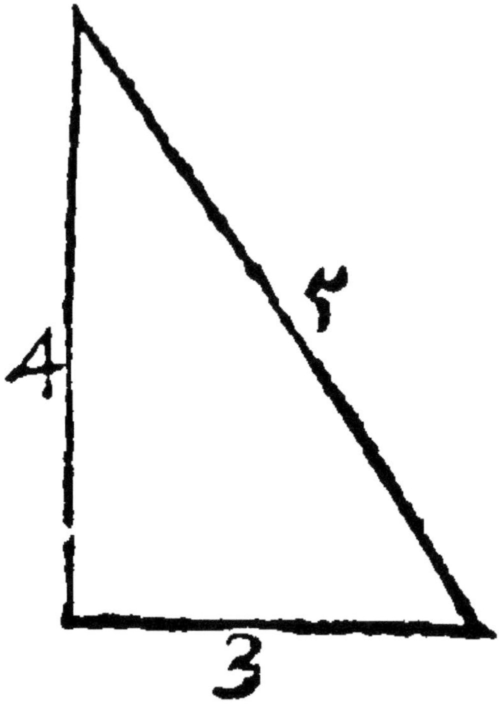

10 If the quadrate of an odde number, given for the first shanke, be made lesse by an unity; the halfe of the remainder shall be the other shanke; increased by an unity it shall be the base.

Or thus: If the square of an odde number given for the first foote, have an unity taken from it, the halfe of the remainder shall be the other foote, and the same halfe increased by an unitie, shall be the base: *H.*

As in the example: The sides are 3, 4, and 5. And 25. the square of 5. the base, is equall to 16. and 9. the squares of the shanks 4. and 3.

Againe, the quadrate or square of 3. the first shanke is 9. and 9 - 1. is 8, whose halfe 4, is the other shanke. And 9 + 1, is 10. whose halfe 5. is the base. *Plato's* way is thus by an even number.

11 If the halfe of an even number given for the first shanke be squared, the square number diminished by an unity shall be the other shanke, and increased by an unitie it shall be the base.

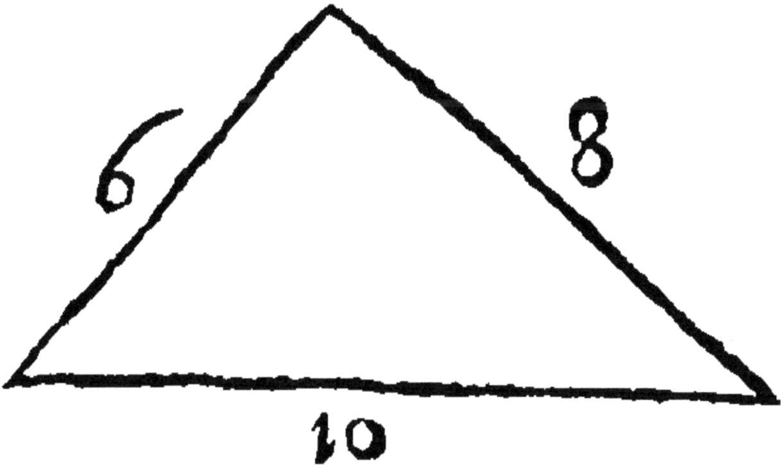

As in this example where the sides are 6, 8. and 10. For 100. the square of 10. the base is equall to 36. and 64. the squares of the shankes 6. and 8.

Againe, the quadrate or square of 3. the halfe of 6, the first shanke, is 9. and 9 - 1, is 8, for the second shanke. And out of this rate of rationall powers (as *Vitruvius*, in the 2. Chapter of his IX. booke) saith *Pythagoras* taught how to make a most exact and true squire, by joyning of three rulers together in the forme of a triangle, which are one unto another as 3, 4. and 5. are one to another.

From hence Architecture learned an Arithmeticall proportion in the parts of ladders and stayres. For that rate or proportion, as in many businesses and measures is very commodious; so also in buildings, and making of ladders or staires, that they may have moderate rises of the steps, it is very speedy. For 9 + 1. is, 10, base.

12. The power of the diagony is twise asmuch, as is the power of the side, and it is unto it also incommensurable.

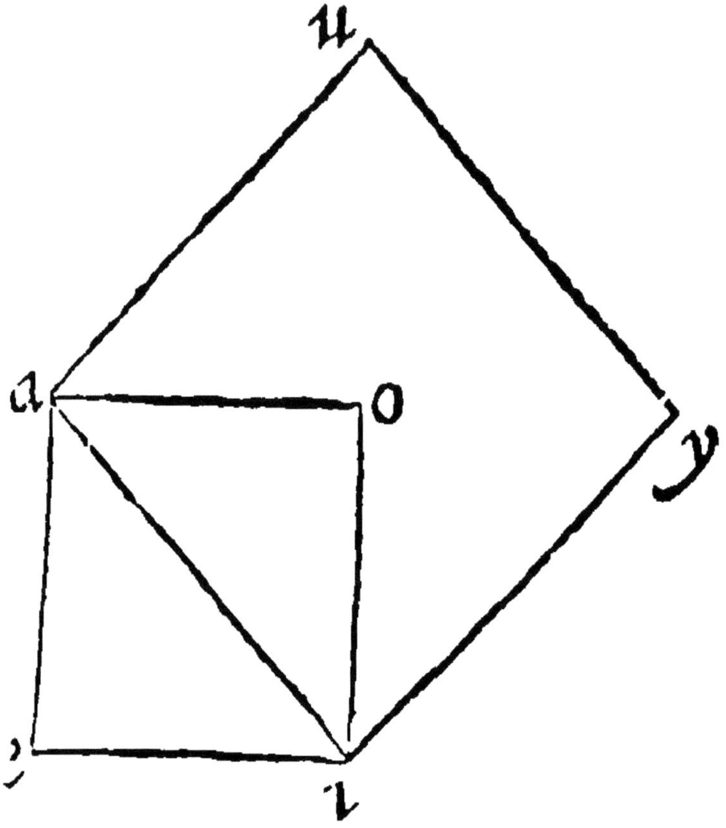

Or thus: The diagonall line is in power double to the side, and is incommensurable unto it, *H.*

As here thou seest, let the first quadrate bee *aeio:* Of whose diagony *ai*, let there be made the quadrate *aiuy:* This, I say, shall be the double of that: seeing that the diagonies power is equall to the power of both the equall shankes. Therefore it is double to the power of one of them.

This is the way of doubling of a square taught by *Plato*, as *Vitruvius* telleth us: Which notwithstanding may be also doubled, trebled, or according to any reason assigned increased, by the 25 e iiij, as there was foretold.

But that the Diagony is incommensurable unto the side it is the 116 p x. The reason is, because otherwise there might be given one quadrate number, double to another quadrate number: Which as *Theon* and *Campanus* teach us, is impossible to be found. But that reason which *Aristotle* bringeth is more cleare which is this; Because otherwise an even number should be odde. For if the Diagony be 4, and the side 3: The square of the Diagony 16, shall be double to the square of the side: And so the square of the side shall be 8. and the same square shall be 9, to wit, the square of 3. And so even shall be odde, which is most absurd.

Hither may be added that at the 42 p x. That the segments of a right line diversly cut; the more unequall they are the greater is their power.

13. If the base of a right angled triangle be cut by a perpendicular from the right angle in a doubled reason, the power of it shall be halfe as much more, as is the power of the greater shanke: But thrise so much as is the power of the lesser. If in a quadrupled reason, it shall be foure times and one fourth so much as is the greater: But five times so much as is the lesser, At the 13, 15, 16 p xiij.

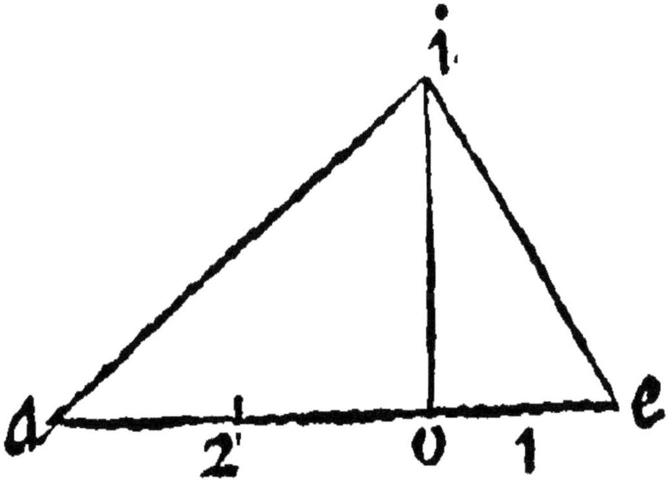

Or thus: If the base of a right angled triangle be cut in double proportion, by a perpendicular comming from the right angle, it is in power sesquialter to the greater foote; and treble to the lesser: But if the base be cut in quadruple proportion, it is sesquiarta to the greater side, and quintuple to the lesser.

As in *aei*, let the base *ae*, be so cut that the segment *ao*, be double to the segment *oe*, to wit, as 2 is to 1. The whole *ae*, shall be unto *ao sesquialtera*, that is, as 2 is to 3. And therefore by the 10 e viij, and 25 e iiij, the square of *ae*, shall be *sesquialterum* unto the square of *ai*.

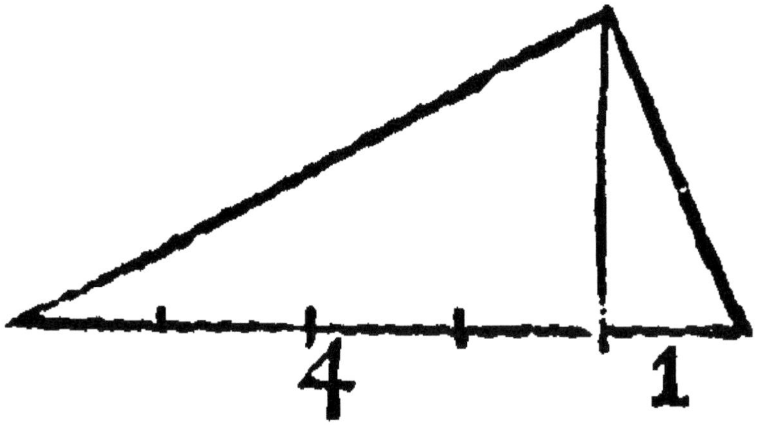

And by the same argument it shall be treble unto the quadrate or square of *ei*.

The other, of the fourefold or quadruple section, are manifest in the figure following, by the like argument.

14 If a right line be cut into how many parts so ever, the power of it is manifold unto the power of segment, denominated of the square of the number of the section.

Or thus: if a right line be cut into how many parts soever it is in power the multiplex of the segment, the square of the number of the section, being denominated thereof: *H*.

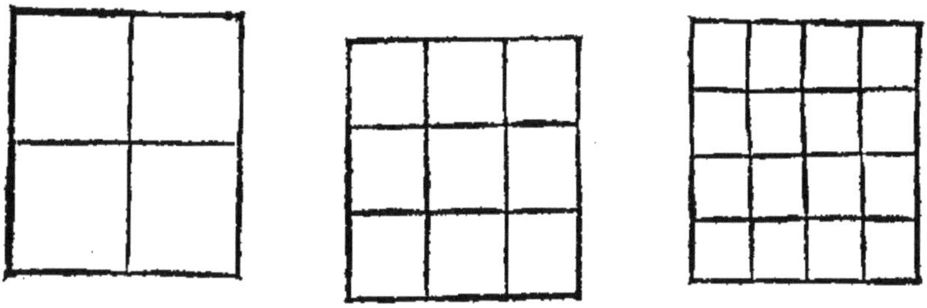

So if it be cut into two equall parts, the power of it shall be foure times so much, as is the power of the halfe, taking demonstration from 4, which is the square of 2, according to which the division was made: If it be cut into three equall parts, the power of it shall be nine fold the power of the third part. If into foure equall parts, it shall be 16 times so much as is the power of the quarter: As here thou seest in these examples.

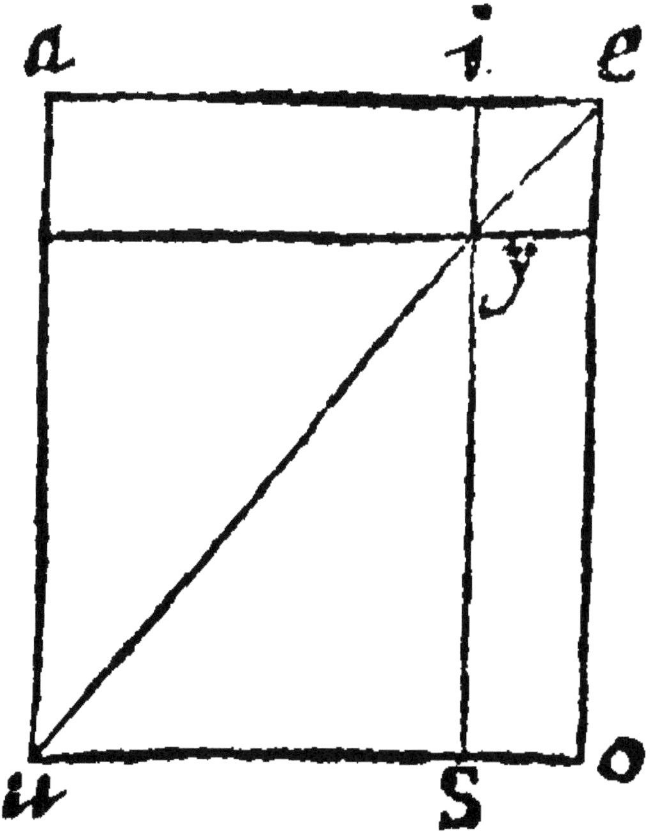

15. If a right line be cut into two segments, the quadrate of the whole is equall to the quadrats of the segments, and a double rectanguled figure, made of them both. 4 p ij.

The third rate of a quadrate is hereafter with two rectangles, and two quadrates, and first of equality.

This is a consectary out of the 22 e x: Because a parallelogramme is equall to his two diagonals and complements. If the right *ae*, be cut in *i*, it maketh the quadrate *aeuo*, greater than *eyi*, and *yus*, the quadrates of the segments, by the two rectangles *ay* and *yo*. This is the rate of a quadrate with a rectangle and a quadrate. But the side of a quadrate proposed in a number is oft times sought. Therefore by the next precedent element and his consectaries, the analysis or finding of the side of a quadrate is made and taught.

Therefore

16. *The side of the first diagonall, is the side of one of the complements; And being doubled, it is the side of them both together: Now the other side of the same complements both together, is the side of the other diagonall.*

The side of a quadrate given is many times in numbers sought. Therefore by the former element and his consectaries the resolution of a quadrates side is framed and performed.

Let therefore the side now of the quadrate number given be sought: And first let the Genesis or making be considered, such as you see here by the multiplication of numbers in the numbers themselves:

10 2		10 2	*The number or*
10 2		10 2	*side divided.*
———		———	
2 4		4	*The lesser diagonall.*
2 0	Or thus	2 0	*One Complement.*
1 0 0		2 0	*Th'other Complement*
———		1 0 0	*The greater diagonall.*
1 4 4		———	
		1 4 4	*The quadrate.*

This is the rate of a quadrate with a rectangle & a quadrate, from whence is had the analisis or resolution of the side of a quadrate expressable by a number. For it is the same way fro *Cambridge* to *London*, that is from *London* to *Cambridge*. And this use of geometricall analysis remaineth, as afterward in a Cube, when as otherwise through the whole booke of *Euclides* Elements there is no other use at all of that.

Here therefore thou shalt note or marke out the severall quadrates, beginning at the right hand and so proceeding towards the left; after this manner, 144. These notes doe signifie that so many severall sides to be found, to make up the whole side of the quadrate given. And here first, it shall not be amisse to warne thee, before thou commest to practice, that for helpe of memory and speed in working, thou know the Quadrats of the nine single numbers of figures; which are these

1.	2.	3.	4.	5.	6.	7.	8.	9.	*Sides.*
1.	4.	9.	16.	25.	36.	49.	64.	81.	*Qu.*

Then beginning at the left hand, as in Division, that is where we left in multiplication, and I seeke amongst the squares the greatest conteined in the first periode, which here is 1; And the side of it, which is also 1, I place with my quotient: Then I square this quotient, that is I multiply it by it selfe, and the product 1, I sect under the same first periode: Lastly, I subtract it from the same periode, and there remaineth not any thing. Then as in division I set up the figures of the next periode one degree higher.

~~44~~
~~1~~, ~~44~~ (12
2) ~~1~~
~~44~~

Or thus:

Secondly double the side now found, and it shall be 2, which I place in manner of a Divisor, on the left hand, within the semicircle: By this I divide the 40, the two complements or Plaines, and I finde the quotient or second side 2; which I place in the quotient by 1, This side I multiply first quadrate like, that is by it selfe; and I make 4, the lesser Diagonall: And therefore I place under the last 4: Then I multiply the said Divisor 2, by the same 2 the quotient, and I make in like manner, 4 which I place under the dividend, or the first 4. Lastly I subtract these products from the numbers above them, and remaineth nothing. Therefore I say first, That 144, the number given is a quadrate: And more-over, That 12 is the true side of it.

~~44~~
~~1~~, ~~44~~ (12
20) ~~1~~
~~40~~
4

Againe, let the side of 15129 be sought. First divide it into imperfect periods as before was taught; in this manner: 15129. Then I seeke amongst the former quadrates, for the side of 1, the quadrate of the first periode; and I finde it to be 1: This side I place within the quotient or lunular on the right side: Lastly I subtract 1 from 1, and nothing remaineth. Then I double the said side found; and I make 2: This 2, I place for my divisor within the lunular or semicircle on the left hand: By which I divide 5; and I finde the quotient 2, which I place by the former quotient: Then I multiply the same 2, first quadratelike by it selfe, and I make 4. Then I multiply the sayd divisour by 2, the quotient, and I make likewise 4: which I place underneath 51. Lastly, I subtract the same 44, from 51, and there remaine 7, over the head of 1; By which I place 29, the last periode remaining.

~~7~~ ~~29~~
~~51~~
~~1~~, ~~51~~, ~~29~~ (123
2) ~~1~~
24) ~~44~~
~~7~~ ~~29~~

Or thus:

Againe I double 12, my whole quotient, and I make 24. By this double I divide 72, the double Complement remaining, and I finde 3 for the side or quotient: First this side I multiply quadratelike by it selfe, and I make 9, which I place underneath 9, the last figure of my dividende. Then againe, by the same quotient, or side 3, I multiply 24: my divisour, and I make 72; which I place under 72, the two figures of my Dividende: Lastly I subtract the under figures, from the upper, and there is likewise nothing remaining: Wherefore I say, as afore; that the figurate 15129 given, is a square: And the side thereof is 123.

~~7~~ ~~29~~
~~51~~
~~1~~, ~~51~~, ~~29~~ (123
20) ~~1~~
~~40~~
4
240) ~~7~~ ~~20~~
9

Sometime after the quadrate now found, in the next places, there is neither any plaine nor square to bee found: Therefore the single side thereof shall be *O*. As in the quadrate 366025, the whole side is 605, consisting of three severall sides, of which the middle one is *o*.

Sometime also the middle plaine doth containe a part of the quadrate next following: Therfore if the other side remaining be greater than the side of the quadrate following, it is to be made equall unto it: As for example, Let the side of the quadrate 784, be sought; The side of the first quadrate shall be 2, and there shall remaine 3, thus: Then the same side doubled is 4 for the quotient; Which is found in 38, the double plaine remaining 9 times, for the other side: But this side is greater than the side of the next following quadrate: Take therefore 1 out of it: And for nine take 8, and place it in your quotient; Which 8 multiplyed by it selfe maketh 64, for the Lesser quadrate: And againe the same multiplyed by 4 the divisour maketh 32; the summe of which two products 384, subtracted from the remaine 384, leave nothing: Therefore 784 is a Quadrate: And the side is 28.

~~384~~
4) ~~784~~ (28
4

Or thus:

~~384~~
40) ~~784~~ (28
4
~~320~~
~~64~~

And from hence the invention of a meane proportionall, betweene two numbers given, (if there be any such to be found) is manifest. For if the product of two numbers given be a quadrate, the side of the quadrate shall be the meane proportionall, betweene the numbers given; as is apparent by the golden rule: As for example, Betweene 4. and 9. two numbers given, I desire to know what is the meane proportion. I multiply therefore 4 and 9. betweene themselves, and the product is 36: which is a quadrate number; as you see in the former; And the side is 6. Therefore I say, the meane proportionall betweene 4. and 9. is 6, that is, As 4. is to 6. so is 6. to 9.

If the number given be not a quadrate, there shall no arithmeticall side, and to be expressed by a number be found: And this figurate number is but the shadow of a Geometricall figure, and doth not indeede expresse it fully, neither is such a quadrate rationall: Yet notwithstanding the numerall side of the greatest square in such like number may be found: As in 148. The greatest quadrate continued is 144 and the side is 12. And there doe remaine 4. Therefore of such kinde of number, which is not a quadrate, there is no true or exact side: Neither shall there ever be found any so neare unto the true one; but there may still be one found more neare the truth. Therefore the side is not to be expressed by a number.

Of the invention of this there are two wayes: The one is by the *Addition of the gnomon*; The other is by the Reduction of the number assigned unto parts of some greater denomination. The first is thus:

17 If the side found be doubled, and to the double a unity be added, the whole shall be the gnomon of the next greater quadrate.

For the sides is one of the complements, and being doubled it is the side of both together. And an unity is the latter diagonall. So the side of 148 is 12.4/25.

The reason of this dependeth on the same proposition, from whence also the whole side, is found. For seeing that the side of every quadrate lesser than the next follower differeth onely from the side of the quadrate next above greater than it but by an 1. the same unity, both twice multiplied by the side of the former quadrate, and also once by it selfe, doth make the *Gnomon* of the greater to be added to the quadrate. For it doth make the quadrate 169. Whereby is understood, that looke how much the numerator 4. is short of the denominatour 25. so much is the quadrate 148. short of the next greater quadrate. For if thou doe adde 21. which is the difference whereby 4 is short of 25. thou shalt make the quadrate 169. whose side is 13. The second is by the reduction, as I said, of the number given unto parts assigned of some great denomination, as 100. or 1000. or some smaller than those, and those quadrates, that their true and certaine may be knowne: Now looke how much the smaller they are, so much nearer to the truth shall the side found be.

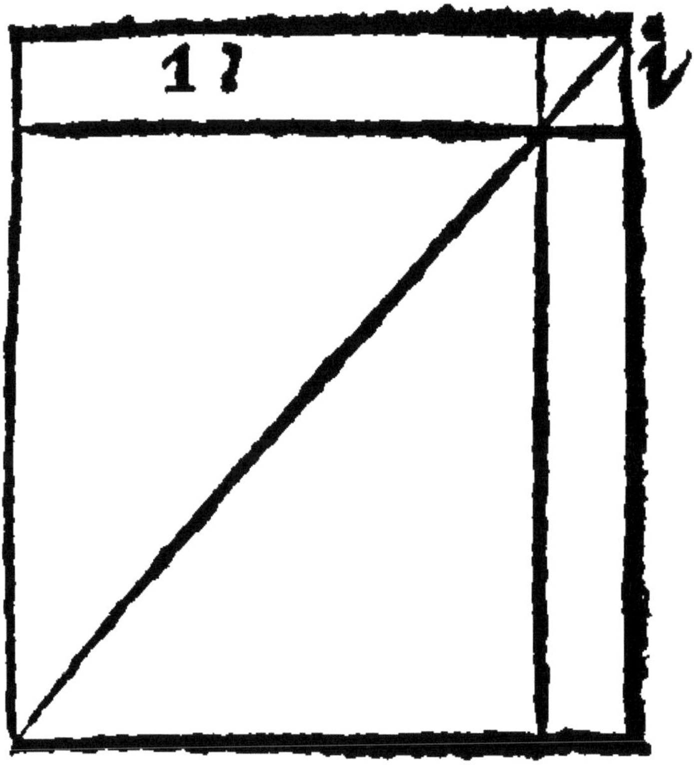

Let the same example be reduced unto hundreds squared parts, thus: 1480000/10000. The side of 10000. by grant is 100. But the side of 1480000. the numerator by the former is 1216. and beside there doe remaine 1344: thus, 1216/100. that is, 12.16/100, or 4/25 which was discovered by the former way. But in the side of the numerator there remained 1344. By which little this second way is more accurate and precise than the first. Yet notwithstanding those remaines are not regarded, because they cannot adde so much as 1/100 part unto the side, found: For neither in deed doe 1344/2426. make one hundred part.

Moreover in lesser parts, the second way beside the other, doth shew the side to be somewhat greater than the side, by the first way found: as in 7. the side by the first way is 3/25. But by the second way the side of 7. reduced unto thousands quadrates, that is unto 7000000/1000000, that is, 2645/1000, and beside there doe remaine 3975. But 645/1000 are greater than 3/5. For 3/5. reduced unto 1000. are but 600/1000. Therefore the second way, in this example, doth exceed the first by 45/1000. those remaines 3975. being also neglected. Therefore this is the Analysis or manner of finding the side of a quadrate, by the first rate of a quadrate, equall to a double rectangle and quadrate.

The Geodesy or measuring of a Triangle.

There is one generall Geodesy or way of measuring any manner of triangle whatsoever in *Hero*, by addition of the sides, halving of the summe, subduction, multiplication, and invention of the quadrates side, after this manner.

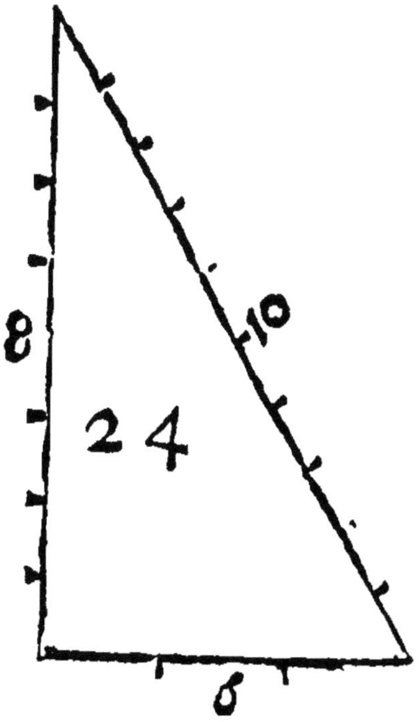

18 If from the halfe of the summe of the sides, the sides be severally subducted, the side of the quadrate continually made of the halfe, and the remaines shall be the content of the triangle.

As for example, Let the sides of the triangle *aei*, be 6. 8. 10: The summe 24. the halfe of the summe 12. From which halfe subduct the sides 6. 8. 10. and let the remaines be 6. 4. 2. Now multiply continually these foure numbers 12. 6. 4. 2. and thou shalt make first of 12. and 6. 72. Of 72. and 4. 288. Lastly, of 288. and 2. 576. And the side of 576. by the 16. e. shall be found to be 24. for the content of the triangle: which also here will be found to be true, by multiplying the sides *ae* and *ei*, containing the right angle, the one by the other; and then taking the halfe of the product.

This generall way of measuring a triangle is most easie and speedy, where the sides are expressed by whole numbers.

The speciall geodesy of rectangle triangle was before taught (at the 9 e xj.) But of an oblique angle it shall hereafter be spoken. But the generall way is farre more excellent than the speciall; For by the reduction of an obliquangle many fraudes and errours doe fall out, which caused the learned *Cardine* merrily to wish, that hee had but as much land as was lost by that false kinde of measuring.

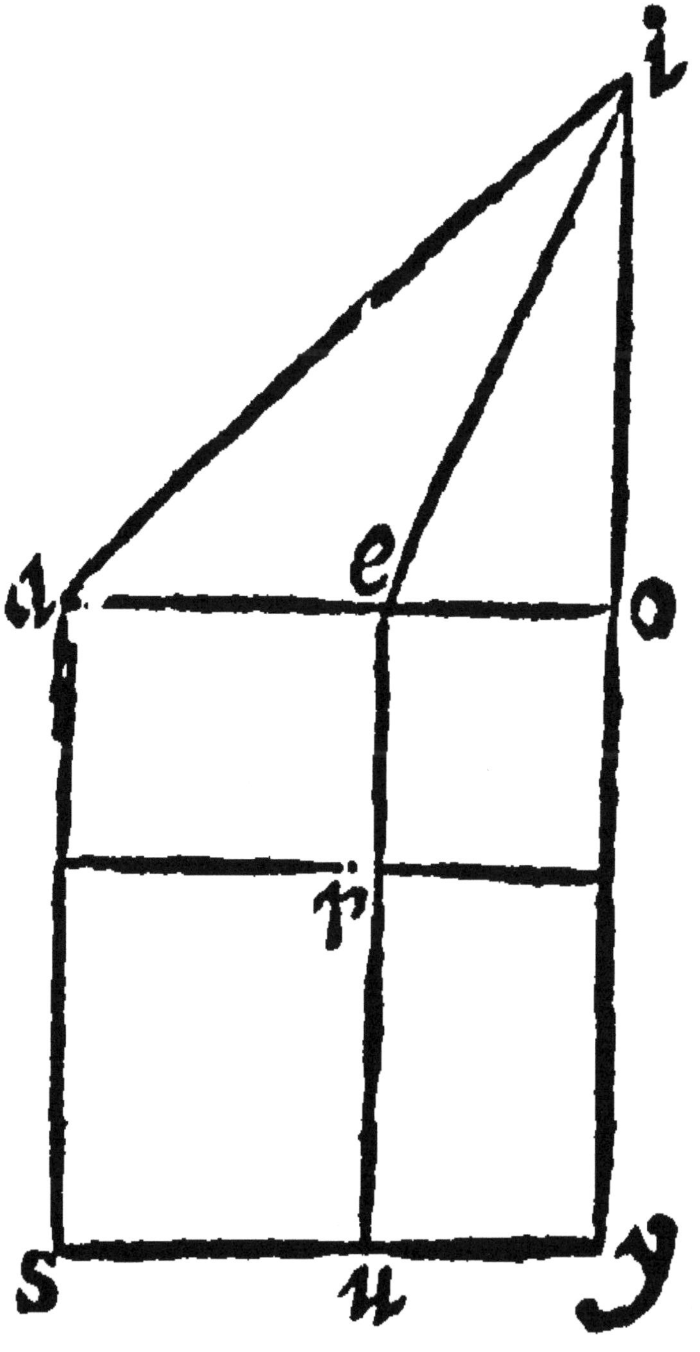
i
e
a
o
p
s
u
y

19 If the base of a triangle doe subtend an obtuse angle, the power of it is more than the power of the shankes, by a double right angle of the one, and of the continuation from the said obtusangle unto the perpendicular of the toppe. 12. p ij.

Or thus: If the base of a triangle doe subtend an obtuse angle, it is in power more than the feete, by the right angled figure twise taken, which is contained under one of the feete and the line continued from the said foote unto the perpendicular drawne from the toppe of the triangle. *H.*

There is a comparison of a quadrate with two in like manner triangles, and as many quadrates, but of unequality.

As in the triangle *aei*, the quadrate of the base *ai*, is greater in power, than the quadrates of the shankes *ae*, and *ei*, by double of the rectangle *ar*, which is made of *ae*, one of the shankes, and of *eo*, the continuation of the same *ae*, unto *o*, the perpendicular of the toppe *i*.

For by 9. e, the quadrate of *ai*, is equall to the quadrates of *ao*, and *oi*, that is, to three quadrates, of *io*, *oe*, *ea*, and the double rectangle aforesaid. But the quadrates of the shankes *ae*, *ei*, are equall to those three quadrates, to wit, of *ai*, his owne quadrate, and of *ei*, two, the first *io*, the second *oe*, by the 9. e. Therefore the excesse remaineth of a double rectangle.

Of Geometry, the thirteenth Booke, Of an Oblong.

1 An Oblong is a rectangle of inequall sides, 31. d j.

Or thus: An Oblong is a rectangled parallelogramme, being not equilater: *H*. As here is *ae, io*.

This second kinde of rectangle is of Euclide in his elements properly named for a definitions sake onely.

The rate of Oblongs is very copious, out of a threefold section of a right line given, sometime rationall and expresable by a number: The first section is as you please, that is, into two segments, equall or unequall: From whence a five-fold rate ariseth.

2 An oblong made of an whole line given, and of one segment of the same, is equall to a rectangle made of both the segments, and the square of the said segment. 3. p ij.

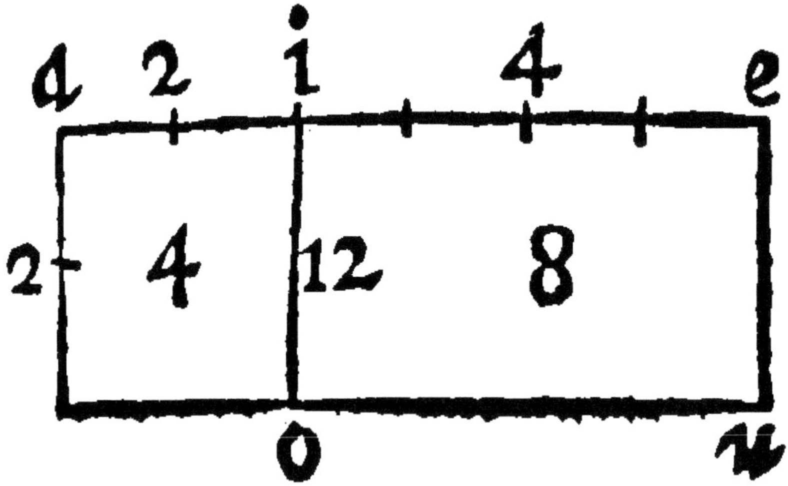

It is a consectary out of the 7 e xj. For the rectangle of the segments, and the quadrate, are made of one side, and of the segments of the other.

As let the right line *ae*, be 6. And let it be cut into two parts *ai*, 2. and *ie*, 4. The rectangle 12. made of *ae*, 6. the whole, and of *ai*, 2. the one segment, shall be equall to *iu*, 8. the rectangle made of the same *ai*, 2. and of *ie*, 4. And also to *ao*, 4. the quadrate of the said segment *ai*, 2.

Now a rectangle is here therefore proposed, because it may be also a quadrate, to wit, if the line be cut into to equall parts.

Secondarily,

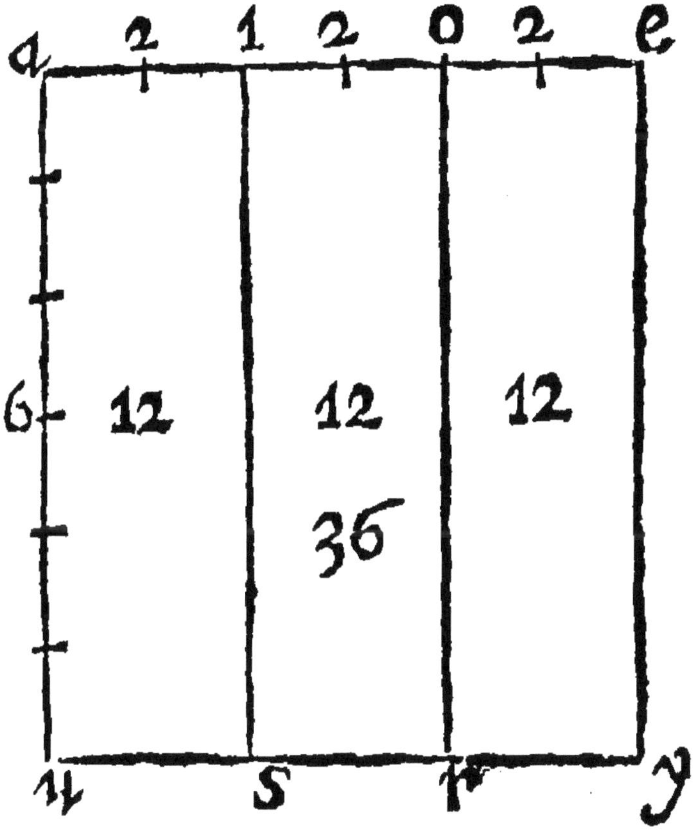

> *3 Oblongs made of the whole line given, and of the segments, are equall to the quadrate of the whole 2 p ij.*

This is also a Consectary out of the 7. e xj.

As let the line *ae*, 6. be cut into *ai*, 2. *io*. 2. and *oe*, 2. The Oblongs *as*, 12. *ir*, 12. and *oy*, 12. made of the whole *ae*, and of those segments, are equall to *ay*, the quadrates of the whole.

Here the segments are more than two, and yet notwithstanding from the first the rest may be taken for one, seeing that the particular rectangle in like manner is equall to them. This proposition is used in the demonstration of the 9. e xviij.

Thirdly,

> *4 Two Oblongs made of the whole line given, and of the one segment, with the third quadrate of the other segment, are equall to the quadrates of the whole, and of the said segment. 7 p ij.*

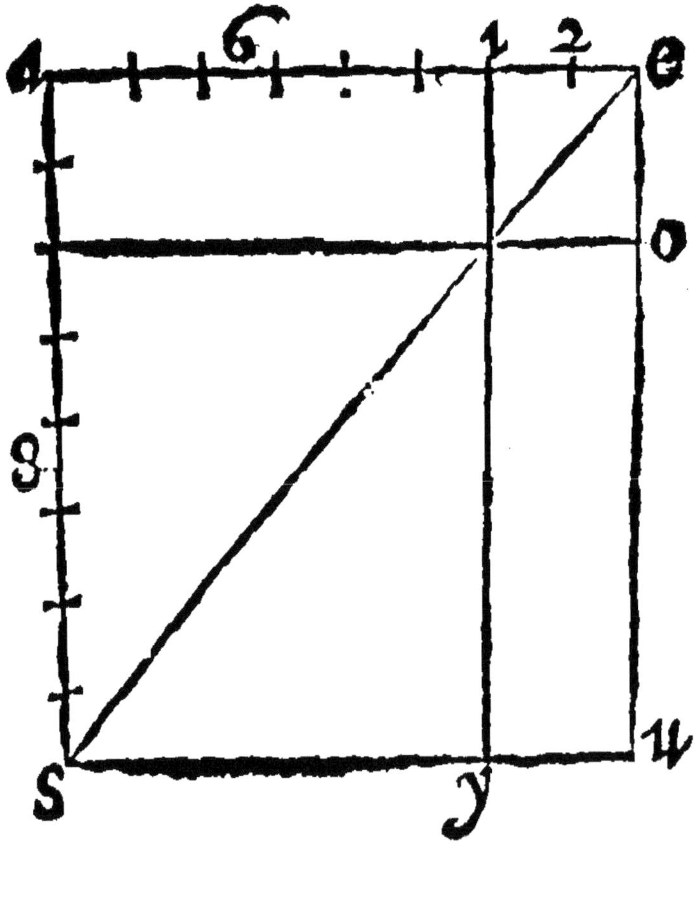

As for example, let the right line *ae*, 8. be cut into *ai*, 6. and *ie*, 2. The oblongs *ao*, and *iu*, of the whole, and 2. the segments, are 32. The quadrate of 6. the other segment 36. And the whole 68. Now the quadrate, of the whole *ae*. 8. is 64. And the quadrate of the said segment 2, is 4. And the summe of these is 68.

5. The base of an acute triangle is of lesse power than the shankes are, by a double oblong made of one of the shankes, and the one segment of the same, from the said angle, unto the perpendicular of the toppe. 13 p. ij.

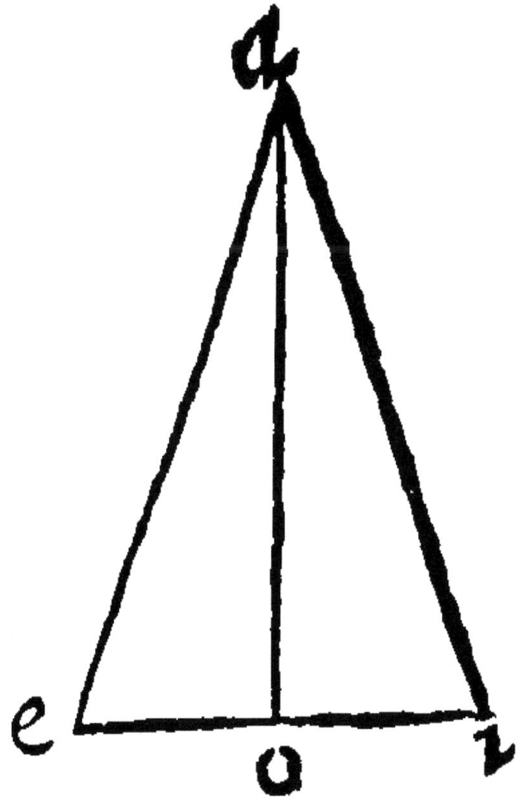

As in the triangle *aei*, let the angle at *i*, be taken for an acute angle. Here by the 4. e, two obongs of *ei*, and *oi*, with the quadrate of *eo*, are equall to the quadrates of *ei*, and *oi*. Let the quadrate of *ao*, be added to both in common. Here the quadrate of *ei*, with the quadrates of *io*, and *oa*, that is the 9 e xij, with the quadrate of *ia*, is equall to two oblongs of *ei*, and *oi*, with two quadrates of *eo* & *oa*, that is by the 9 e xij, with the quadrate of *ea*. Therefore two oblongs with the quadrate of the base, are equall to the quadrates of the shankes: And the base is exceeded of the shankes by two oblongs.

And from hence is had the segment of the shanke toward the angle, and by that the perpendicular in a triangle.

Therefore

6. If the square of the base of an acute angle be taken out of the squares of the shankes, the quotient of the halfe of the remaine, divided by the shanke, shall be the segment of the dividing shanke from the said angle unto the perpendicular of the toppe.

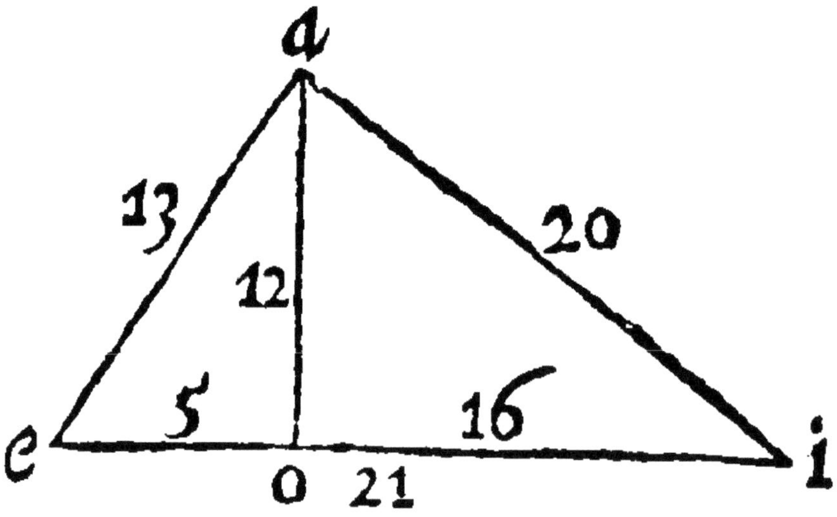

As in the acute angled triangle *aei*, let the sides be 13, 20, 21. And let *ae* be the base of the acute angle. Now the quadrate or square of 13 the said base is 169: And the quadrate of 20, or *ai*, is 400: And of 21, or *ei*, is 441. The summe of which is 841. And 841, 169, are 672: Whose halfe is 336. And the quotient of 336, divided by 21, is 16, the segment of the dividing shanke *ei*, from the angle *aei*, unto *ao*, the perpendicular of the toppe. Now 21, 16, are 5. Therefore the other segment or portion of the said *ei*, is 5.

Now againe from 169, the quadrate of the base 13, take 25, the quadrate of 5, the said segment: And the remaine shall be 144, for the quadrate of the perpendicular *ao*, by the 9 e xij.

Here the perpendicular now found, and the sides cut, are the sides of the rectangle, whose halfe shall be the content of the Triangle: As here the Rectangle of 21 and 12 is 252; whose halfe 126, is the content of the triangle.

The second section followeth from whence ariseth the fourth rate or comparison.

7. If a right line be cut into two equall parts, and otherwise; the oblong of the unequall segments, with the quadrate of the segment betweene them, is equall to the quadrate of the bisegment. 5 p ij.

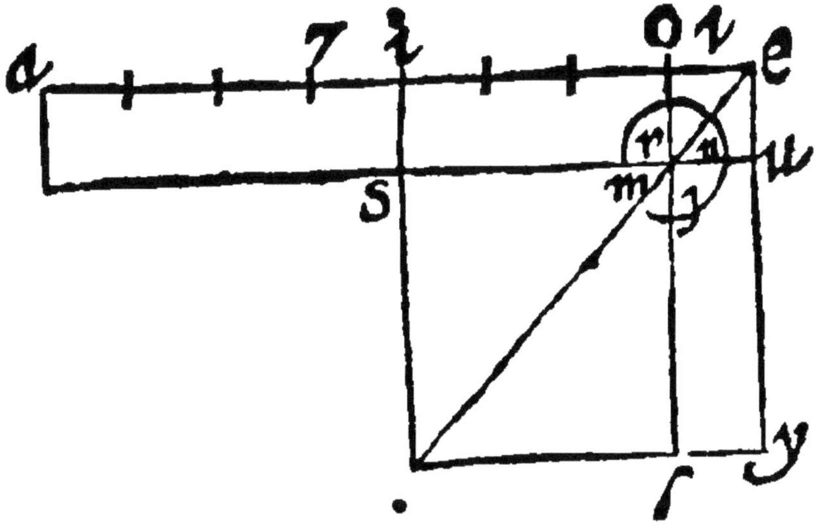

As for example, Let the right line *ae* 8, be cut into two equall portions, *ai* 4, and *ie* 4. And otherwise that is into two unequall portions, *ao* 7, and *oe* 1: The oblong of 7 and 1, with 9, the quadrate of 3, the intersegment, (or portion cut betweene them) that is 16; shall bee equall to the quadrate of *ie* 4, which is also 16. Which is also manifest by making up the diagramme as here thou seest. For as the parallelogramme *as* is by the 26 e x, equall to the parallelogramme *iu*; And therefore by the 19 e x, it is equall to *oy*. For *ou*, is common to both the equall complements, Therefore if *so* be added in common to both; the *ar*, shall be equall to the gnomon *mni*: Now the quadrate of the segment betweene them is *sl*. Wherefore *ar*, the oblong of the unequall segments, with *s* the quadrate of the intersegment, is equall to *iy* the quadrate of the said bisegment.

The third section doth follow, from whence the fifth reason ariseth.

8. If a right line be cut into equall parts; and continued; the oblong made of the continued and the continuation, with the quadrate of the bisegment or halfe, is equall to the quadrate of the line compounded of the bisegment and continuation. 6 p ij.

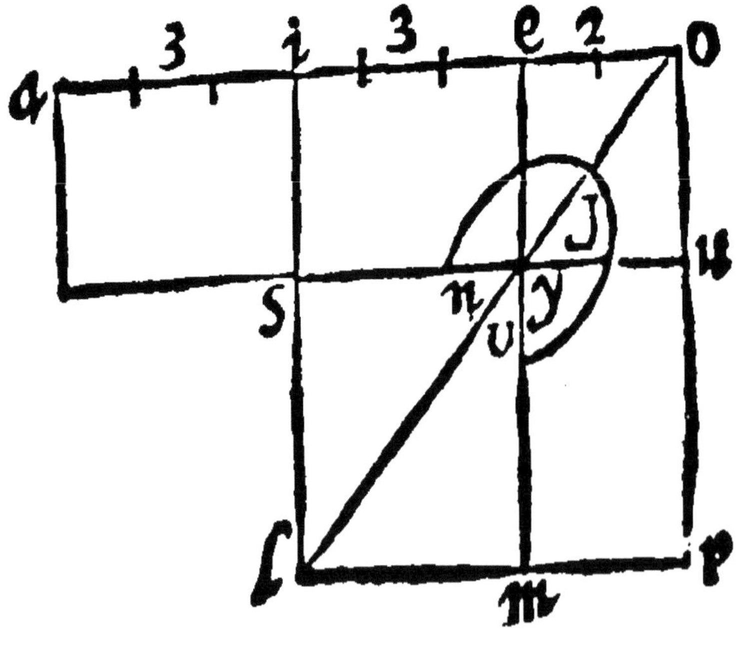

As for example, let the line *ae* 6, be cut into two equall portions, *ai* 3, and *ie* 3: And let it be continued unto *eo* 2: The oblong 16, made of 8 the continued line, and of 2, the continuation; with 9 the quadrate of 3, the halfe, (that is 25.) shall be equall to 25, the quadrate of 3, the halfe and 2, the continuation, that is 5. This as the former, may geometrically, with the helpe of numbers be expressed. For by the 26 e x, *as* is equall to *iy*: And by the 19 e x, it is equall to *yr*, the complement. To these equalls adde *so*. Now the oblong *au*, shall be equall to the gnomon *nju*. Lastly, to the equalls adde the quadrate of the bisegment or halfe. The Oblong of the continued line and of the continuation, with the quadrate of the bisegment, shall be equall to the quadrate of the line compounded of the bisegment and continuation. These were the rates of an oblong with a rectangle.

From hence ariseth the Mesographus or Mesolabus of *Heron* the mechanicke; so named of the invention of two lines continually proportionall betweene two lines given. Whereupon arose the Deliacke probleme, which troubled *Apollo* himselfe. Now the Mesographus of *Hero* is an infinite right line, which is stayed with a scrue-pinne, which is to be moved up and downe in riglet. And it is as *Pappus* saith, in the beginning of his III booke, for architects most fit, and more ready than the Plato's mesographus. The mechanicall handling of this mesographus, is described by *Eutocius* at the 1 Theoreme of the II booke of the speare; But it is somewhat more plainely and easily thus layd downe by us.

9. If the Mesographus, touching the angle opposite to the angle made of the two lines given, doe cut the said two lines given, comprehending a right angled parallelogramme, and infinitely continued, equally distant from the center, the intersegments shall be the meanes continually proportionally, betweene and two lines given.

Or thus: If a Mesographus, touching the angle opposite to the angle made of the lines given, doe cut the equall distance from the center, the two right lines given, conteining a right angled parallelogramme, and continued out infinitely, the segments shall be meane in continuall proportion with the line given: *H*.

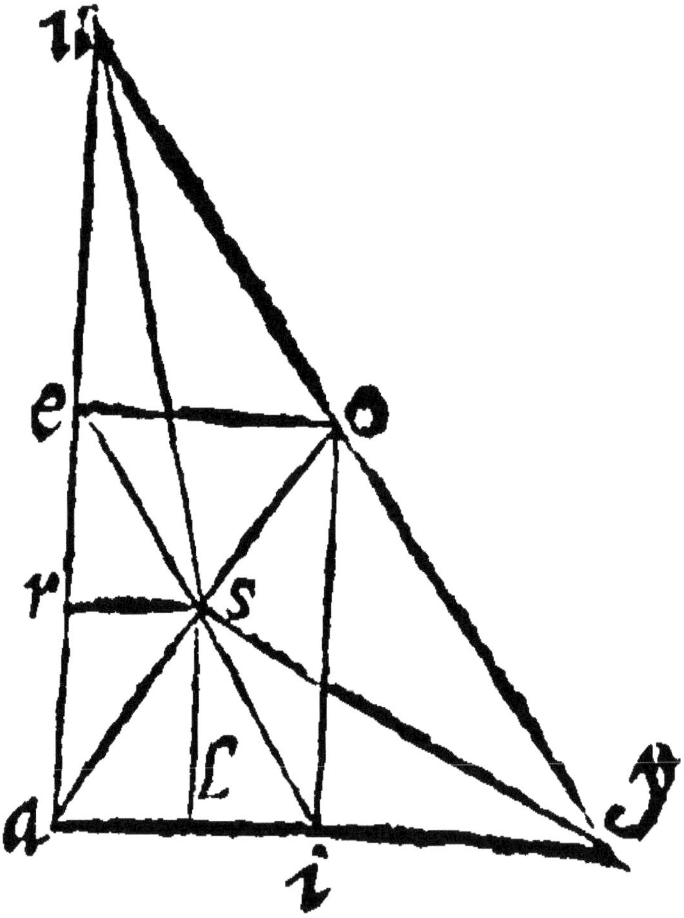

As let the two right-lines given be *ae*, and *ai*: And let them comprehend the rectangled parallelogramme *ao*: And let the said right lines given be continued infinitely, *ae* toward *u*; and *ai* toward *y*. Now let the Mesographus *uy*, touch *o*, the angle opposite to *a*: And let it cut the sayd continued lines equally distant from the Center.

(The center is found by the 8 e iiij, to wit, by the meeting of the diagonies: For the equidistance from the center the Mesographus is to be moved up or downe, untill by the Compasses, it be found.)

Now suppose the points of equidistancy thus found to be *u*, and *y*. I say, That the portions of the continued lines thus are the meane proportionalls sought: And as *ae* is to *iy*: so is *iy* to *eu*, so is *eu*, to *ai*.

First let from *s*, the center, *sr* be perpendicular to the side *ae*: It shall therefore cut the said *ae*, into two parts, by the 5 e xj: And therefore againe, by the 7 e, the oblong made of *au*, and *ue*, with the quadrate of *re*, is equall to the quadrate of *ru*: And taking to them in common *rs*, the oblong with two quadrates *er*, and *rs*, that is, by the 9 e xij, with the quadrate *se* is equall to the quadrates *ru* and *rs*, that is by the 9 e xij, to the quadrate *su*. The like is to be said of the oblong of *ay*, and *yi*, by drawing the perpendicular *sl*, as afore. For this oblong with the quadrates *li*, and *sl*, that is, by the 9 e xij, with the quadrate *is*, is equall to the quadrates *yl*, and *ls*, that is, by the 9 e 12, to *ys*. Therefore the oblongs equall to equalls, are equall betweene themselves: And taking from each side of equall rayes, by the 11 e x, equall quadrates *se* and *si*, there shall remaine equalls. Wherefore by the 27 e x, the sides of equall rectangles are reciprocall: And as *au* is to *ay*: so by the 13 e vij, *oi*, that is, by the 8 e x, *ea*, to *iy*: And so therefore by the concluded, *yi* is to *ue*; And so by the 13 e vij, is *ue* to *eo*, that is, by the 8 e x, unto *ai*. Therefore as *ea* is to *yi*: so is *yi* to *ue*; and so is *ue*, to *ai*. Wherefore *eu*, *iy*, the intersegments or portions cut, are the two meane proportionals betweene the two lines given.

The fourteenth Booke, of P. Ramus *Geometry: Of a right line proportionally cut: And of other Quadrangles, and Multangels.*

Thus farre of the threefold section, from whence we have the five rationall rates of equality: There followeth of the third section another section, into two segments proportionall to the whole. The section it selfe is first to be defined.

1. A right line is cut according to a meane and extreame rate, when as the whole shall be to the greater segment; so the greater shall be unto the lesser. 3 d vj.

This line is cut so, that the whole line it selfe, with the two segments, doth make the three bounds of the proportion: And the whole it selfe is first bound: The greater segment is the middle bound: The lesser the third bound.

2. If a right line cut proportionally be rationall unto the measure given, the segments are unto the same, and betweene themselves irrationall è 6 p xiij.

Euclide calleth each of these segments Ἀποτομὴ that is, *Residuum*, a Residuall or remaine: And surely these cannot otherwise be expressed, then by the name *Residuum*; As if a line of 7 foote should thus be given or put downe: The greater segment shall be called a line of 7 foote, from whence the lesser is substracted: Neither may the lesser otherwise be expressed, but by saying, It is the part residuall or remnant of the line of 7 foote, from which the greater segment was subtracted or taken.

A Triangle, and all Triangulates, that is figures made of triangles, except a Rightangled-parallelogramme, are in Geometry held to be irrationalls. This is therefore the definition of a proportionall section: The section it selfe followeth, which is by the rate of an oblong with a quadrate.

3. If a quadrate be made of a right line given, the difference of the right line from the middest of the conterminall side of the said quadrate made, above the same halfe, shall be the greater segment of the line given proportionally cut: 11 p ij.

Or thus: If a square be made of a right line given, the difference of a right line drawne from the angle of the square made unto the middest of the next side, above the halfe of the side, shall be the greater segment of the line given, being proportionally cut: *H*.

Let the right line gived be *ae*. The quadrate of the same let it be *aeio*: And from the angle *e*, unto *u*, the middest of the conterminal side, let the right line *eu*, be drawne; Then compare or lay it to the halfe *ua*; The difference of it above the said halfe shall be *ay*, This *ay*, say 1, is the greater segment of *ae*, the line given, proportionally cut.

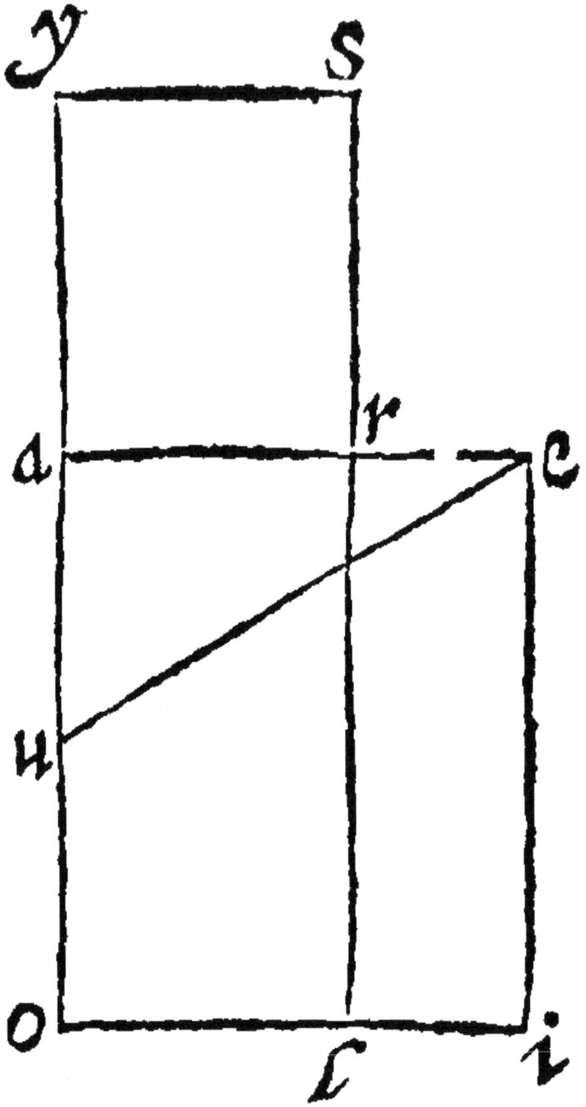

For of *ya*, let the quadrate *aysr*, be made: And let *sr*, be continued unto *l*. Now by the 8 e, xiij. the oblong of *oy*, and *ay*, with the quadrate of *ua*, is equall to the quadrate of *uy*, that is by the construction of *ue*: And therefore, by the 9 e xij. it is equall to the quadrates *ea*, and *au*: Take away from each side the common oblong *al*, and the quadrate *yr*, shall be equall to the oblong *ri*. Therefore the three right lines, *ea*, *ar*, and *re*, by the 8 e xij. are continuall proportionall. And the right line *ae*, is cut proportionally.

Therefore

4. If a right line cut proportionally, be continued with the greater segment, the whole shall be cut proportionally, and the greater segment shall be the line given. 5 p xiij.

As in the same example, the right line *oy*, is continued with the greater segment, and the oblong of the whole and the lesser segment is equall to the quadrate of the greater. And thus one may by infinitely proportionally cutting increase a right line; and againe decrease it. The lesser segment of a right line proportionally cut, is the greater segment, of the greater proportionally cut. And from hence a decreasing may be made infinitely.

5. The greater segment continued to the halfe of the whole, is of power quintuple unto the said halfe, that is, five times so great as it is: and if the power of a right line be quintuple to his segment, the remainder made the double of the former is cut proportionally, and the greater segment, is the same remainder. 1. and 2. p xiij.

This is the fabricke or manner of making a proportionall section. A three-fold rate followeth: The first is of the greater segment.

Let therefore the right line *ae*, be cut proportionally in *i*: And let the greater segment be *ia*: and let the line cut be continued unto *io*, so that *oa*, be the halfe of the line cut. I say, the quadrate of *io*, is in power five times so great, as *ys*, the power of the quadrate of *ao*. Let therefore of *ao*, be made the quadrate *iosr*: We doe see the quadrate *ua*, to be once contained in the quadrate *si*. Let us now teach that it is moreover foure times comprehended in *lmn*, the gnomon remaining: Let therefore the quadrate *aeiu*, be made of the line given: And let *ri*, be continued unto *f*. Here the quadrate *ae*, is (14. e xij.) foure times so much as is that *au*, made of the halfe: and it is also equall to the gnomon *lmn*: For the part *iu*, is equall to *ry*; first by the grant, seeing that *ai*, is the greater segment, from whence *ry*, is made the quadrate, because the other Diagonall is also a quadrate:

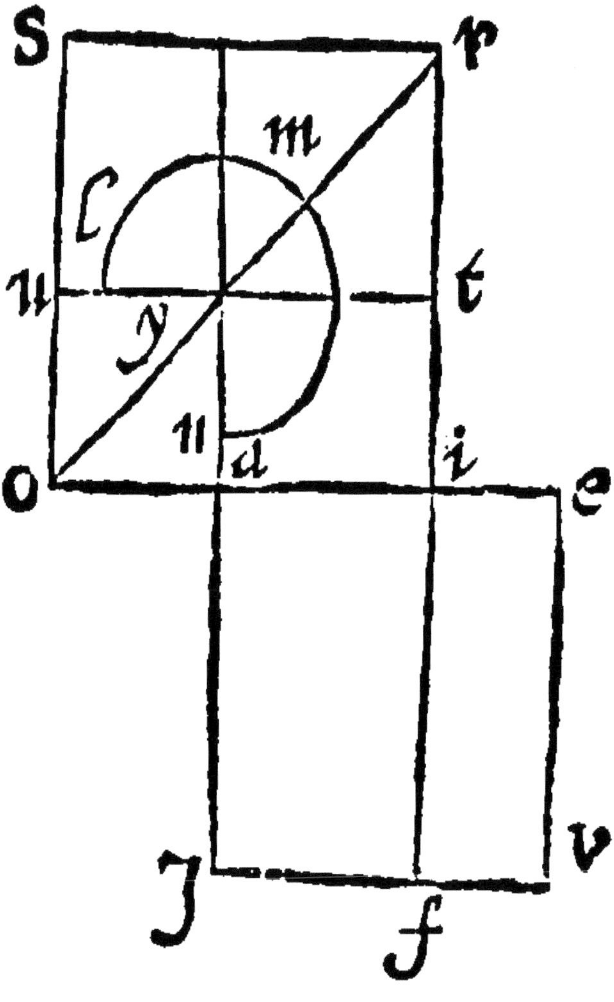

Secondarily the complements *sy*, and *yi*, by the 19. e x, are equall: And to them is equall *af*. For by the 23. e x. and by the grant, it is the double of the complement *yi*. Therefore it is equall to them both. Wherefore the gnomon *lmn*, is equall to the quadruple quadrate of the said little quadrate: And the greater segment continued to the halfe of the right line given is of power five fold to the power of *ao*.

The converse is apparent in the same example: For seeing that *io,* is of power five times so much as is *ao*; the gnomon *lmn,* shall be foure times so much as is *ua*: Whose quadruple also, by the 14. e xij, is *av*. Therefore it is equall to the gnomon. Now *aj,* is equall to *ae*: Therefore it is the double also of *ao,* that is of *ay*: And therefore by the 24. e x. it is the double of *at*: And therefore it is equall to the complements *iy,* and *ys*: Therefore the other diagonall *yr,* is equall to the other rectangle *iv*. Wherefore, by the 8 e xij. as *ev,* that is, *ae,* is to *yt,* that is *ai*: so is *ai,* unto *ie*; Wherefore by the 1 e, *ae,* is proportionall cut: And the greater segment is *ai,* the same remaine.

The other propriety of the quintuple doth follow.

6 The lesser segment continued to the halfe of the greater, is of power quintuple to the same halfe è 3 p xiij.

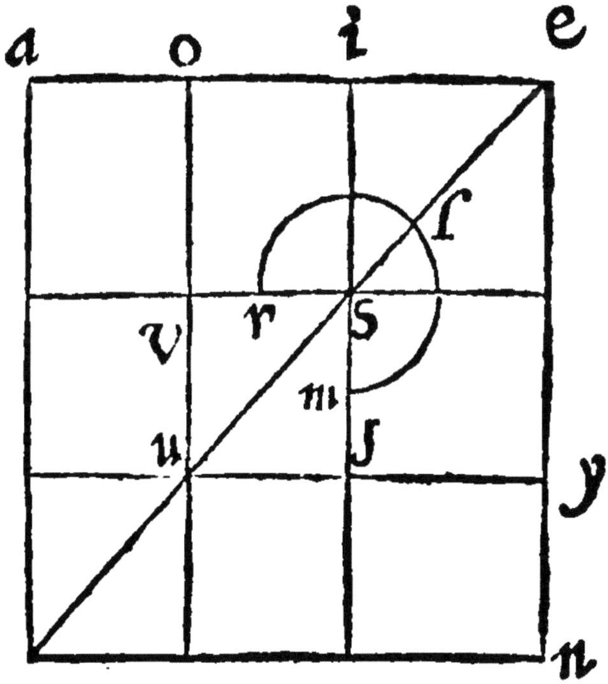

As here, the right line *ae,* let it be cut proportionally in *i*: And the lesser *ie,* let it be continued even unto *o,* the halfe of the greater *ai*. I say, that the power of *oe,* shall be five times as much as is the power of *io*. Let a quadrate therefore be made of *ae*: And let the figure be made up (as you see:) And let the quadrate of the halfe be noted with *su*: And the gnomon *rlm*. Here the first quadrate *oy*, is five times as great, as the second *su*. For it doe containe it once: And the gnomon *rlm*, remaining containeth it foure times. For it is equall to the Oblong *in*; because *os*, the complement is equall to *sy*, by the 19 e x; And therefore also it is equall to *in*; seeing the whole complement *as*, is equall to the whole complement *sn*: And *av*, is equall to *os*, by the construction, and 23. e x: And adding to both the common oblong *iy*, the whole gnomon is equal to the whole oblong. But the oblong *in*, is equall to the quadrate *ai*, by the grant, & 8 e xij. which by the 14. e xij. is foure times as great, as the quadrate *su*. Wherefore the lesser segment *ie*, continued to *io*, the halfe of the greater segment, is of power five times as much as is the halfe of the same.

The rate of the triple followeth.

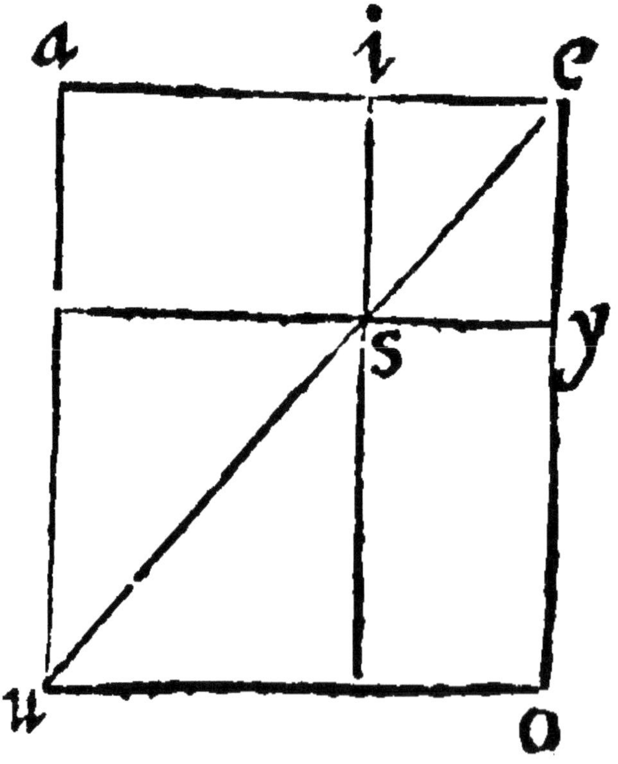

7 The whole line and the lesser segment are in power treble unto the greater. è 4 p xiij.

Let the right line *ae* be proportionally cut in *i*, and let the figure be made up: The oblong *ay*, and *io*, with the quadrate *su*, by the 4 e xiij, are equall to the quadrates of *ae*, and *ie*, whose power is treble to that of *ai*. For they doe once containe the quadrate *su*; And each of the oblongs is equall to the same quadrate *su*, by the grant, and 8 e xij. Therefore they doe containe it thrise.

8 An obliquangled parallelogramme is either a Rhombus, or a Rhomboides.

9 A Rhombus is an obliquangled equilater parallelogramme 32 d j.

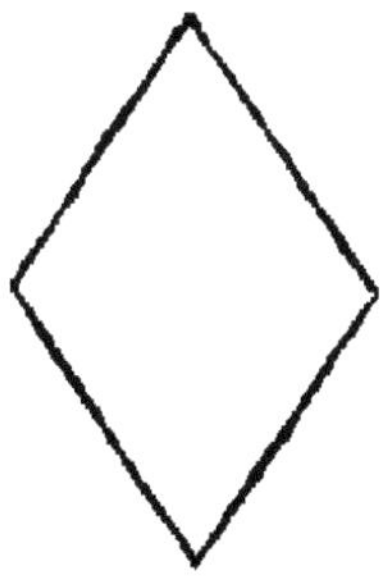

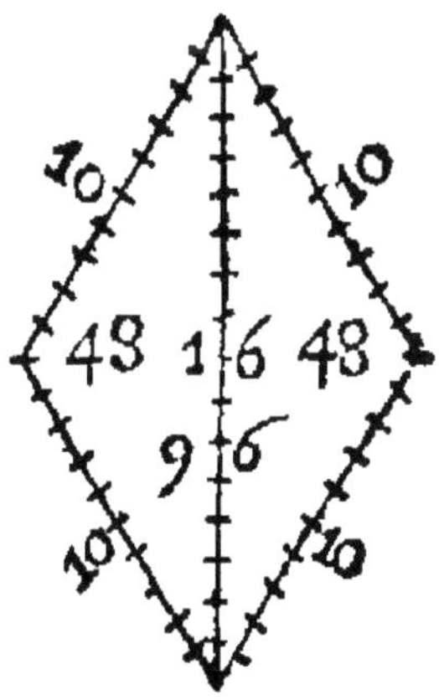

Whereupon it is apparant that a Rhombus is a square having the angles as it were pressed, or thrust nearer together, by which name, both the Byrt or Turbot, a Fish; and a Wheele or Reele, which Spinners doe use; and the quarrels in glasse windowes, because they are cut commonly of this forme, are by the Greekes and Latines so called.

It is otherwise of some called a Diamond.

10 A Rhomboides is an obliquangled parallelogramme not equilater 33. d j.

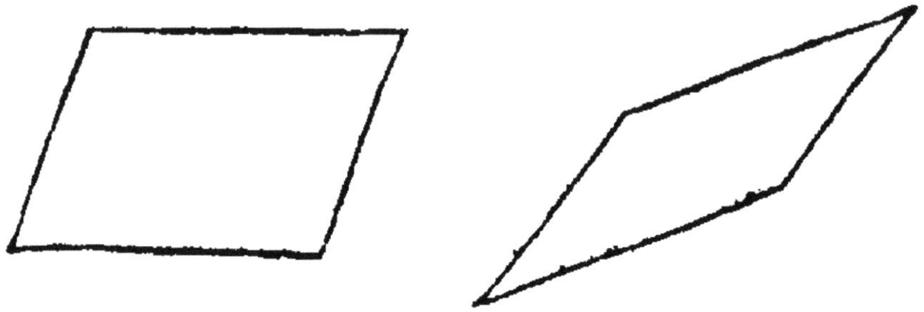

And a Rhomboides is so opposed to an oblong, as a Rhombus is to a quadrate.

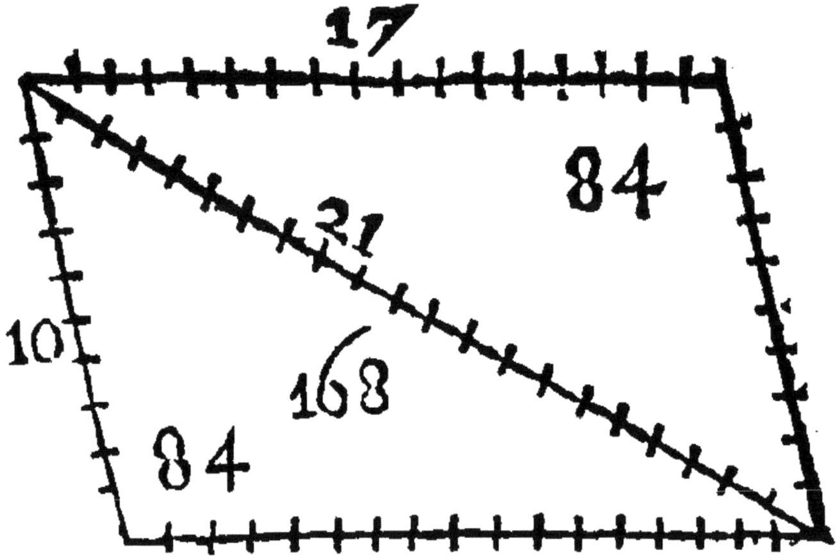

So also looke how much the straightening or pressing together is greater, so much is the inequality of the obtuse and acute angles the greater. As here.

And the Rhomboides is so called as one would say Rhombuslike, although beside the inequality of the angles it hath nothing like to a Rhombus. An example of measuring of a Rhomboides is thus.

11 A Trapezium is a quadrangle not parallelogramme. 34. d j.

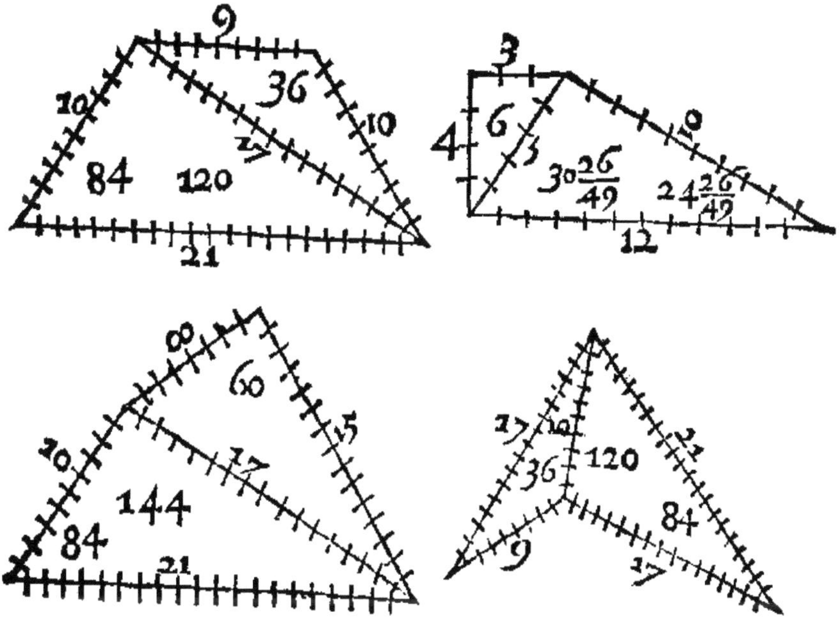

Of the quadrangles the Trapezium remaineth for the last place: *Euclide* intreateth this fabricke to be granted him, that a Trapezium may be called as it were a little table: And surely Geometry can yeeld no reason of that name.

The examples both of the figure and of the measure of the same let these be.

Therefore triangulate quadrangles are of this sort.

12 A multangle is a figure that is comprehended of more than foure right lines. 23. d j.

By this generall name, all other sorts of right lined figures hereafter following, are by Euclide comprehended, as are the quinquangle, sexangle, septangle, and such like inumerable taking their names of the number of their angles.

In every kinde of multangle, there is one ordinate, as we have in the former signified, of which in this place we will say nothing, but this one thing of the quinquangle. The rest shall be reserved untill we come to Adscription.

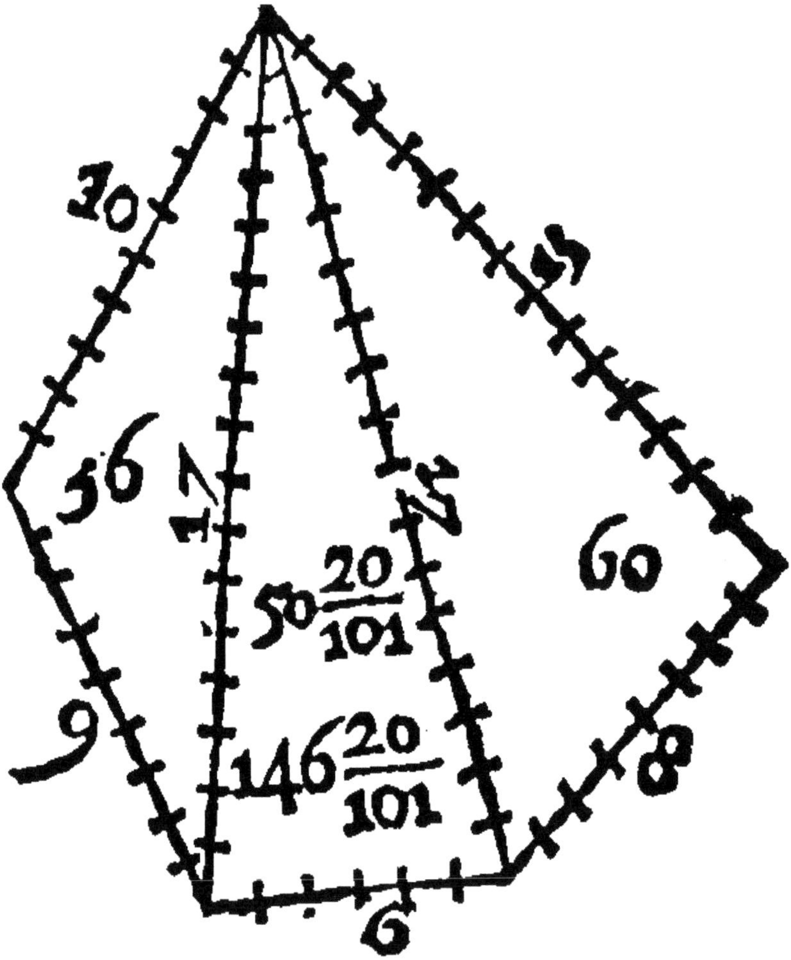

13 Multangled triangulates doe take their measure also from their triangles.

As here, this quinquangle is measured by his three triangles. The first triangle, whose sides are 9. 10. and 17. by the 18. e xij. is 36. The second, whose sides are 6, 17, and 17. by the same e, is 50.20/101. The third, whose sides are 17, 15. and 8. by the same, is 60. And the summe of 36. 50.20/101. and 60. is 146.20/101, for the whole content of the Quinquangle given.

14 If an equilater quinquangle have three sides equall, it is equiangled. 7 p 13.

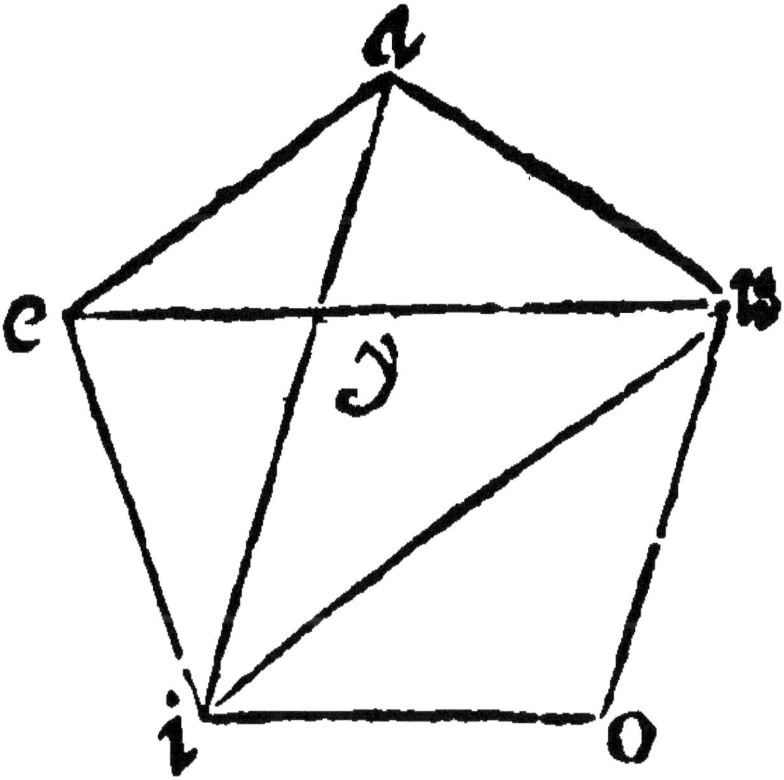

This of some, from the Greeke is called Pentagon; of others a Pentangle, by a name partly Greeke partly Latine.

As in the Quinquangle *aeiou*, the three angles at *a*, *e*, and *i*, are equall: Therefore the other two are equall: And they are equall unto these. For let *eu*, *ai*, *ia*, be knit together with right lines. Here the triangles *aei*, and *eau*. by the grant, and by the 2 and 1 e vij. are equilaters and equiangles: And the Bases *ai*, and *eu*, are equall: And the Angles, *eai*, and *aue*, are equall: Item *aeu*, and *eia*. Therefore *ay*, and *ye*, are equall, by the 17 e vj. Item the remainder *uy*, is equall to the remainder *yi*, when from equals equals be subtracted. Moreover by the grant, and by the 17 e vj, *oui*, and *oiu*, are equall. Wherefore three are equall; And therefore the whole angle is equall at *u*, to the whole angle at *i*. And therefore it is equall to those which are equall to it.

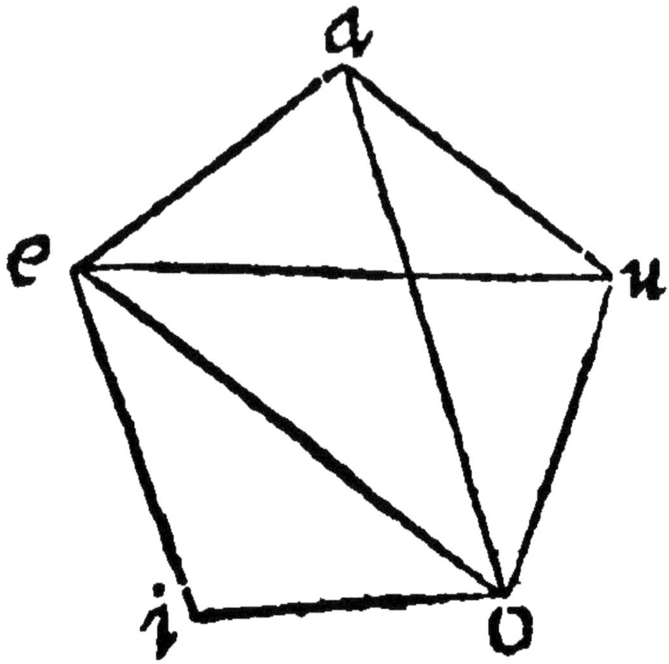

I say moreover that the angle at *o*, is likewise equall, if *ao*, and *oe*, be knit together with a right line, as here: For three in like manner do come to be equall.

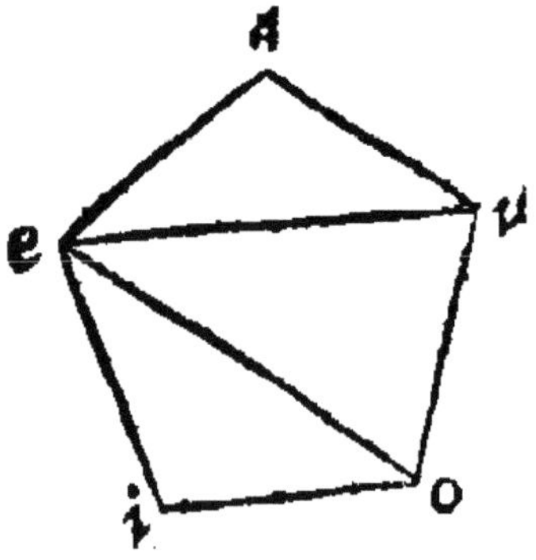

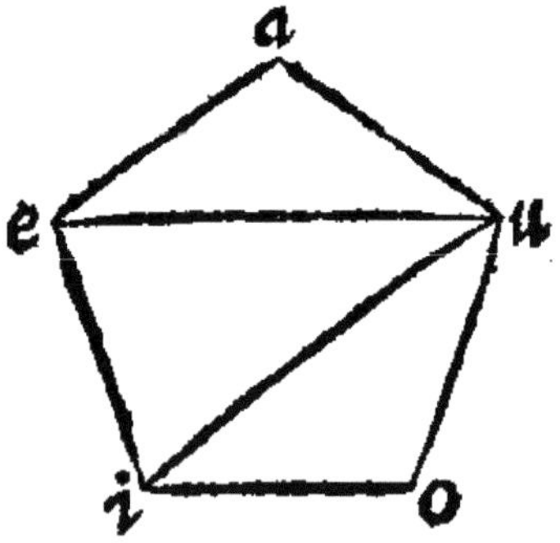

But if the three angles *non deinceps* not successively following be equall, as *aio*, the businesse will yet be more easie, as here: For the angles *eua*, and *eoi*, are equall by the grant: And the inner also *eou*, and *euo*. Therefore the wholes of two are equall. Of the other at *e*, the same will fall out, if *iu*, be knit together with a right line *iu*, as here: For the wholes of two shall be equall.

The fifteenth Booke of Geometry, *Of the Lines in a Circle.*

As yet we have had the Geometry of rectilineals: The Geometry of Curvilineals, of which the Circle is the chiefe, doth follow.

1. A Circle is a round plaine. è 15 d j.

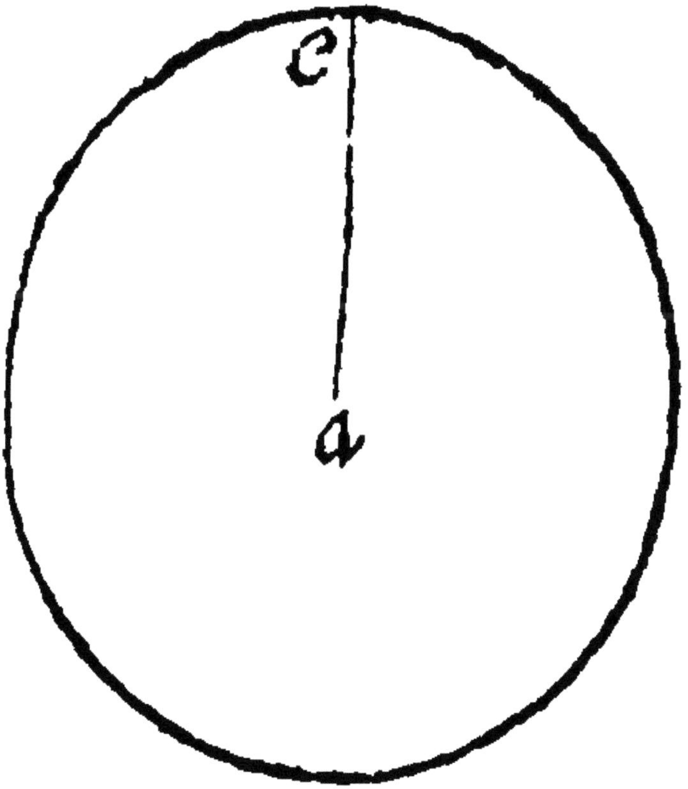

As here thou seest. A Rectilineall plaine was at the 3 e vj, defined to be a plaine comprehended of right lines. And so also might a circle have beene defined to be a plaine comprehended of a periphery or bought-line, but this is better.

The meanes to describe a Circle, is the same, which was to make a Periphery: But with some difference: For there was considered no more but the motion, the point in the end of the ray describing the periphery: Here is considered the motion of the whole ray, making the whole plot conteined within the periphery.

A Circle of all plaines is the most ordinate figure, as was before taught at the 10 e iiij.

2 Circles are as the quadrates or squares made of their diameters 2 p. xij.

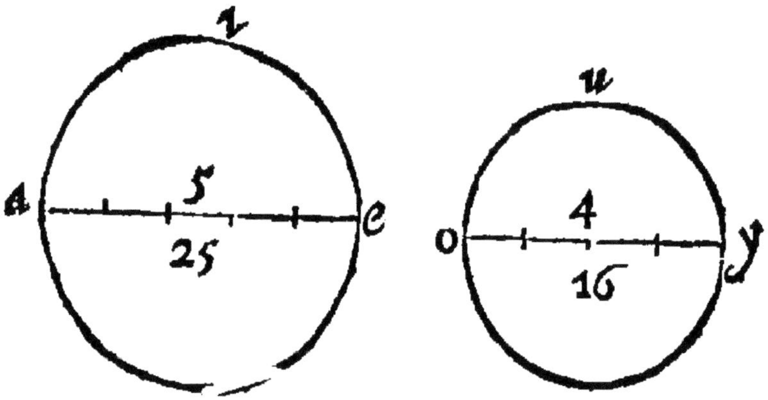

For Circles are like plaines. And their homologall sides are their diameters, as was foretold at the 24 e iiij. And therefore by the 1 e vj, they are one to another, as the quadrates of their diameters are one to another, which indeed is the double reason of their homologall sides. As here the Circle *aei*, is unto the Circle *ouy* as 25, is unto 16, which are the quadrates of their Dieameters, 5 and 4.

Therefore

3. The Diameters are, as their peripheries Pappus, 5 l. xj, and 26th. 18.

As here thou seest in *ae*, and *io*.

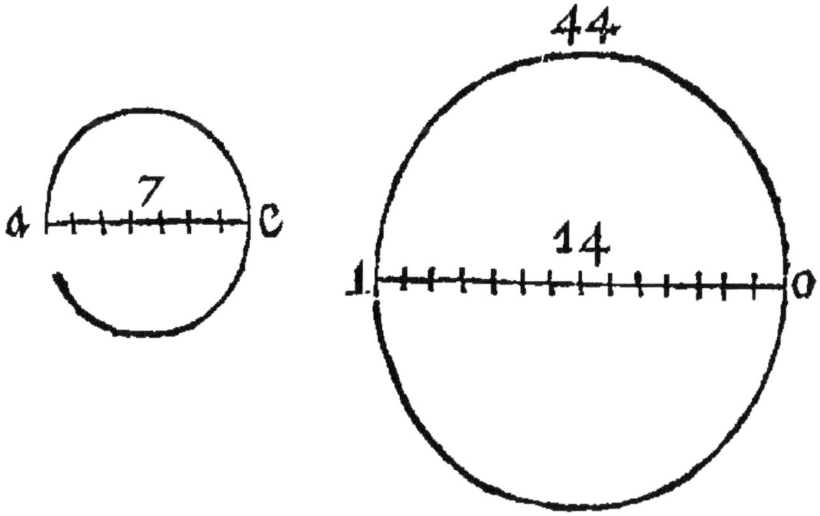

4. Circular Geometry is either in Lines, or in the segments of a Circle.

This partition of the subject matters howsoever is taken for the distinguishing and severing with some light a matter somewhat confused; And indeed concerning lines, the consideration of secants is here the foremost, and first of Inscripts.

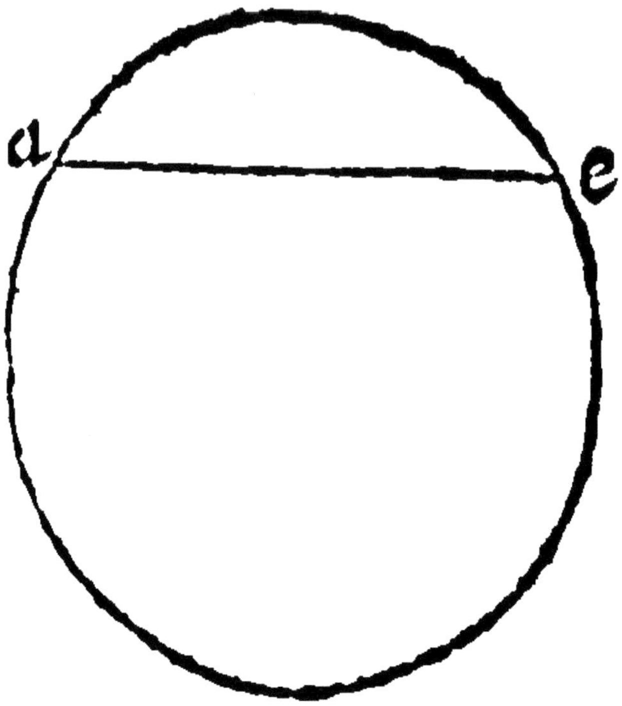

5. If a right line be bounded by two points in the periphery, it shall fall within the Circle. 2 p iij.

As here *ae*, because the right within the same points is shorter, than the periphery is, by the 5 e ij.

From hence doth follow the Infinite section, of which we spake at the 6 e j.

This proposition teacheth how a Rightline is to be inscribed in a circle, to wit, by taking of two points in the periphery.

6. If from the end of the diameter, and with a ray of it equal to the right line given, a periphery be described, a right line drawne from the said end, unto the meeting of the peripheries, shall be inscribed into the circle, equall to the right line given. 1 p iiij.

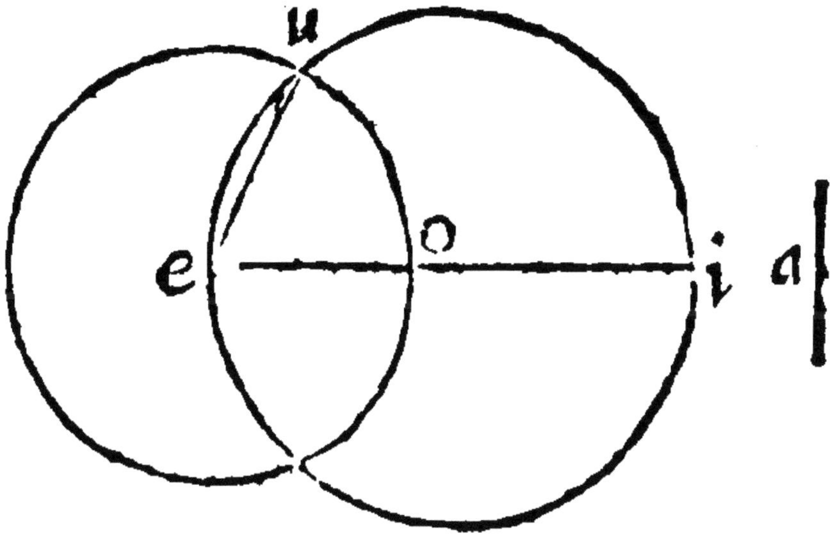

As let the right line given be *a*: And from *e*, the end of the diameter *ei*: And with *eo*, a part of it equall to *a*, the line given, describe the circle *eu*: A right line *eu*, drawne from the end *e*, unto *u*, the meeting of the two peripheries, shall be inscribed in the circle given, by the 5 e, equall to the line given; because it is equall to *eo*, by the 10 e v, seeing it is a ray of the same Circle.

And this proposition teacheth, How a right line given is to be inscribed into a Circle, equall to a line given.

Moreover of all inscripts the diameter is the chiefe: For it sheweth the center, and also the reason or proportion of all other inscripts. Therefore the invention and making of the diameter of a Circle is first to be taught.

7. If an inscript do cut into two equall parts, another inscript perpendicularly, it is the diamiter of the Circle, and the middest of it is the center. 1 p iij.

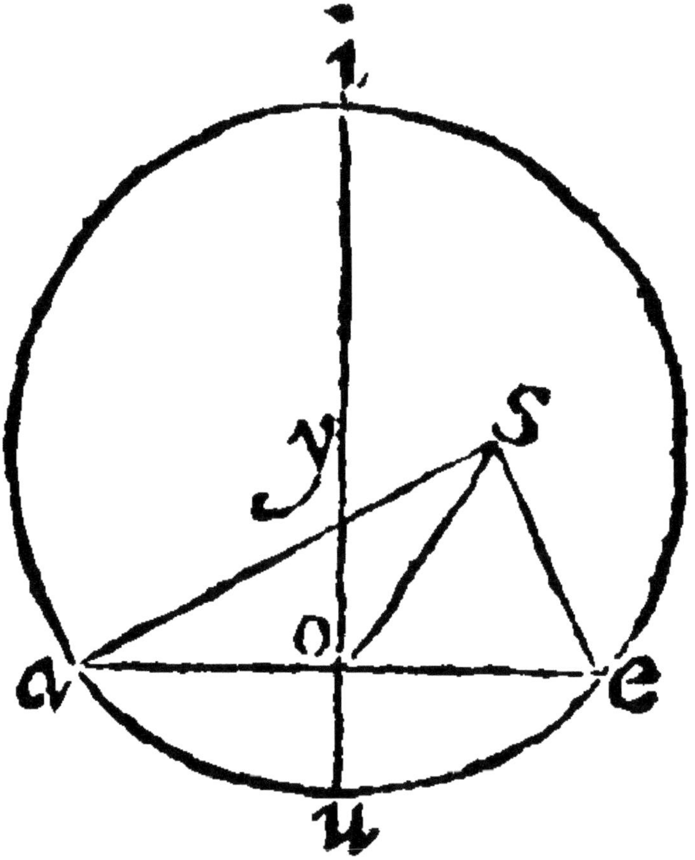

As let the Inscript *ae*, cut the inscript *iu* perpendicularly: dividing it into two equall parts in *o*. I say that the one inscript thus halfing the other, is the diameter of the Circle: And that the middest of it is the center thereof: As in the circle, let the Inscript *is*, cut the inscript *ae*, and that perpendicularly dividing into two equall parts in *o*. I say that *iu*, thus dividing *ae*, is the Diameter of the Circle: And *y*, the middest of the said *iu*, is the Center of the same.

The cause is the same, which was of the 5 e xj. Because the inscript cut into halfes is for the side of the inscribed rectangle, and it doth subtend the periphery cut also into two parts; By the which both the Inscript and Periphery also were in like manner cut into two equall parts: Therefore the right line thus halfing in the diameter of the rectangle: But that the middle of the circle is the center, is manifest out of the 7 e v, and 29 e iiij.

Euclide, thought better of *Impossibile*, than he did of the cause: And thus he forceth it. For if *y* be not the Center, but *s*, the part must be equall to the whole: For the Triangle *aos*, shall be equilater to the triangle *eos*. For *ao, oe,* are equall by the grant: Item *sa,* and *se,* are the rayes of the circle: And *so,* is common to both the triangles. Therefore by the 1 e vij, the angles on each side at *o* are equall; And by the 13 e v, they are both right angles. Therefore *soe* is a right angle; It is therefore equall by the grant, to the right angle *yoe,* that is, the part is equall to the whole, which is impossible. Wherefore *s* is not the Center. The same will fall out of any other points whatsoever out of *y*.

Therefore;

> *8. If two right lines doe perpendicularly halfe two inscripts, the meeting of these two bisecants shall be the Center of the circle è 25 p iij.*

As here *ae*, and *io*, let them cut into halfes the right lines *uy*, and *ys*. And let them meete, that they cut one another in *r*. I say *r* is the center of the circle *ayoseiu*. For before, at the 6, and 7 e, it was manifest that the Center was in the Diameter. And in the meeting of the diameters. [Therefore two manner of wayes is the Center found; First by the middle of the diameter: And then againe by the concourse, or meeting of the diameters, in the middest of the lines halfed or cut into two equall portions.] Here is no neede of the meeting of many diameters, one will serve well enough.

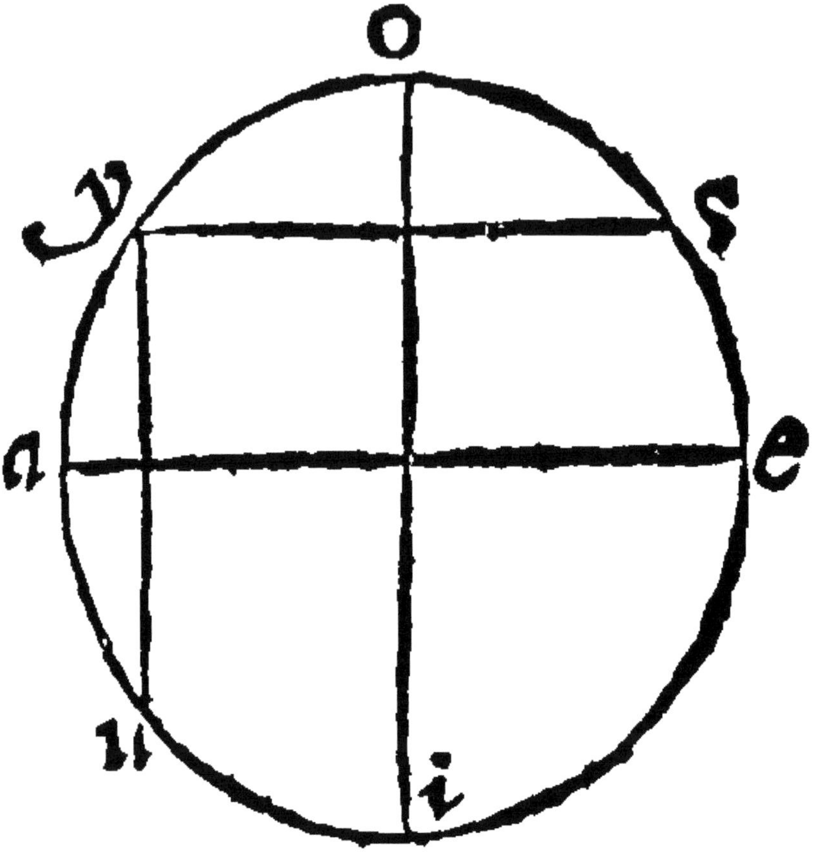

And one may

9. Draw a periphery by three points, which doe not fall in a right line.

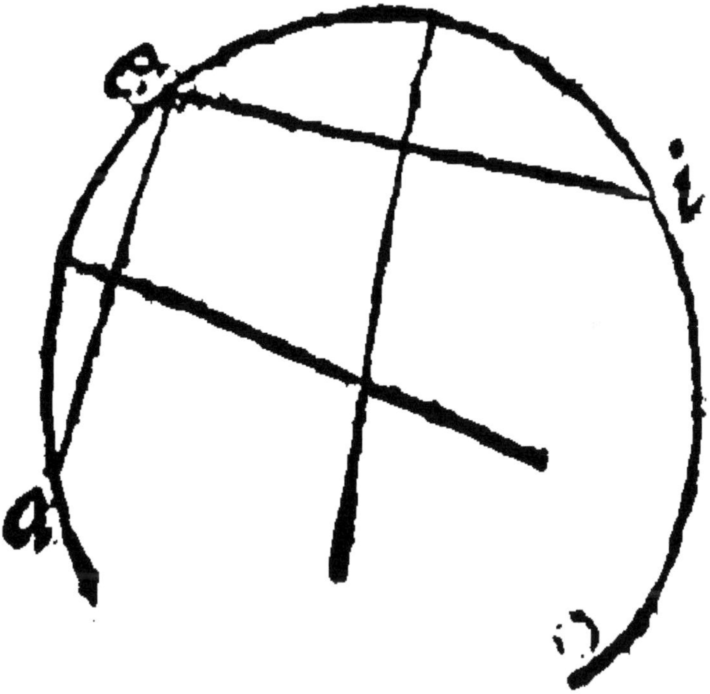

As here, by *aei*, (First from *a*, to *e*, let a right line be drawne; And likewise from *e* to *i*. Then, by the 12 e v, let both these lines be cut into equall parts, by two infinite right lines: These halfing lines also shall meete: And in their meeting shall be the Center, by the 8 e. And therefore from that meeting unto any of the sayd points given is the ray of the periphery desired.)

10. If a diameter doe halfe an inscript, that is, not a diameter, it doth cut it perpendicularly: And contrariwise: 3 p iij.

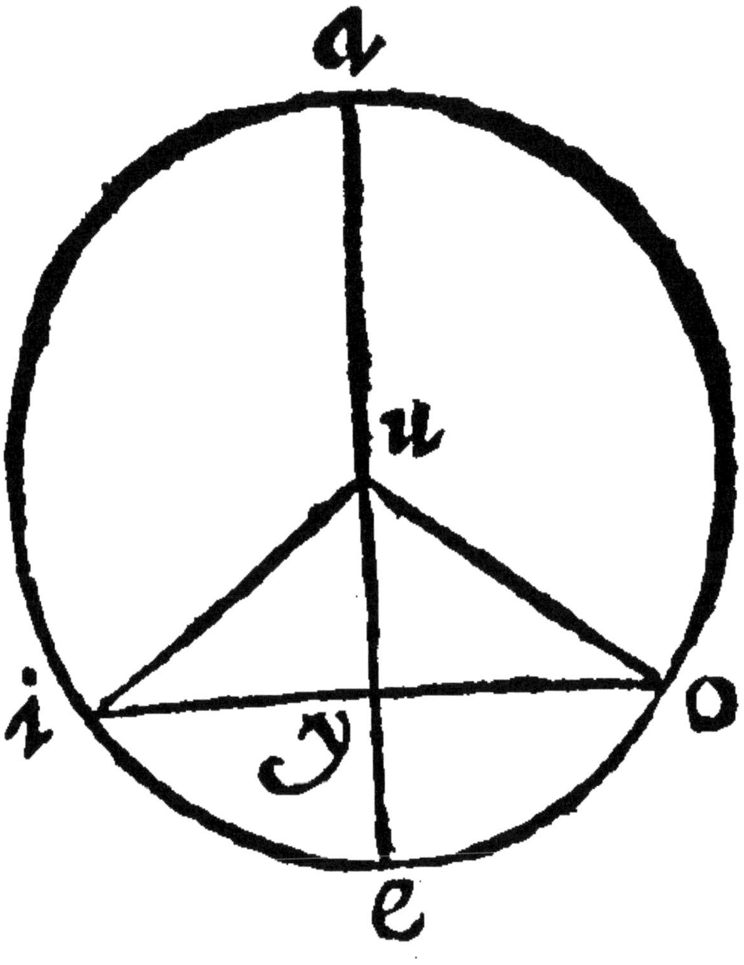

As let the diameter *ae*, halfe the inscript *io*, which is not a diameter: And let the raies of the circle bee *ui*, and *uo*. The cause in all is the same, which was of the 5 e xj.

11. If inscripts which are not diameters doe cut one another, the segments shall be unequall. 4 p iij.

This is a consectary drawne out of the 28 e iiij. For if the inscripts were halfed, they should be diameters, against the grant.

But rate hath beene hitherto in the parts of inscripts: Proportion in the same parts followeth.

12 If two inscripts doe cut one another, the rectangle of the segments of the one is equall to the rectangle of the segments of the other. 35 p iij.

If the inscripts thus cut be diameters, the proportion is manifest, as in the first figure. For the Rectangle of the segments, of the one is equall to the rectangle of the segments of the other, seeing they be both quadrates of equall sides. If they be not diameters let them otherwise as *ae*, and *io*: I say the Oblong of *au*, and *ue*, is equall to the Oblong of *ou*, and *ui*. For let the raies from the Center *y*, be *ye*, and *yi*. To the quadrate of each of these both the rectangles of the segments shall be equall. For by the 7 e, let the diameter *yu*, fall upon the point of the common section *u*; And let *ys*, and *sr*, be perpendiculars. Here by the 5 e xj. the inscripts are cut equally in the points *r* and *s*: And unequally in the point *u*: Therefore by the 7 e xiij, the oblong, of *ou*, and *ui*, with the quadrate *su*, is equall to the quadrates *si*; And adding *ys*, the same oblong, with the quadrates *us* and *sy*, that is, by the 9 e xij, with the quadrate *yu*, is equall to the quadrates *is* and *sy*, that is, by the 9 e xij, to the quadrate *iy*, that is, by the 5 e xij, to *ye*, to the which by the same cause it is manifest the other oblong with the quadrate *yu* is equall. Let the quadrate *yu*, bee taken from each of them: And then the oblongs shall be equall to the same: And therefore betweene themselves.

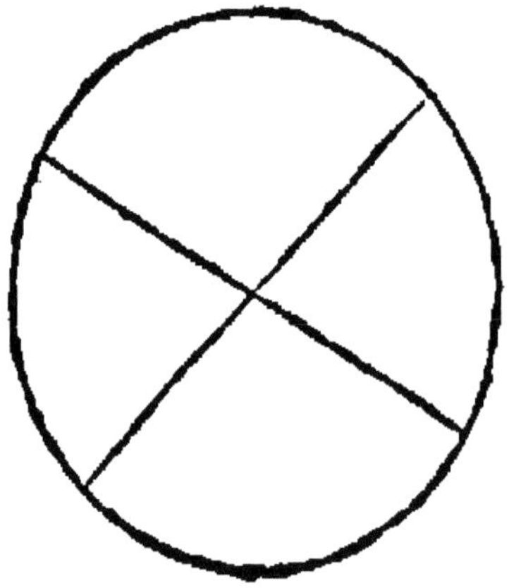

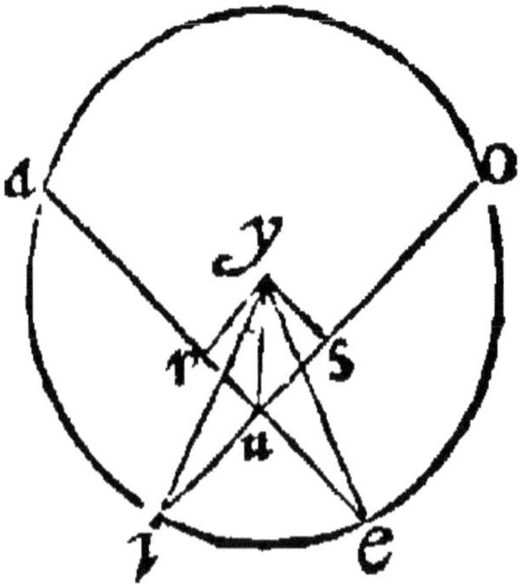

And this is the comparison of the parts inscripts. The rate of whole inscripts doth follow, the which whole one diameter doth make:

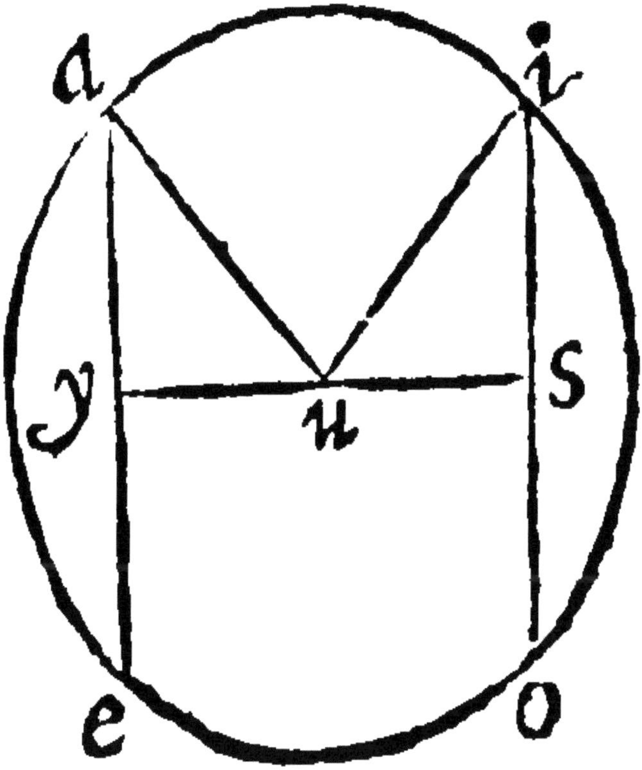

13 Inscripts are equall distant from the center, unto which the perpendiculars from the center are equall 4 d iij.

As it appeareth in the next figure, of the lines *ae* and *io*, unto which the perpendiculars *uy* and *us*, from the Center *u*, are equall.

14. If inscripts be equall, they be equally distant from the center: And contrariwise. 13 p iij.

The diameters in the same circle, by the 28 e iiij, are equall: And they are equally distant from the center, seeing they are by the center, or rather are no whit at all distant from it: Other inscripts are judged to be equall, greater, or lesser one than another, by the diameter, or by the diameters center.

Euclide doth demonstrate this proposition thus: Let first *ae* and *io* be equall; I say they are equidistant from the center. For let *uy*, and *us*, be perpendiculars: They shall cut the assigned *ae*, & *io*, into halfes, by the 5 e xj: And *ya* and *si* are equall, because they are the halfes of equals. Now let the raies of the circle be *ua*, and *ui*: Their quadrates by the 9 e xij, are equall to the paire of quadrates of the shankes, which paires are therefore equall betweene themselves. Take from equalls the quadrates *ya*, and *si*, there shall remaine *yu*, and *us*, equalls: and therefore the sides are equall, by the 4 e 12.

The converse likewise is manifest: For the perpendiculars given do halfe them: And the halfes as before are equall.

15 Of unequall inscripts the diameter is the greatest: And that which is next to the diameter, is greater than that which is farther off from it: That which is farthest off from it, is the least: And that which is next to the least, is lesser than that which is farther off: And those two onely which are on each side of the diameter are equall è 15 e iij.

This proposition consisteth of five members: The first is, The diameter is the greatest inscript: The second, That which is next to the diameter is greater than that which is farther off: The third, That which is farthest off from the diameter is the least: The fourth, That next to the least is lesser, than that farther off: The fifth, That two onely on each side of the diameter are equall betweene themselves. All which are manifest, out of that same argument of equalitie, that is the center the beginning of decreasing, and the end of increasing. For looke how much farther off you goe from the center, or how much nearer you come unto it, so much lesser or greater doe you make the inscript.

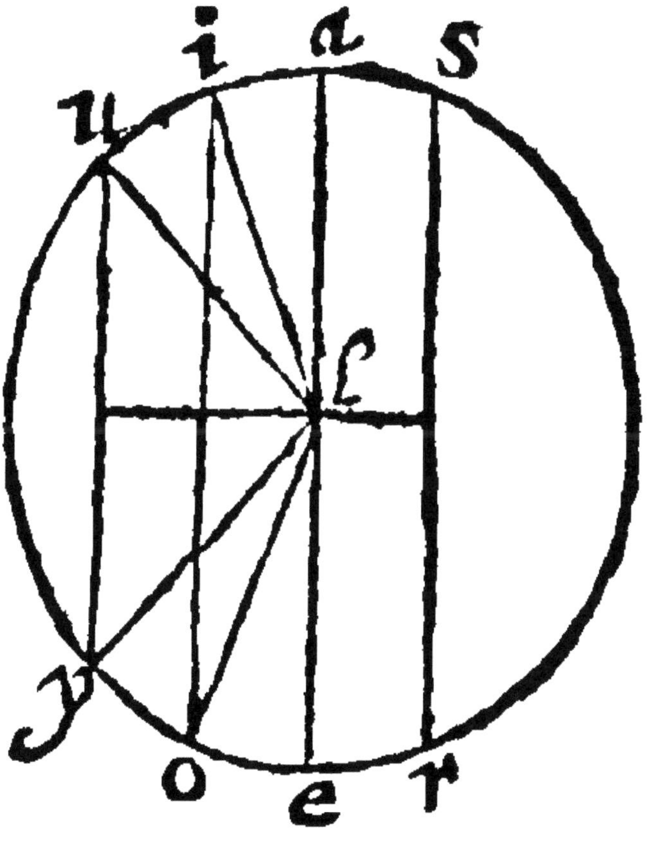

Let there be in a circle; many inscripts, of which one, to wit, *ae*, let it be the diameter: I say, that it is of them all the greatest or longest. But let *io*, be nearer to the diameter, (or as in the former Elements was said) nearer to the center, than *uy*. I say that *io*, is longer than *uy*. Moreover, let *uy*, be the farthest off from the same diameter or center; I say the same *uy*, is the shortest of them all. Now to this shortest *uy*, let *io*, be nearer than *ae*; I say therefore that *io*, also is lesser than *ae*. Let at length *io*, be not the diameter: I say that beyond the diameter *ae*, there may onely a line be inscribed equall unto it, such as is *sr*. And those equal betweene themselves on each side of the diametry may only be given, not three, nor more. And after the same manner also, onely one beyond the diameter, may possibly be equall to *uy*, to wit, that which is as farre off from the diameter as it is; and so in others.

But Euclides conclusion is by triangles of two sides greater than the other; and of the greater angle.

The first part is plaine thus: Because the diameter *ae*, is equall to *il*, and *lo*, *viz*. to the raies; And to those which are greater than *io*, the base by the 9. e vj &c.

The second part of the nearer, is manifest by the 5 e vij. because of the triangle *ilo*, equicrurall to the triangle *uly*, is greater in angle: And therefore it is also greater in base.

The third and fourth are consectaries of the first.

The fifth part is manifest by the second: For if beside *io*, and *sr*, there be supposed a third equall, the same also shall be unequall, because it shall be both nearer and farther off from the diameter.

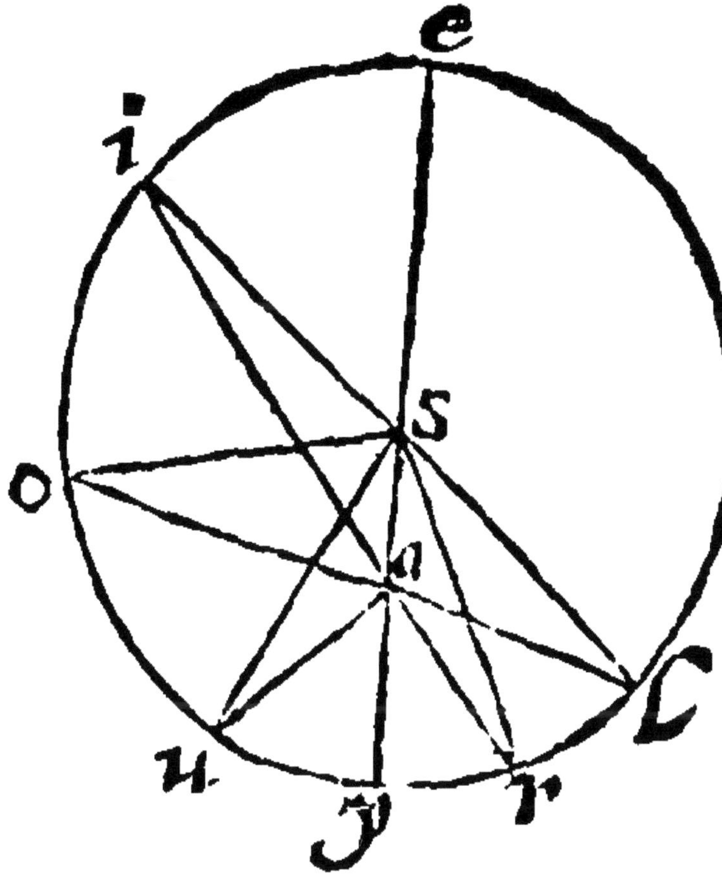

16 Of right lines drawne from a point in the diameter which is not the center unto the periphery, that which passeth by the center is the greatest: And that which is nearer to the greatest, is greater than that which is farther off: The other part of the greatest is the left. And that which is nearest to the least, is lesser than that which is farther off: And two on each side of the greater or least are only equall. 7 p iij.

The first part of *ae*, and *ai*, is manifest, as before, by the 9 e vj. The second of *ai*, and *ao*; Item of *ao*, and *au*, is plaine by the 5 e vij.

The third, that *ay*, is lesser than *au*, because *sy*, which is equall to *su*, is lesser than the right lines *sa*, and *au*, by the 9 e vj: And the common *sa*, being taken away, *ay* shall be left, lesser than *au*.

The fourth part followeth of the third.

The fifth let it be thus: *sr*, making the angle *asr*, equall to the angle *asu*, the bases *au*, and *ar*, shall be equall by the 2 e vij. To these if the third be supposed to be equall, as *al*, it would follow by the 1 e vij. that the whole angle *sa*, should be equall to *rsa*, the particular angle, which is impossible. And out of this fifth part issueth this Consectary.

Therefore

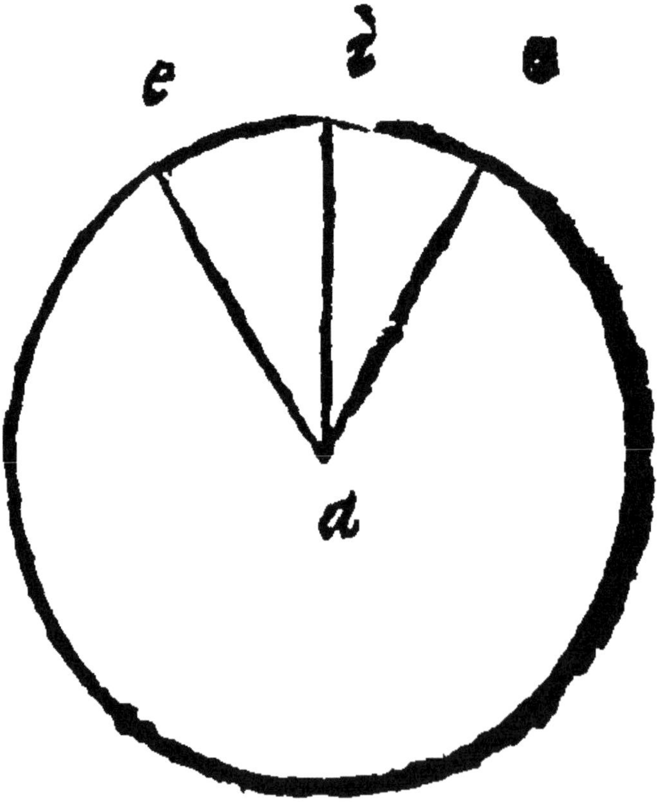

17 If a point in a circle be the bound of three equall right lines determined in the periphery, it is the center of the circle. 9 p iij.

Let the point *a*, in a circle be the common bound of three right lines, ending in the periphery and equall betweene themselves, be *ae*, *ai*, *au*. I say this point is the center of the Circle.

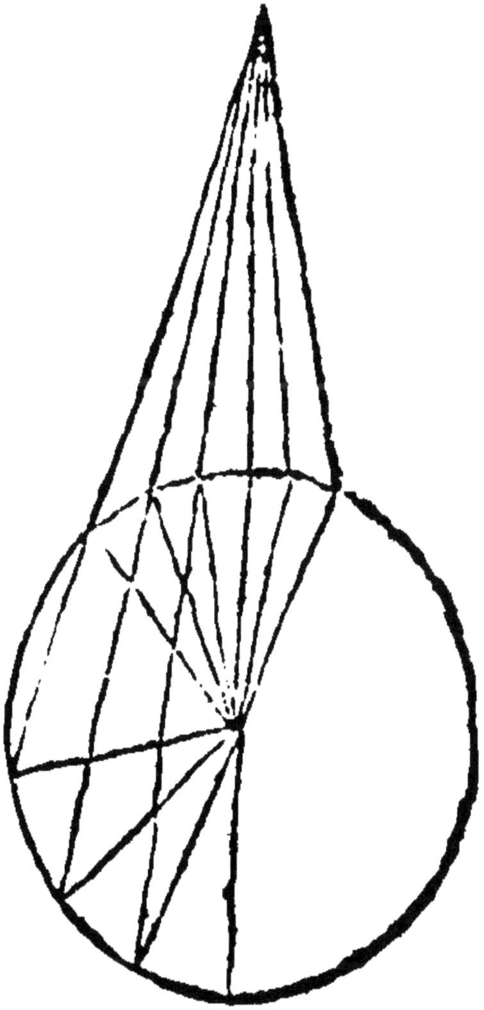

Otherwise from a point of the diameter which is not the center, not onely two right lines on each side should be equall. For by any point whatsoever the diameter may be drawne. Such was before observed in a quinquangle; If three angles be equall, all are equall; so in a Circle: If three right lines falling from the same point unto the perephery be equall, all are equall.

18 Of right lines drawne from a point assigned without the periphery, unto the concavity or hollow of the same, that which is by the center is the greatest; And that next to the greatest, is greater than that which is farther off: But of those which fall upon the convexitie of the circumference, the segment of the greatest is least. And that which is next unto the least is lesser than that is farther off: And two on each side of the greatest or least are onely equall. 8 p iij.

The demonstration of this is very like unto the above mentioned, of five parts. And thus much of the secants, the Tangents doe follow.

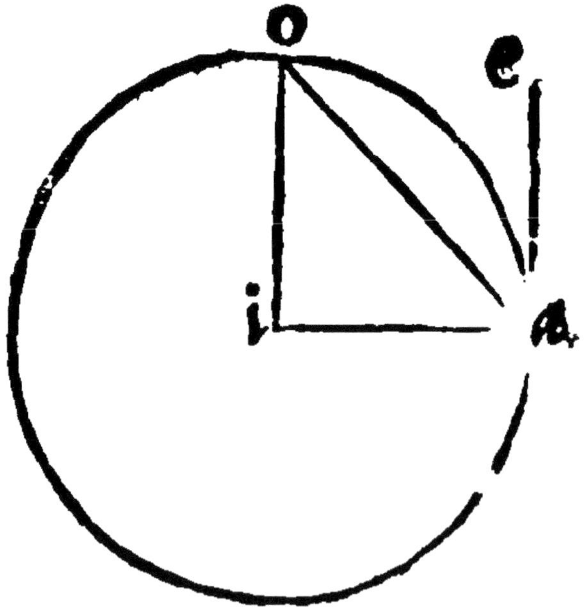

19 If a right line be perpendicular unto the end of the diameter, it doth touch the periphery: And contrariwise è 16 p iij.

As for example, Let the circle given *ae*, be perpendicular to the end of the diameter, or the end of the ray, in the end *a*, as suppose the ray be *ia*: I say, that *ea*, doth touch, not cut the periphery in the common bound *a*.

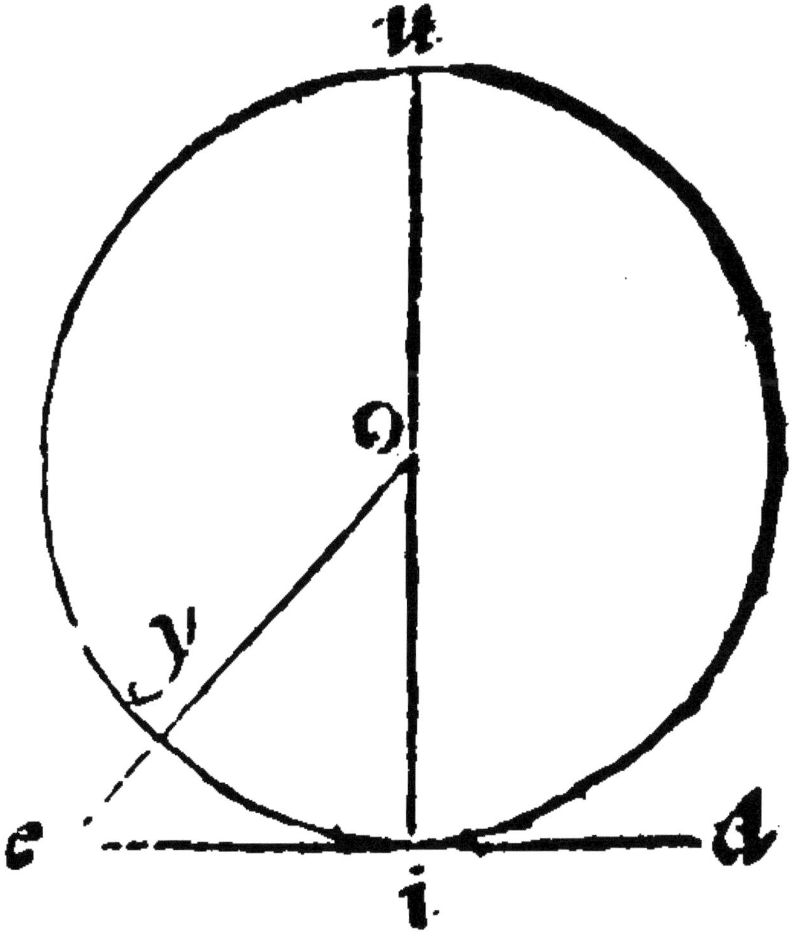

This was to have beene made a *postulatum* out of the definition of a perpendicle: Because if this should leane never so little, it should cut the periphery, and should not be perpendicular: Notwithstanding *Euclide* doth force it thus: Otherwise let the right line *ae*, be perpendicular to the diameter *ai*. And a right line from *o*, with the center *i*, let it fall within the circle at *o*, and let *oi*, joyned together. Here in the triangle *aoi*, two angles, contrary to the 13 e vj, should be right angles at *a*, by the grant: And at *o*, by the 17 e vj.

The demonstration of the converse is like unto the former. For if the tangent, or touch-line *ae*, be not perpendicular to the diameter *iou*, let *oe*, from the center *o*, be drawne perpendicular; Then shall the angle *oei*, be right angle: And *oie* an acutangle: And therefore by the 22 e vj, *oi*, that is *oy*, shall be greater then *oye*, that is the part, then the whole.

Therefore

20 If a right line doe passe by the center and touch-point, it is perpendicular to the tangent or touch-line. 18 p iij.

Or thus, as *Schoner* amendeth it: If a right line be the diameter by the touch point, it is perpendicular to the tangent.

21 If a right line be perpendicular unto the tangent, it doth passe by the center and touch-point. 19. p iij.

Or thus: if it be perpendicular to the tangent, it is a diameter by the touch point: *Schoner*.

For a right line either from the center unto the touch-point; or from the touch point unto the center is radius or semidiameter.

And

22 The touch-point is that, into which the perpendicular from the center doth fall upon the touch line.

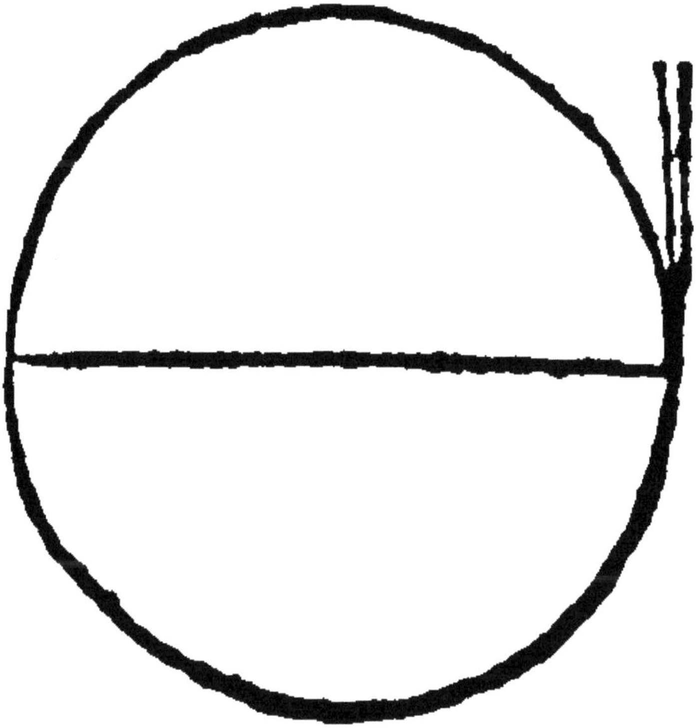

23 A tangent on the same side is onely one.

Or touch line is but one upon one, and the same side: *H*. Or. A tangent is but one onely in that point of the periphery *Schoner*.

It is a consectary drawne out of the xiij. e ij. Because a tangent is a very perpendicular.

Euclide propoundeth this more specially thus; that no other right line may possibly fall betweene the periphery and the tangent.

And

24 A touch-angle is lesser than any rectilineall acute angle, è 16 p ij.

Angulus contractus, A touch angle is an angle of a straight touch-line and a periphery. It is commonly called *Angulus contingentiæ*: Of *Proclus* it is named *Cornicularis*, an horne-like corner; because it is made of a right line and periphery like unto a horne. It is lesse therefore than any acute or sharpe right-lined angle: Because if it were not lesser, a right line might fall between the periphery and the perpendicular.

And

25 All touch-angles in equall peripheries are equall.

But in unequall peripheries, the cornicular angle of a lesser periphery, is greater than the Cornicular of a greater.

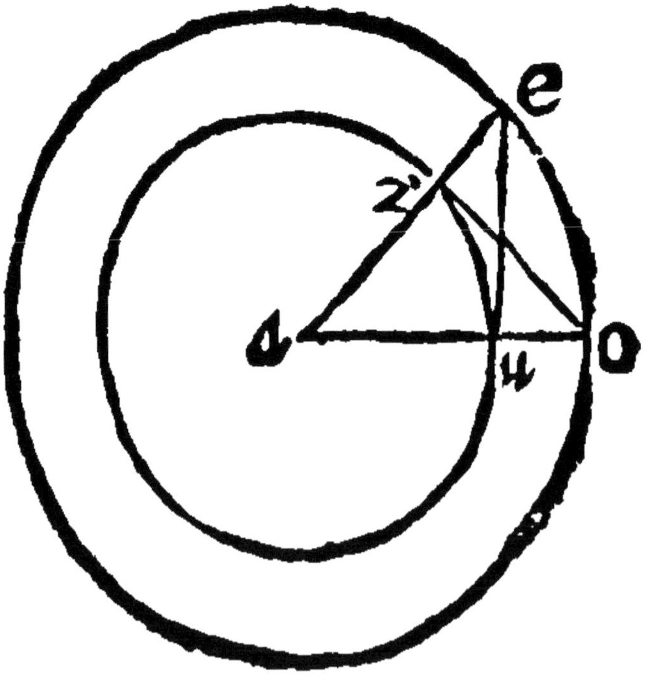

26 If from a ray out of the center of a periphery given, a periphery be described unto a point assigned without, and from the meeting of the assigned and the ray, a perpendicular falling upon the said ray unto the now described periphery, be tied by a right line with the said center, a right line drawne from the point given unto the meeting of the periphery given, and the knitting line shall touch the assigned periphery 17 p iij.

As with the ray *ae*, from the center *a*, of the periphery assigned, unto the point assigned *e*, let the periphery *eo*, be described: And let *io*, be perpendicular to the ray unto the described periphery. This knit by a right line unto the center *a*, let *eu*, be drawne. I say, that *eu*, doth touch the periphery *iu*, assigned: Because it shall be perpendicular unto the end of the diameter. For the triangles *eau*, and *oai*, by the 2 e vij, seeing they are equicrurall; And equall in shankes of the common angle; they are equall in the angles at the base. But the angle *aio*, is a right angle: Therefore the angle *eua*, shall be a right angle. And therefore the right line *eu*, by the 13 e ij, is perpendicular to *ao*.

Thus much of the Secants and Tangents severally: It followeth of both kindes joyntly together.

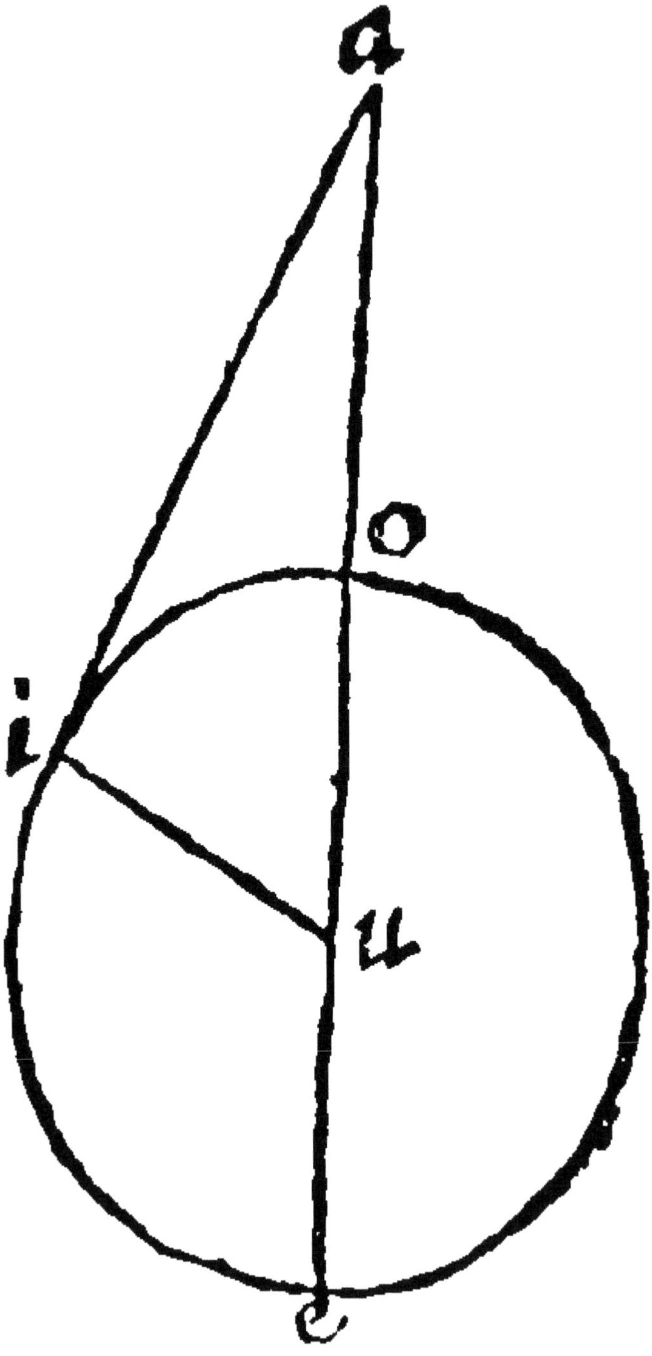
a
o
i
u
e

27 If of two right lines, from an assigned point without, the first doe cut a periphery unto the concave, the other do touch the same; the oblong of the secant, and of the outter segment of the secant, is equall to the quadrate of the tangent: and if such a like oblong be equall to the quadrate of the other, that same other doth touch the periphery: 36, and 37. p iij.

If the secant or cutting line do passe by the center, the matter is more easie and as here, Let *ae*, cut; And *ai*, touch: The outter segment is *ao*, and the center *u*, Now *ui*, shall be perpendicular to the tangent *ai*, by the 20. e: Then by 8 e xiij, the oblong of *ea*, and *ao*, with the quadrate of *au*, that is, of *iu*, is equall to the quadrate of *au*, that is, by the 9 e xij. to the quadrates of *ai*, and *iu*. Take *iu*, the common quadrate: The Rectangle shall be equall to the quadrate of the tangent.

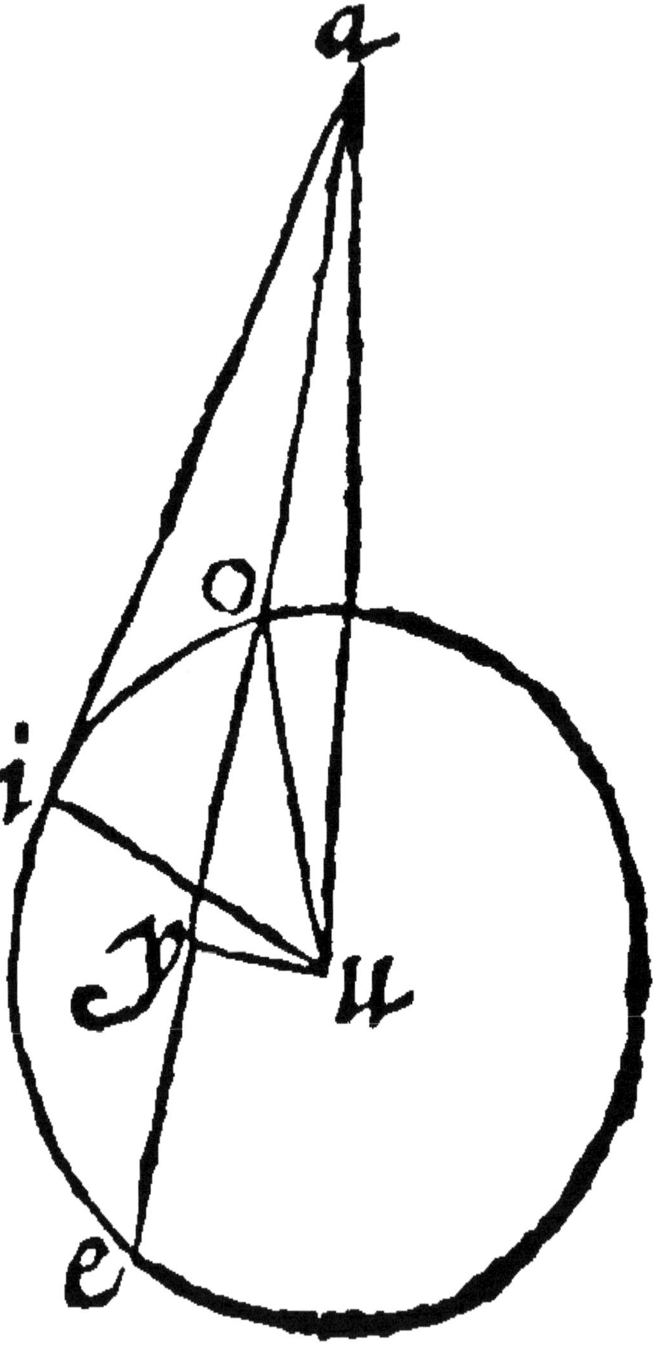

If the secant doe not passe by the center, as in this figure, the center *u*, found by the 7 e, *iu*, shall be by the 20 e perpendicular unto the tangent *ai*;

then draw *ua*, and *uo*, and the perpendicular halving *oe*, by the 10 e. Here by the 8 e xiij, the oblong of *ae*, and *ao*, with the quadrate *oy*, is equal to the quadrate *ay*: Therefore *yu*, the common quadrate added, the same oblong, with the quadrates *oy*, and *yu*, that is by the 9 e xij. with the quadrate *ou*, is equall to the quadrates *ay*, and *uy*, that is, by the 9 e xij, to *au*, that is, againe, to *ai*, and *iu*. Lastly, let *ur*, and *iu*, two equall quadrates be taken from each, and there wil remaine the oblong equall to the quadrate of the tangent.

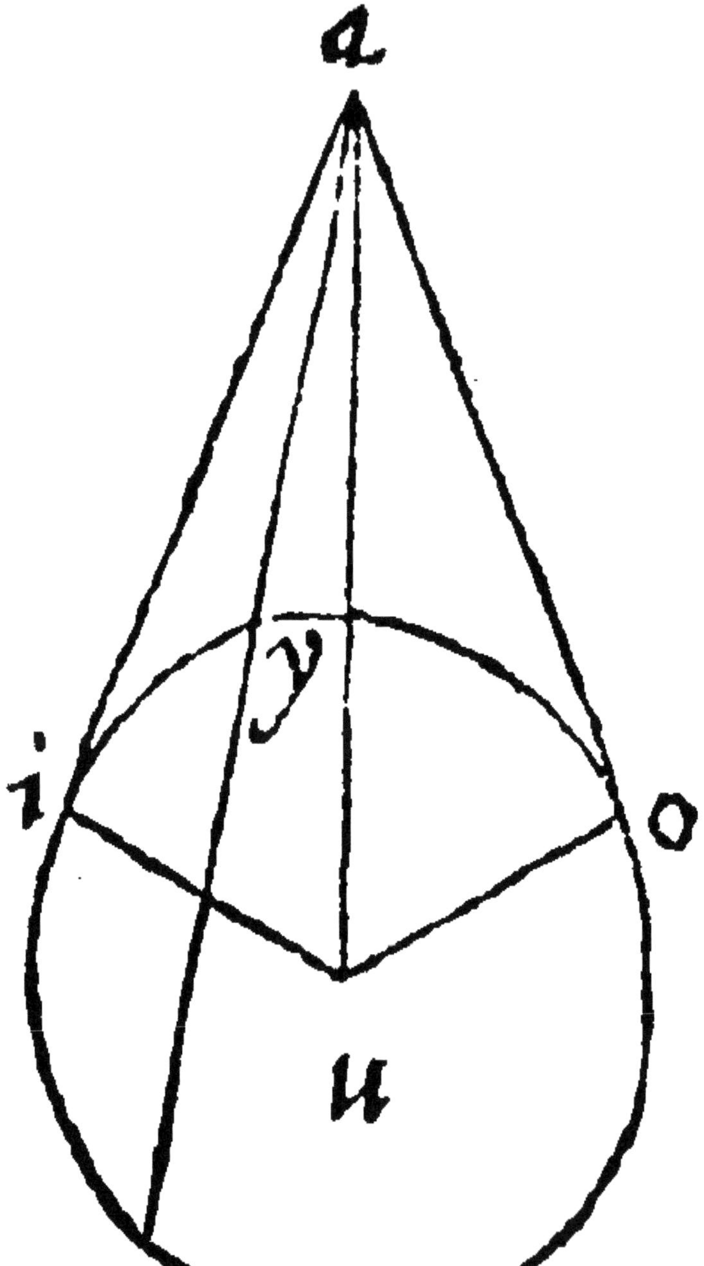

The converse is likewise demonstrated in this figure. Let the Rectangle of *ae*, and *ay*, be equall to the quadrate of *ai*. I say, that *ai* doth touch the circle.

For let, by the 26 e, *ao* the tangent be drawne: Item let *au*, *ui*, and *uo* bee drawne. Here the oblong of *ea*, and *ay*, is equall to the quadrate of *ao*, by the 27 e: And to the quadrate of *ai*, by the grant. Therefore *ai*, and *ao*, are equall. Then is *uo*, by the 20 e, perpendicular to the tangent. Here the triangles *auo*, and *aui*, are equilaters: And by the 1 e vij, equiangles. But the angle at *o* is a right angle: Therefore also a right angle and equall to it is that at *i*, by the 13 e iij, wherefore *ai* is perpendicular to the end of the diameter: And, by the 19 e, it toucheth the periphery.

Therefore

28. All tangents falling from the same point are equall.

Or, Touch lines drawne from one and the same point are equall: *H.*

Because their quadrates are equall to the same oblong.

And

29. The oblongs made of any secant from the same point, and of the outter segment of the secant are equall betweene themselves. Camp. 36 p iij.

The reason is because to the same *thing*.

And

30. To two right lines given one may so continue or joyne the third, that the oblong of the continued and the continuation may be equall to the quadrate remaining. Vitellio 127 p j.

As in the first figure, if the first of the lines given be *eo*, the second *ia*, the third *oa*.

Now are we come to Circular Geometry, that is to the Geometry of Circles or Peripheries cut and touching one another: And of Right lines and Peripheries.

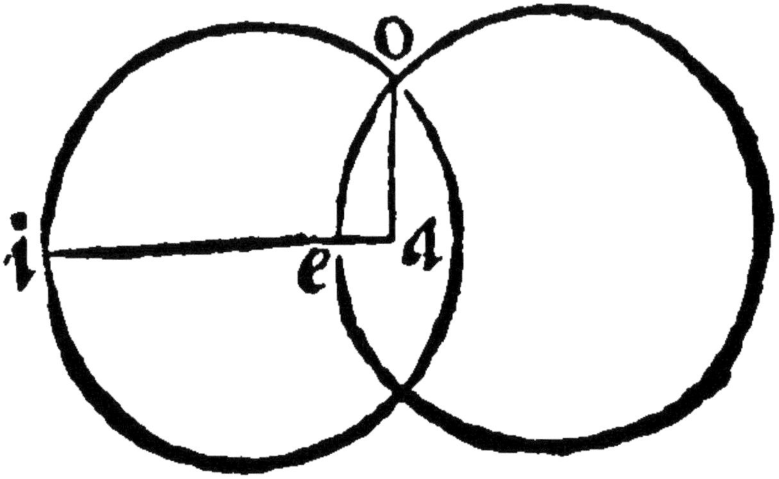

31. If peripheries doe either cut or touch one another, they are eccentrickes: And they doe cut one another in two points onely, and these by the touch point doe continue their diameters, 5. 6. 10, 11, 12 p iij.

All these might well have beene asked: But they have also their demonstrations, *ex impossibili*, not very difficult.

The first part is manifest, because the part should be equall to the whole, if the Center were the same to both, as *a*. For two raies are equall to the common raie *ao*: And therefore *ae* and *ai*, that is, the part and the whole, are equall one to another.

The second part is demonstrated as the first: For otherwise the part must be equall to the whole, as here ae *and* ai, *the raies of the lesser periphery; And* ae, *and* ao, *the raies of the greater are equall. Wherefore* ai, *should be equall to* ao *the Part to the whole.*

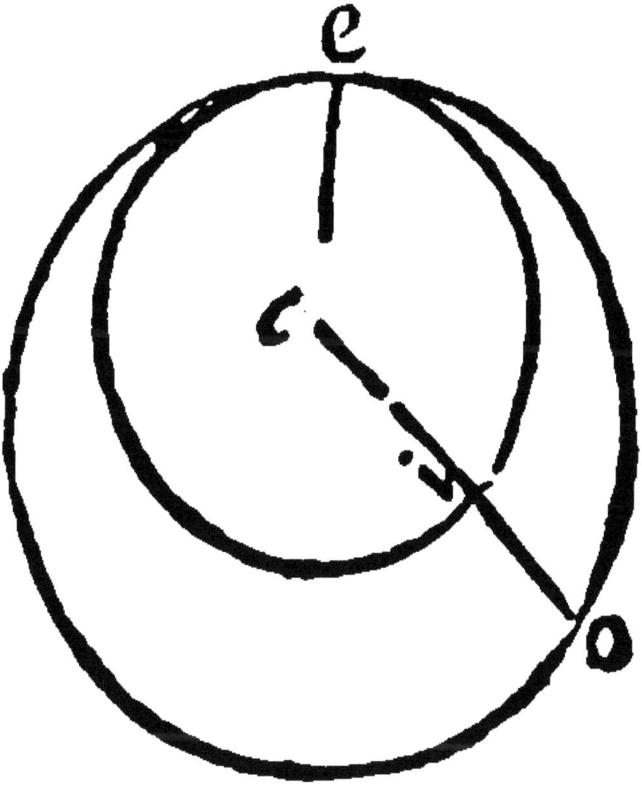

If the Peripheries be outwardly contiguall, the matter is more easie, and by the judgement of *Euclide* it deserved not a demonstration, as here.

The third part is apparent out of the first: Otherwise those which cut one another should be concentrickes. For, by the 7 e, the center being found: And by the 9 e, three right lines being drawne from the center unto three points of the sections, the three raies must be equall, as here.

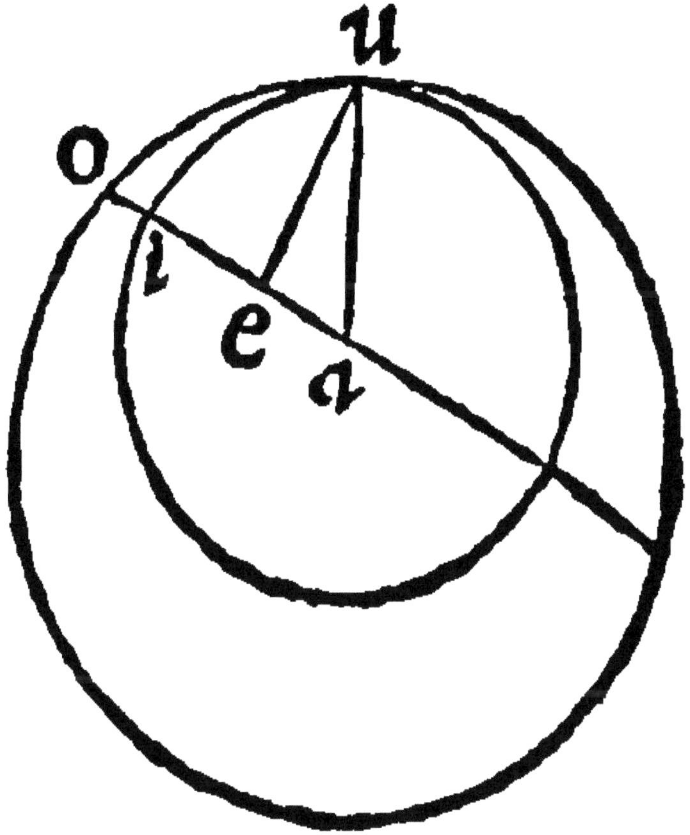

The fourth part is demonstrated after the same manner: Because otherwise the Part must be greater then the whole. For let the right line *aeio*, be drawne by the centers *a* and *e*: And let the particular raies be *eu*, and *au*. Here two sides *ue*, and *ea*, of the triangle *uea*, by the 9 e vj, are greater than *ua*: And therefore also then *ao*; Take away *ae*, the remainder *ue*, shall be greater than *eo*. But *ei* is equall to *eu*. Wherefore *ei* is greater than *eo*, the part, than the whole.

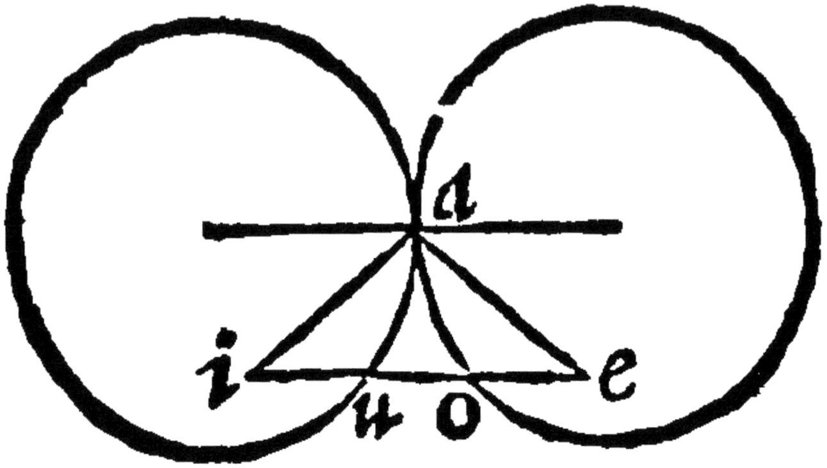

The same will fall out, if the touch be without, as here: For, by the 9 e vj, *ea* and *ia*, are greater than *ie*. But *eo* and *iu*, are equall to *ea*, and *ia*. Wherefore *eo*, and *iu*, are greater than *ie*, the parts than the whole.

Of right lines and Peripheries joyntly the rate is but one.

32. If inscripts be equall, they doe cut equall peripheries: And contrariwise, 28, 29 p iij.

Or thus: If the inscripts of the same circle or of equall circles be equall, they doe cut equall peripheries: And contrariwise *B*.

Or thus: If lines inscribed into equall circles or to the same be equall, they cut equall peripheries: And contrariwise, if they doe cut equall peripheries, they shall themselves be equall: *Schoner*.

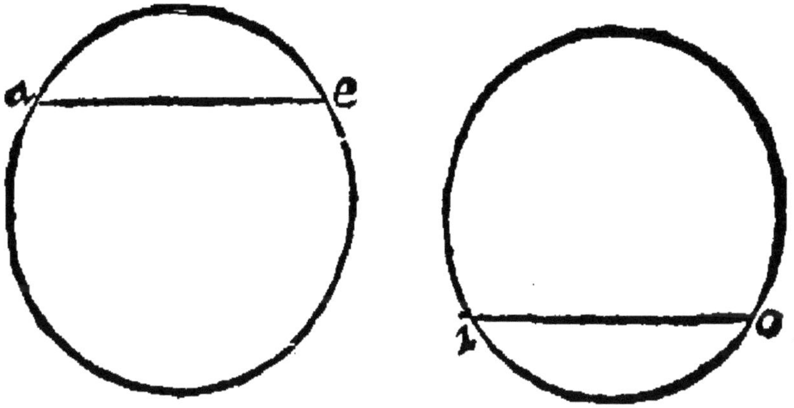

The matter is apparent by congruency or application: as here in this example. For let the circles agree, and then shall equall inscripts and peripheries agree.

Except with the learned *Rodulphus Snellius*, you doe understand aswell two equall peripheries to be given, as two equall right lines, you shall not conclude two equall sections, and therefore we have justly inserted *of the same, or of equall Circles*; which we doe now see was in like manner by *Lazarus Schonerus*.

The sixteenth Booke of Geometry, Of the Segments of a Circle.

1. A Segment of a Circle is that which is comprehended outterly of a periphery, and innerly of a right line.

The Geometry of Segments is common also to the spheare: But now this same generall is hard to be declared and taught: And the segment may be comprehended within of an oblique line either single or manifold. But here we follow those things that are usuall and commonly received. First therefore the generall definition is set formost, for the more easie distinguishing of the species and severall kindes.

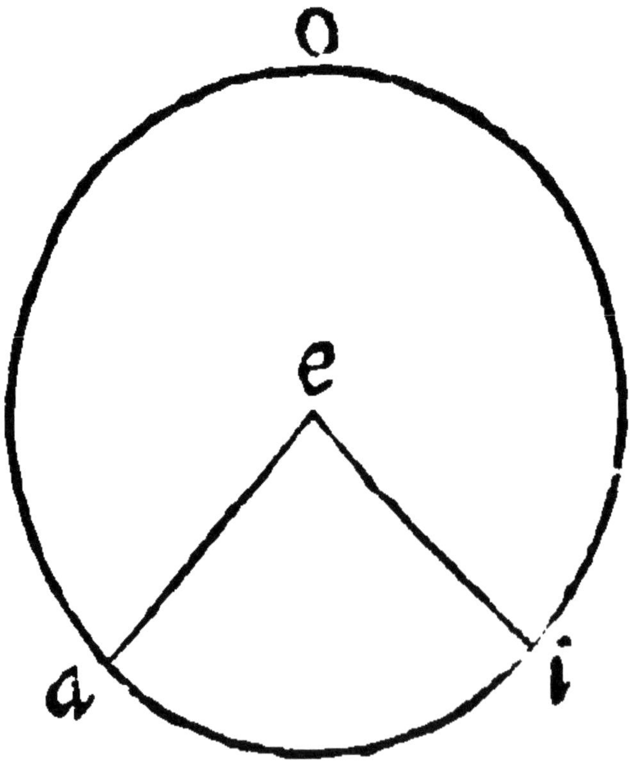

2. A segment of a Circle is either a sectour, or a section.

Segmentum a segment, and *Sectio* a section, and *Sector* a sectour, are almost the same in common acceptation, but they shall be distinguished by their definitions.

3. A Sectour is a segment innerly comprehended of two right lines, making an angle in the center; which is called an angle in the center: As the periphery is, the base of the sectour, 9 d iij.

As *aei* is a sectour. Here a sectour is defined, and his right lined angle, is absolutely called *The greater Sectour* which notwithstanding may be cut into two sectours by drawing of a semidiameter, as after shall be seene in the measuring of a section.

4. An angle in the Periphery is an angle comprehended of two right lines inscribed, and jointly bounded or meeting in the periphery. 8 d iij.

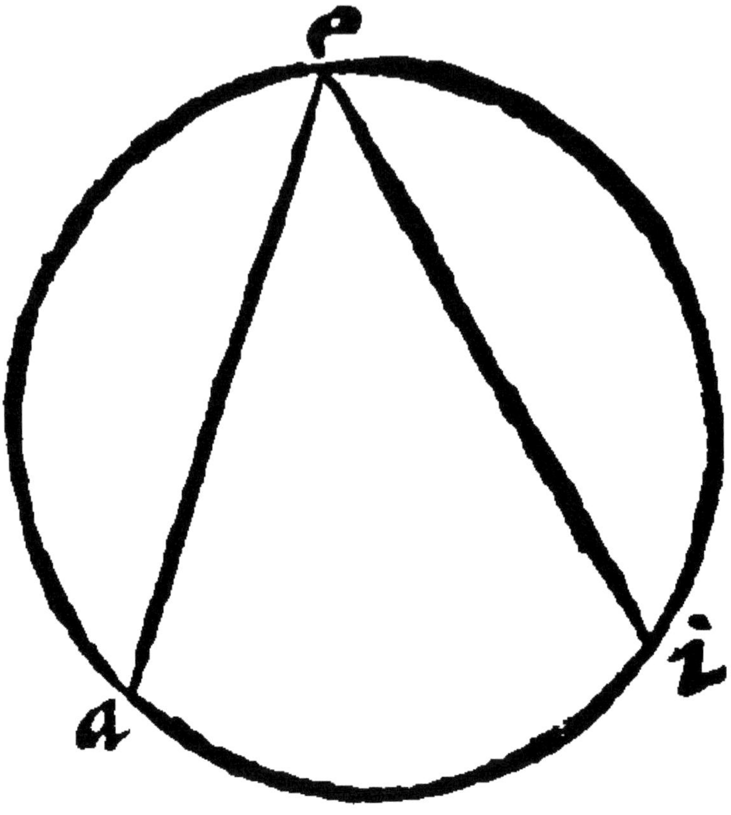

This might have beene called *The Sectour in the Periphery*, to wit, comprehended innerly of two right lines joyntly bounded in the periphery; as here *aei*.

5. The angle in the center, is double to the angle of the periphery standing upon the same base, 20 p iij.

The variety or the example in *Euclide* is threefold, and yet the demonstration is but one and the same: As here *eai*, the angle in the center, shall be prooved to be double to *eoi*, the angle in the periphery, the right line *ou* cutting it into two triangles on each side equicrurall; And, by the 17 e vj, at the base equiangles: Whose doubles severally are the angles, *eau*, of *eoa*: And *iau*, of *ioa*, For seeing it is equall to the two inner equall betweene themselves by the 15 e vj; it shall be the double of one of them. Therefore the whole *eai*, is the double of the whole *eoi*.

The second example is thus of the angle in the center *aei*: And in the periphery *aoi*. Here the shankes *eo*, and *ei*, by the 28 e iiij, are equall: And by the 17 e vj, the angles at *o* and *i* are equall: To both which the angle in the center is equall, by the 15 e vj. Therefore it is double of the one.

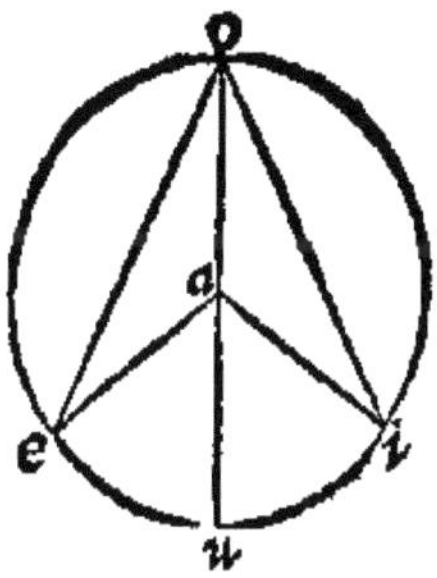

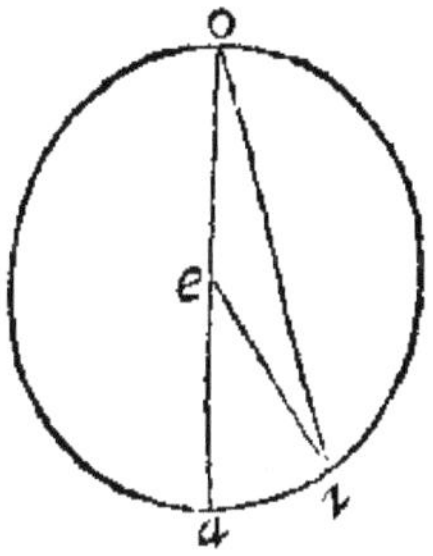

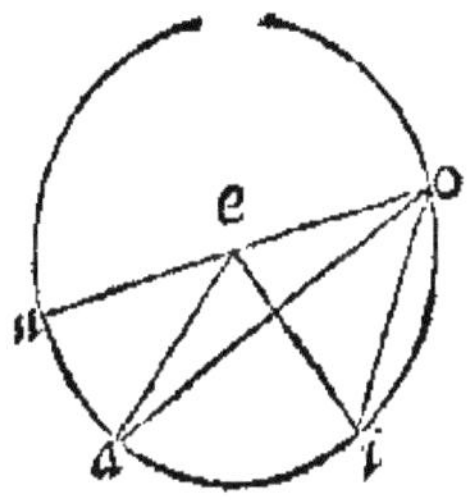

The third example is of the angle in the center, *aei*, And in the periphery *aoi*, Let the diameter be *oeu*. Here the whole angle *ieu*, by the 15 e vj, is equall to the two inner angles *eoi*, and *eio*, which are equall one to another, by the 17 e vj: And therefore it is double of the one. Item the particular angle *aeu*, is equall by the 15 e vj, to the angles *eoa*, and *eao*, equall also one to another, by the 17 e vj. Therefore the remainder *aei*, is the double of the other *aoi*, in the periphery.

Therefore

6. If the angle in the periphery be equall to the angle in the center, it is double to it in base. And contrariwise.

This followeth out of the former element: For the angle in the center is double to the angle in the periphery standing upon the same base: Wherefore if the angle in the periphery be to be made equall to the angle in the center, his base is to be doubled, and thence shall follow the equality of them both: *S.*

7. The angles in the center or periphery of equall circles, are as the Peripheries are upon which they doe insist: And contrariwise. è 33 p vj, and 26, 27 p iij.

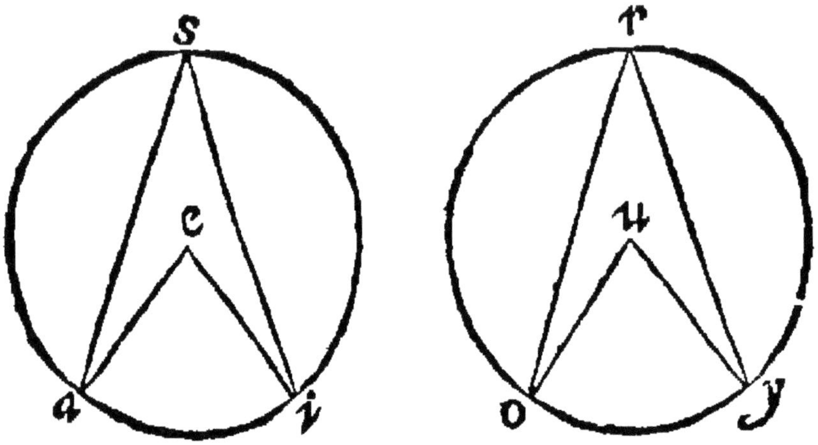

Here is a double proportion with the periphery underneath, of the angles in the center: And of angles in the periphery. But it shall suffice to declare it in the angles in the center.

First therefore let the Angles in the center *aei*, and *ouy* be equall: The bases *ai*, and *oy*, shall be equall, by the 11 e vij: And the peripheries, *ai*, and *oy*, by the 32 e xv, shall likewise be equall. Therefore if the angles be unequall, the peripheries likewise shall be unequall.

The same shall also be true of the Angles in the Periphery. The Converse in like manner is true: From whence followeth this consectary:

Therefore

8. As the sectour is unto the sectour, so is the angle unto the angle: And Contrariwise.

And thus much of the Sectour.

9. A section is a segment of a circle within cōprehended of one right line, which is termed the base of the section.

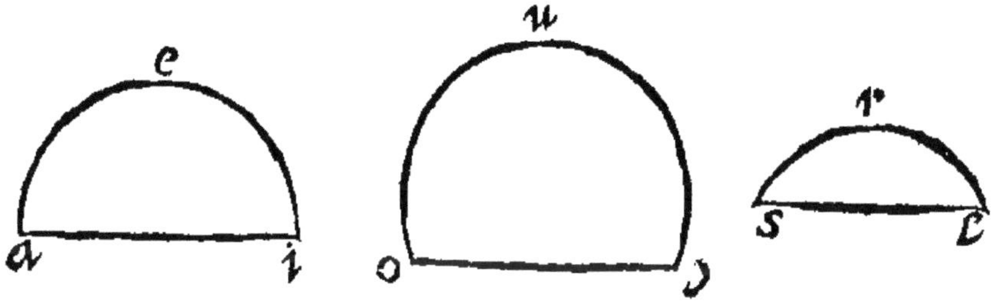

As here, *aei*, and *ouy*, and *srl*, are sections.

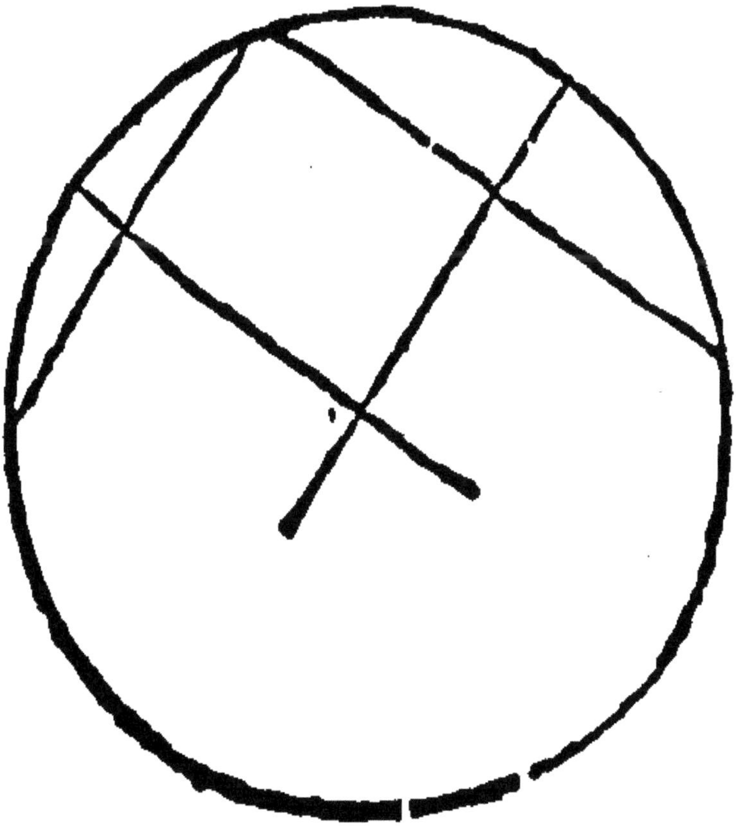

10. A section is made up by finding of the center.

The Invention of the center was manifest at the 7 e xv: And so here thou seest a way to make up a Circle, by the 8 e xv.

11 The periphery of a section is divided into two equall parts by a perpendicular dividing the base into two equall parts. 20. p iij.

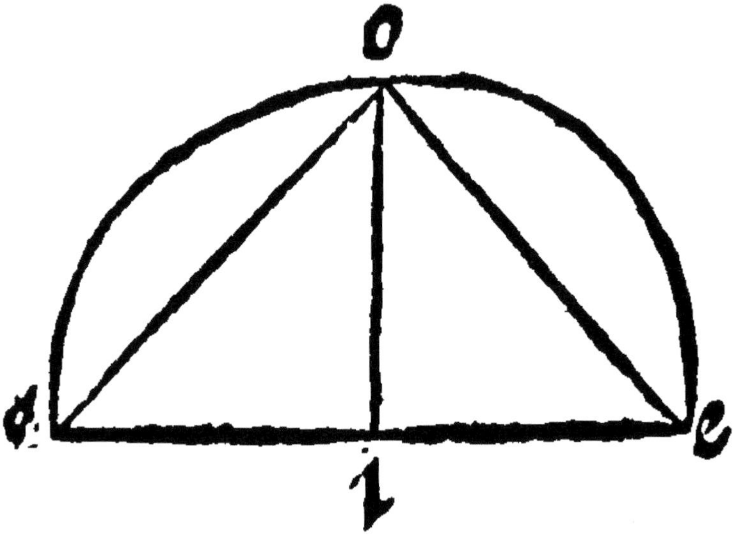

Let the periphery of the section *aoe*, to be halfed or cut into two equall parts. Let the base *ae*, be cut into two equall parts by the pendicular *io*, which shall cut the periphery in *o*, I say, that *ao*, and *oe*, are bisegments. For draw two right lines *ao*, and *oe*, and thou shalt have two triangles *aio*, and *eio*, equilaters by the 2 e vij. Therefore the bases *ao*, and *oe*, are equall: And by the 32. e xv. equall peripheries to the subtenses.

Here *Euclide* doth by congruency comprehende two peripheries in one, and so doe we comprehend them.

12 An angle in a section is an angle comprehended of two right lines joyntly bounded in the base and in the periphery joyntly bounded 7 d iij.

Or thus: An angle in the section, is an angle comprehended under two right lines, having the same tearmes with the bases, and the termes with the circumference: *H*. As *aoe*, in the former example.

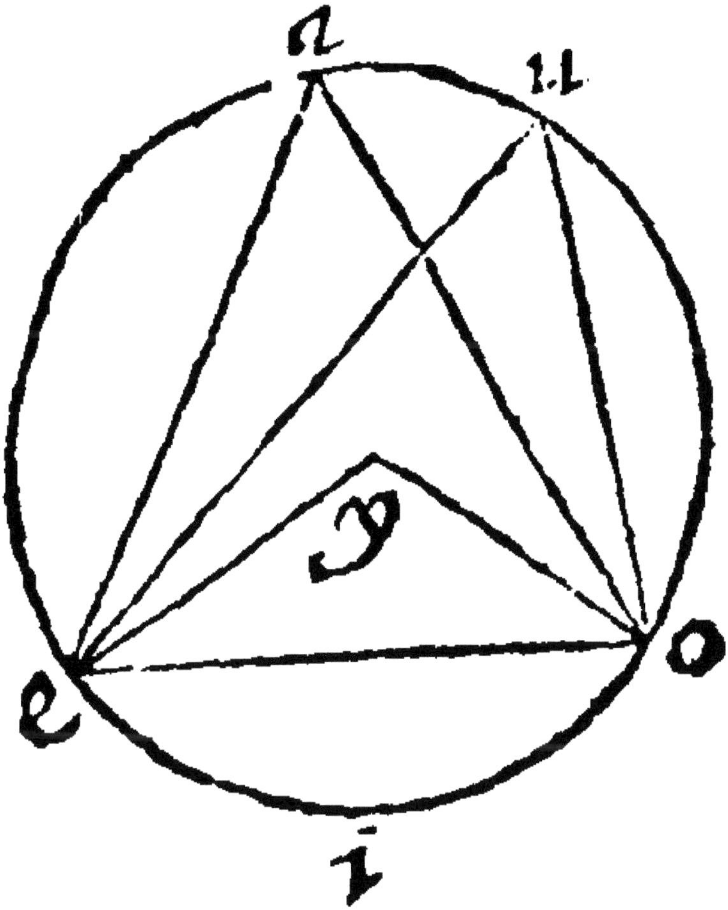

13 The angles in the same section are equall. 21. p iij.

Let the section be *eauo*, And in it the angles at *a*, & *u*: These are equall, because, by the 5 e, they are the halfes of the angle *eyo*, in the center: Or else they are equall, by the 7 e, because they insist upon the same periphery.

Here it is certaine that angles in a section are indeed angles in a periphery, and doe differ onely in base.

14 The angles in opposite sections are equall to two right angles. 22. p iij.

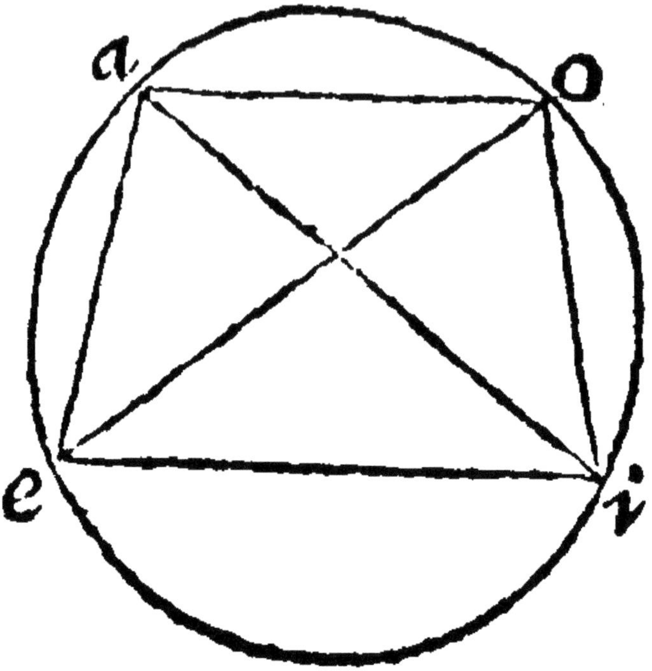

For here the opposite angles at *a*, and *i*, are equall to the three angles of the triangle *eoi*, which are equall to two right angles, by the 13 e vj. For first *i*, is equall to it selfe: Then *a*, by parts is equall to the two other. For *eai*, is equall to *eoi*, and *iao*, to *oei*, by the 13 e. Therefore the opposite angles are equall to two right angles.

The reason or rate of a section is thus: The similitude doth follow.

15 If sections doe receive [or containe] equall angles, they are alike è 10. d iij.

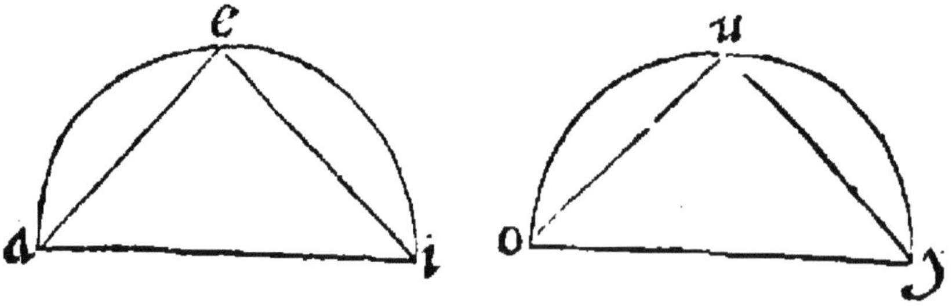

As here *aei*, and *ouy*. The triangle here inscribed, seeing they are equiangles, by the grant; they shall also be alike, by the 12 e vij.

16 If like sections be upon an equall base, they are equall: and contrariwise. 23, 24. p iij.

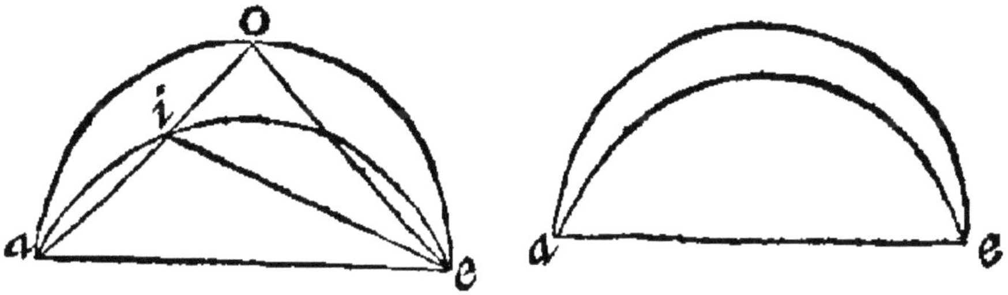

In the first figure, let the base be the same. And if they shall be said to unequall sections; and one of them greater than another, the angle in that *aoe*, shall be lesse than the angle *aie*, in the lesser section, by the 16 e vj. which notwithstanding, by the grant, is equall.

In the second figure, if one section be put upon another, it will agree with it: Otherwise against the first part, like sections upon the same base, should not be equall. But congruency is here sufficient.

By the former two propositions, and by the 9 e xv. one may finde a section like unto another assigned, or else from a circle given to cut off one like unto it.

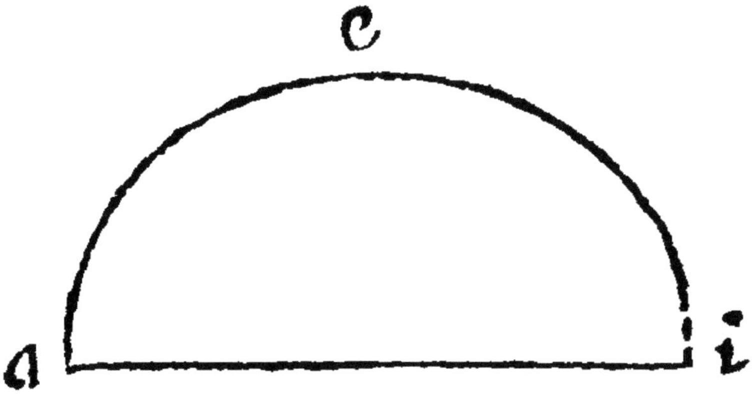

17 Angle of a section is that which is comprehended of the bounds of a section.

As here *eai*: And *eia*.

18 A section is either a semicircle: or that which is unequall to a semicircle.

A section is two fold, a semicircle, to wit, when it is cut by the diameter: or unequall to a semicircle, when it is cut by a line lesser than the diameter.

19 A semicircle is the half section of a circle.

Or it is that which is made the diameter.

Therefore

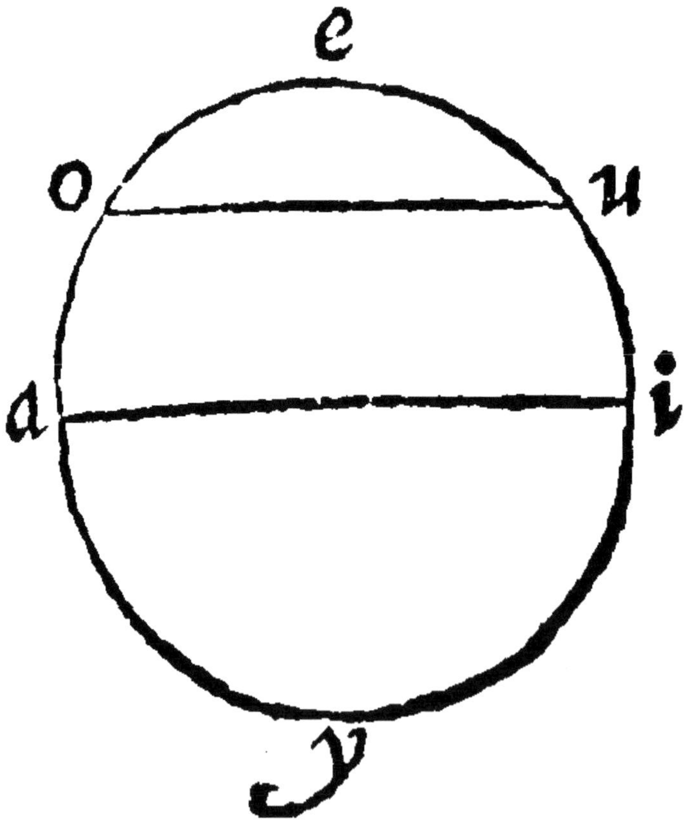

20 A semicircle is comprehended of a periphery and the diameter 18 d j.

As *aei,* is a semicircle: The other sections, as *oyu,* and *oeu,* are unequall sections: that greater; this lesser.

21 The angle in a semicircle is a right angle: The angle of a semicircle is lesser than a rectilineall right angle: But greater than any acute angle: The angle in a greater section is lesser than a right angle: Of a greater, it is a greater. In a lesser it is greater: Of a lesser, it is lesser, è 31. and 16. p iij.

Or thus: The angle in a semicircle is a right angle, the angle of a semicircle is lesse than a right rightlined angle, but greater than any acute angle: The angle in the greater section is lesse than a right angle: the angle of the greater section is greater than a right angle: the angle in the lesser section is greater than a right angle, the angle of the lesser section, is lesser than a right angle: *H.*

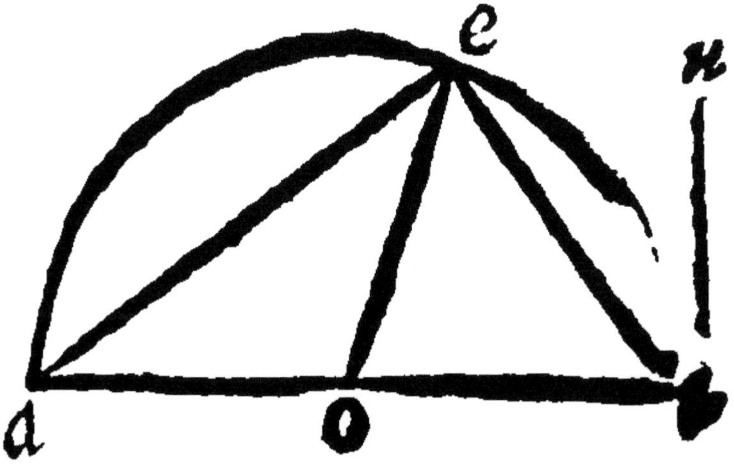

There are seven parts of this Element: The first is that *The angle in a semicircle is a right angle*: as in *aei*: For if the ray *oe*, be drawne, the angle *aei*, shall be divided into two angles *aeo*, and *oei*, equall to the angles *eao*, and *eio*, by the 17 e vj. Therefore seeing that one angle is equall to the other two, it is a right angle, by the 6 e viij. *Aristotle* saith that the angle in a semicircle is a right angle, because it is the halfe of two right angles, which is all one in effect.

The second part, *That the angle of a semicircle is lesser than a right angle*; is manifest out of that, because it is the part of a right angle. For the angle of the semicircle *aie*, is part of the rectilineall right angle *aiu*.

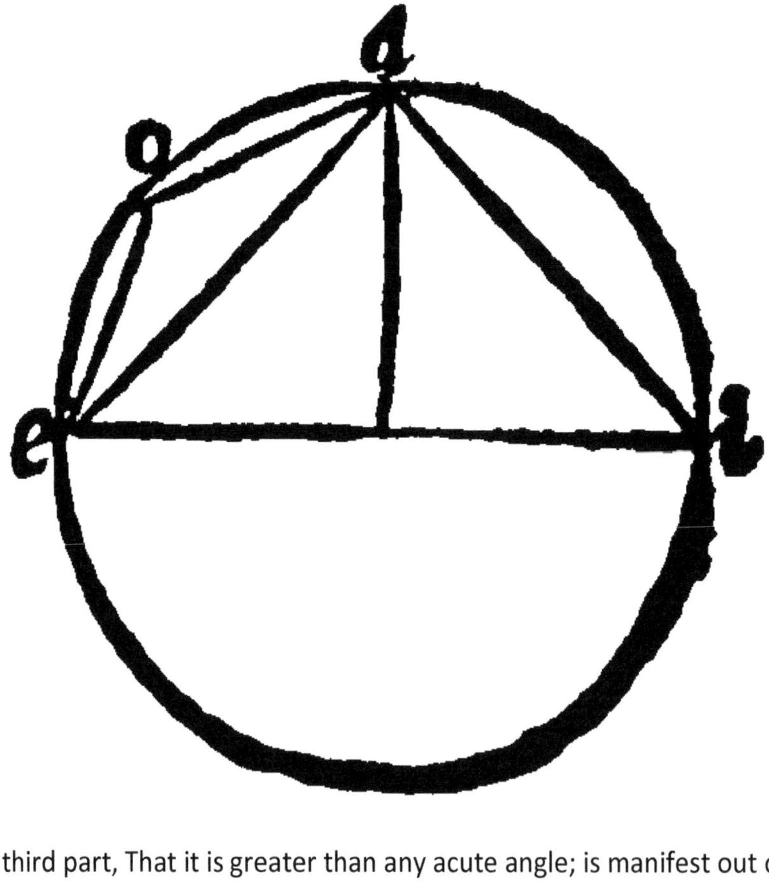

The third part, That it is greater than any acute angle; is manifest out of the 23. e xv. For otherwise a tangent were not on the same part one onely and no more.

The fourth part is thus made manifest: The angle at *i*, in the greater section *aei*, is lesser than a right angle; because it is in the same triangle *aei*, which at *a*, is a right angle. And if neither of the shankes be by the center, not withstanding an angle may be made equall to the assigned in the same section.

The fifth is thus: The angle of the greater section *eai*, is greater than a right angle: because it containeth a right-angle.

The sixth is thus, the angle *aoe*, in a lesser section, is greater than a right angle, by the 14 e xvj. Because that which is in the opposite section, is lesser than a right angle.

The seventh is thus. The angle *eao*, is lesser than a right-angle: Because it is part of a right angle, to wit of the outter angle, if *ia*, be drawne out at length.

And thus much of the angles of a circle, of all which the most effectuall and of greater power and use is the angle in a semicircle, and therefore it is not without cause so often mentioned of *Aristotle*. This Geometry therefore of *Aristotle*, let us somewhat more fully open and declare. For from hence doe arise many things.

Therefore

22 If two right lines jointly bounded with the diameter of a circle, be jointly bounded in the periphery, they doe make a right angle.

Or thus; If two right lines, having the same termes with the diameter, be joyned together in one point, of the circomference, they make a right angle. *H.*

This corollary is drawne out of the first part of the former Element, where it was said, that an angle in a semicircle is a right angle.

And

23 If an infinite right line be cut of a periphery of an externall center, in a point assigned and contingent, and the diameter be drawne from the contingent point, a right line from the point assigned knitting it with the diameter, shall be perpendicular unto the infinite line given.

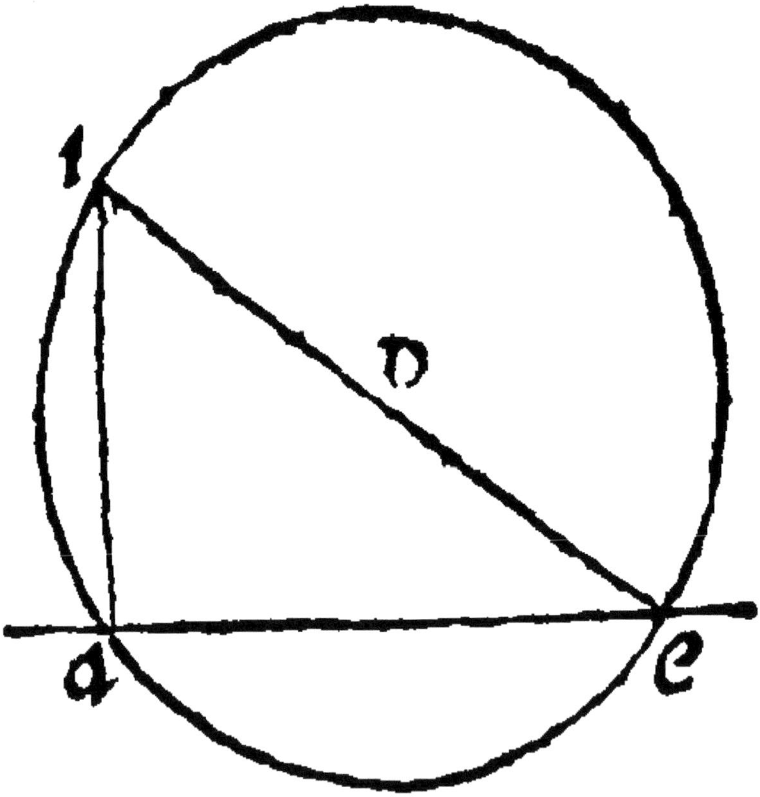

Let the infinite right line be *ae*, from whose point *a*, a perpendicular is to be raised.

The right line *ae*, let it be cut by the periphery *aei*, (whose center *o*, is out of the assigned *ae*,) and that in the point *a*, and a contingent point, as in *e*: And from *e*, let the diamiter be *eoi*: The right line *ai*, from *a*, the point given, knitting it with the diameter *ioe*, shall be perpendicular upon the infinite line *ae*; Because with the said infinite, it maketh an angle in a semicircle.

And

24 If a right line from a point given, making an acute angle with an infinite line, be made the diameter of a periphery cutting the infinite, a right line from the point assigned knitting the segment, shall be perpendicular upon the infinite line.

As in the same example, having an externall point given, let a perpendicular unto the infinite right line *ae* be sought: Let the right line *ioe*, be made the diameter of the peripherie; and withall let it make with the infinite right line given an acute angle in *e*, from whose bisection for the center, let a periphery cut the infinite, &c.

And

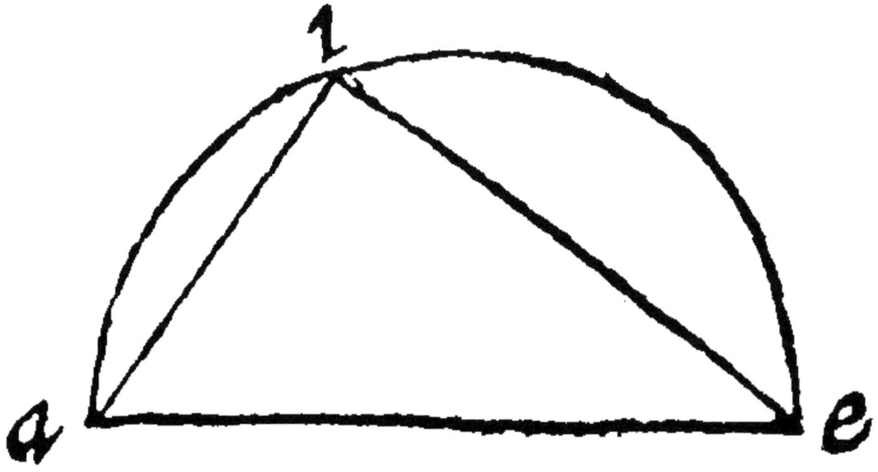

25 If of two right lines, the greater be made the diameter of a circle, and the lesser jointly bounded with the greater and inscribed, be knit together, the power of the greater shall be more than the power of the lesser by the quadrate of that which knitteth them both together. ad 13 p. x.

As in this example; The power of the diameter *ae*, is greater than the power of *ei*, by the quadrate of *ai*. For the triangle *aei*, shall be a rectangle; And by the 9 e xij. *ae*, the greater shall be of power equall to the shankes. Out of an angle in a semicircle Euclide raiseth two notable fabrickes; to wit, the invention of a meane proportionall betweene two lines given: And the Reason or rate in opposite sections. The *genesis* or invention of the meane proportionall, of which we heard at the 9 e viij. is thus:

26 If a right line continued or continually made of two right lines given, be made the diameter of a circle, the perpendicular from the point of their continuation unto the periphery, shall be the meane proportionall betweene the two lines given. 13 p vj.

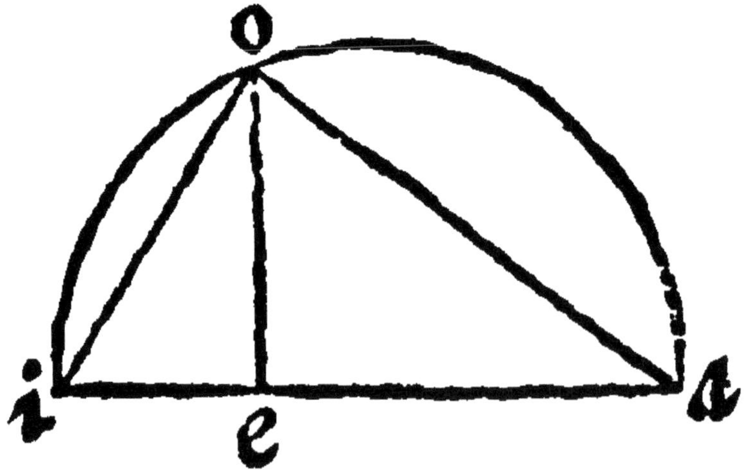

As for example, let the assigned right lines be *ae*, and *ei*, of the which *aei*, is continued. And let *eo*, be perpendicular from the periphery *aoi*, unto *e*, the point of continuation or joyning together of the lines given. This *eo*, say I, shall be the meane proportionall: Because drawing the right lines *ao*, and *io*, you shall make a rectangled triangle, seeing that *aoi*, is an angle in a semicircle: And, by the 9 e viij. *oe*, shall be proportionall betweene *ae*, and *ei*.

So if the side of a quadrate of 10. foote content, were sought; let the sides 1. foote and 10. foote an oblong equall to that same quadrate, be continued; the meane proportionall shall be the side of the quadrate, that is, the power of it shall be 10. foote. The reason of the angles in opposite sections doth follow.

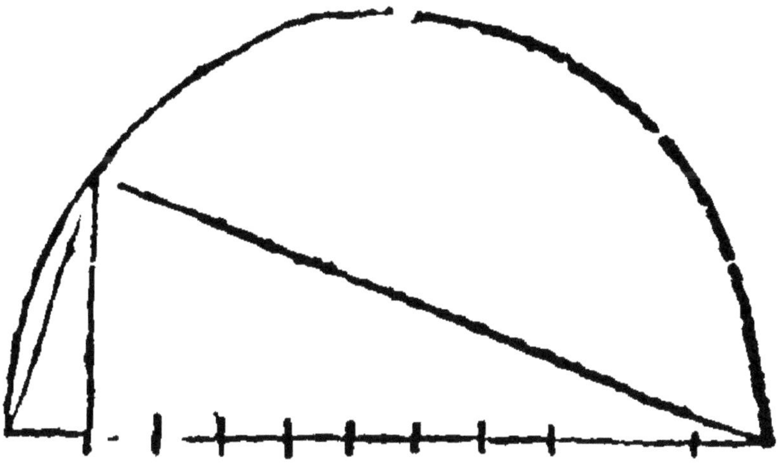

27 The angles in opposite sections are equall in the alterne angles made of the secant and touch line. 32. p iij.

If the sections be equall or alike, then are they the sections of a semicircle, and the matter is plaine by the 21 e. But if they be unequall or unlike the argument of demonstration is indeed fetch'd from the angle in a semicircle, but by the equall or like angle of the tangent and end of the diameter.

As let the unequall sections be *eio*, and *eao*: the tangent let it be *uey*: And the angles in the opposite sections, *eao*, and *eio*. I say they are equall in the alterne angles of the secant and touch line *oey*, and *oeu*. First that which is at *a*, is equall to the alterne *oey*: Because also three angles *oey*, *oea*, and *aeu*, are equall to two right angles, by the 14 e v. Unto which also are equall the three angles in the triangle *aeo*, by the 13 e vj. From three equals take away the two right angles *aue*, and *aoe*: (For *aoe*, is a right angle, by the 21 e; because it is in a semicircle:) Take away also the common angle *aeo*: And the remainders *eao*, and *oey*, alterne angles, shall be equall.

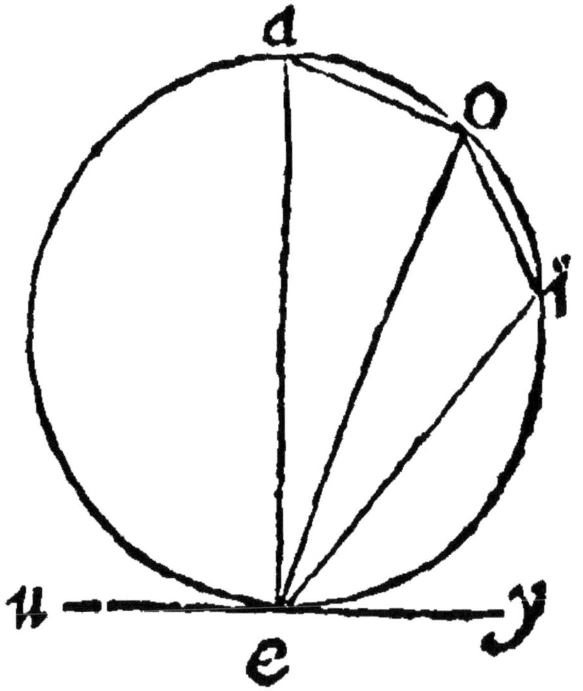

Secondarily, the angles at *a*, and *i*, are equall to two right angles, by the 14, e: To these are equall both *oey*, and *oeu*. But *eao*, is equall to the alterne *oey*. Therefore that which is at *i*, is equall to, the other alterne *oeu*. Neither is it any matter, whether the angle at *a*, be at the diameter or not: For that is onely assumed for demonstrations sake: For wheresoever it is, it is equall, to wit, in the same section. And from hence is the making of a like section, by giving a right line to be subtended.

Therefore

28 If at the end of a right line given a right lined angle be made equall to an angle given, and from the toppe of the angle now made, a perpendicular unto the other side do meete with a perpendicular drawn from the middest of the line given, the meeting shall be the center of the circle described by the equalled angle, in whose opposite section the angle upon the line given shall be made equall to the assigned è 33 p iij.

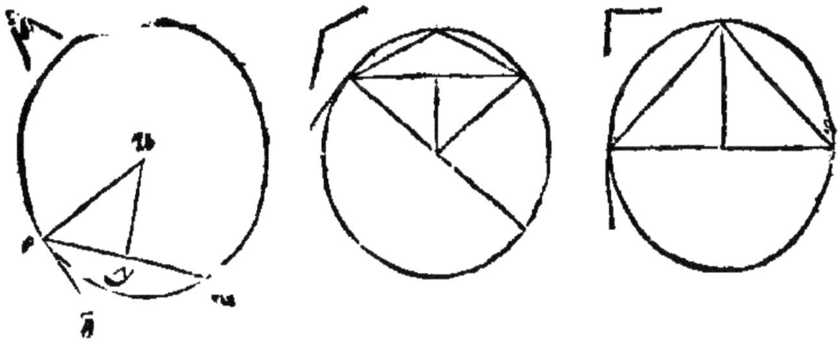

This you may make triall of in the three kindes of angles, all wayes by the same argument: as here the angle given is *a*: The right line given *ei*: at the end *e*, the equalled angle, *ieo*: The perpendicular to the side *eo*, let it be *eu*: But from the middest of the line given let it be *yu*. Here *u*, shall be the center desired. And from hence one may make a section upon a right line given, which shall receive a rectilineall angle equall to an angle assigned.

And

29 If the angle of the secant and touch line be equall to an assigned rectilineall angle, the angle in the opposite section shall likewise be equall to the same. 34. p iij.

As in this figure underneath. And from hence one may from a circle given cut off a section, in which there is an angle equall to the assigned. As let the angle given be *a*: And the circle *eio*. Thou must make at the point *e*, of the secant *eo*, and the tangent *yu*, an angle equall to the assigned, by the 11 e iij. such as here is *oeu*: Then the section *oei*, shall containe an angle equall to the assigned.

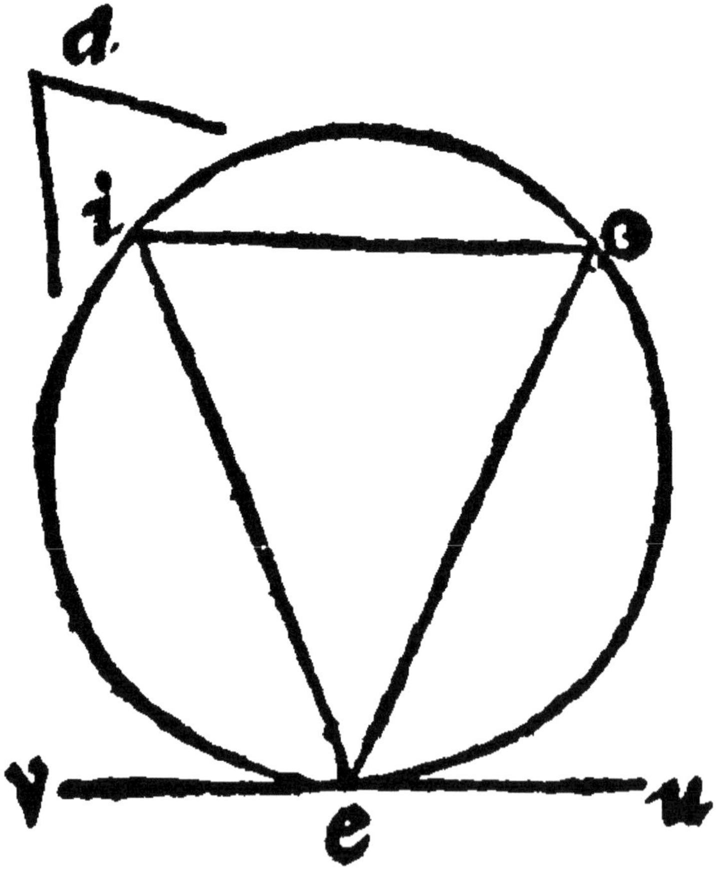

Of Geometry *the seventeenth Booke, Of the Adscription of a Circle and Triangle.*

Hitherto we have spoken of the Geometry of Rectilineall plaines, and of a circle: Now followeth the Adscription of both: This was generally defined in the first book 12 e. Now the periphery of a circle is the bound therof. Therefore a rectilineall is inscribed into a circle, when the periphery doth touch the angles of it 3 d iiij. It is circumscribed when it is touched of every side by the periphery; 4 d iij.

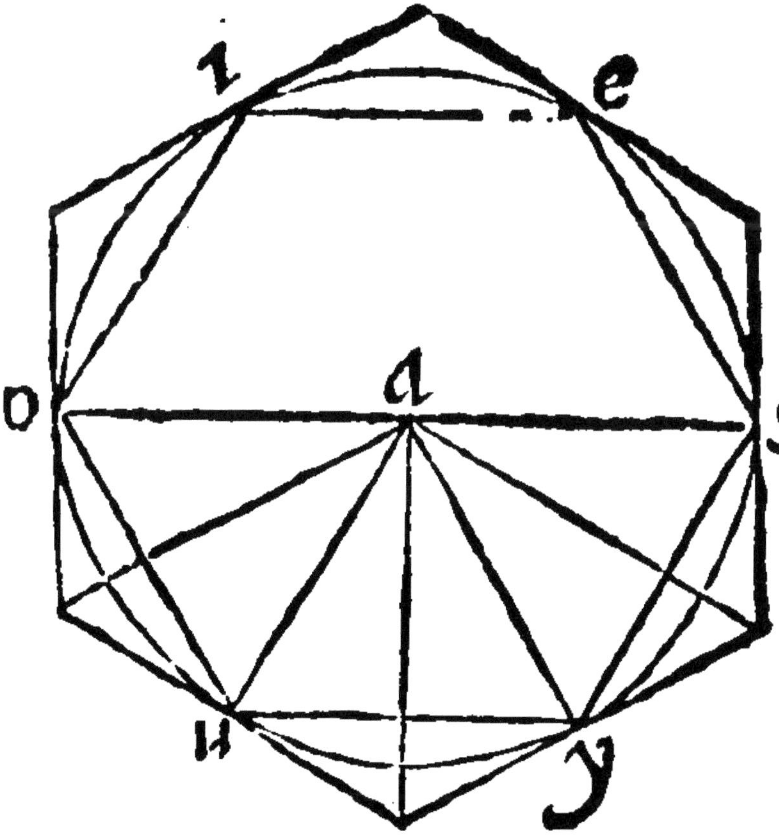

1. If rectilineall ascribed unto a circle be an equilater, it is equiangle.

Of the inscript it is manifest; And that of a Triangle by it selfe: Because if it be equilater, it is equiangle, by the 19 e vj. But in a Triangulate the matter is to be prooved by demonstration. As here, if the inscripts *ou*, and *sy*, be equall, then doe they subtend equall peripheries, by the 32 e xv. Then if you doe omit the periphery in the middest betweene them both, as here *uy*, and shalt adde *oies* the remainder to each of them, the whole *oiesy*, subtended to the angle at *u*: And *uoies*, subtended to the angle at *y*, shall be equall. Therefore the angles in the periphery, insisting upon equall peripheries are equall.

Of the circumscript it is likewise true, if the circumscript be understood to be a circle. For the perpendiculars from the center *a*, unto the sides of the circumscript, by the 9 e xij, shal make triangles on each side equilaters, & equiangls, by drawing the semidiameters unto the corners, as in the same exāple.

2. It is equall to a triangle of equall base to the perimeter, but of heighth to the perpendicular from the center to the side.

As here is manifest, by the 8 e vij. For there are in one triangle, three triangles of equall heighth.

The same will fall out in a Triangulate, as here in a quadrate: For here shal be made foure triangles of equall height.

Lastly every equilater rectilineall ascribed to a circle, shall be equall to a triangle, of base equall to the perimeter of the adscript. Because the perimeter conteineth the bases of the triangles, into the which the rectilineall is resolved.

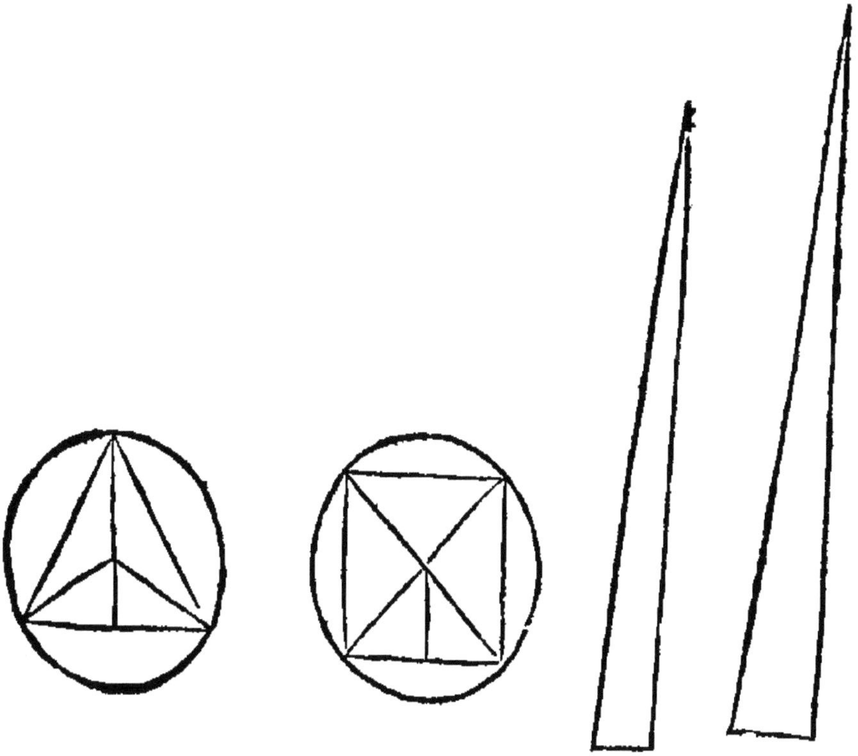

> *3. Like rectilinealls inscribed into circles, are one to another as the quadrates of their diameters, 1 p. xij.*

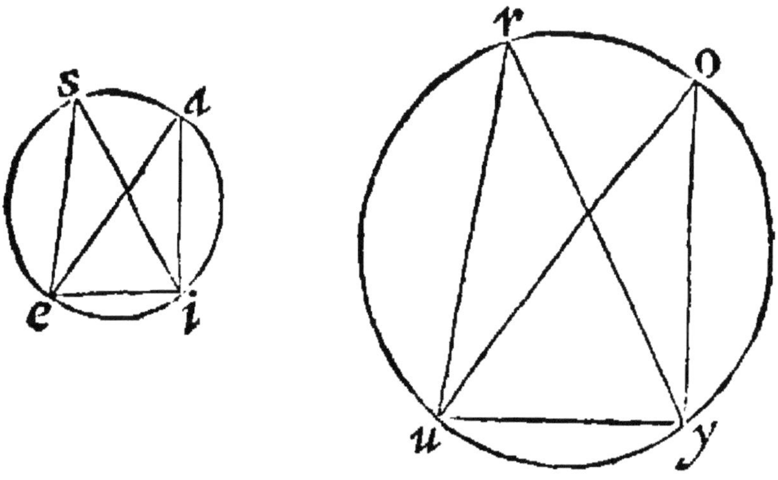

Because by the 1 e vj, like plains have a doubled reasó of their homologall sides. But in rectilineals inscribed the diameters are the homologall sides, or they are proportionall to their homologall sides. As let the like rectangled triangles be *aei*, and *ouy*; Here because *ae* and *ou*, are the diameters, the matter appeareth to be plaine at the first sight. But in the Obliquangled triangles, *sei*, and *ruy*, alike also, the diameters are proportionall to their homologall sides, to wit, *ei* and *uy*. For by the grant, as *se* is to *ru*: so is *ei* to *uy*, And therefore, by the former, as the diameter *ea* and *uo*.

In like Triangulates, seeing by the 4 e x, they may be resolved into like triangles, the same will fall out.

Therefore

4. If it be as the diameter of the circle is unto the side of rectilineall inscribed, so the diameter of the second circle be unto the side of the second rectilineall inscribed, and the severall triangles of the inscripts be alike and likely situate, the rectilinealls inscribed shall be alike and likely situate.

This *Euclide* did thus assume at the 2 p xij, and indeed as it seemeth out of the 18 p vj. Both which are conteined in the 23 e iiij. And therefore we also have assumed it.

Adscription of a Circle is with any triangle: But with a triangulate it is with that onely which is ordinate: And indeed adscription of a Circle is common *to all.*

5. If two right lines doe cut into two equall parts two angles of an assigned rectilineall, the circle of the ray from their meeting perpendicular unto the side, shall be inscribed unto the assigned rectilineall. 4 and 8. p. iiij.

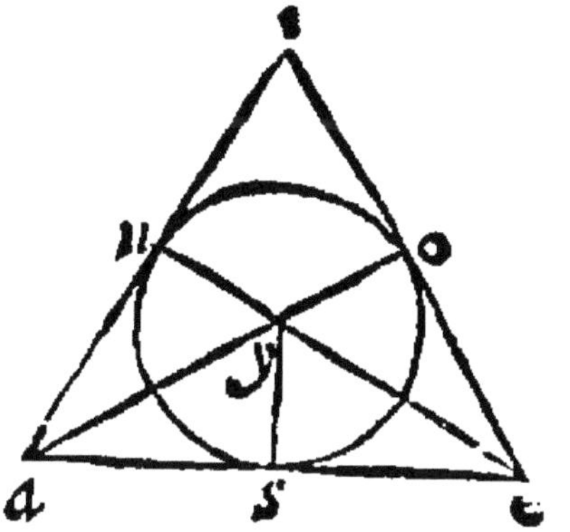

As in the Triangle *aei*, let the right lines *ao*, and *eu*, halfe the angles *a* and *e*: And from *y*, their meeting, let the perpendiculars unto the sides be *yo*, *yu*, *ys*; I say that the center *y*, with the ray *yo*, or *ya*, or *ys*, is the circle inscribed, by the 17 e xv. Because the halfing lines with the perpendiculars shall make equilater triangles, by the 2 e vij. And therefore the three perpendiculars, which are the bases of the equilaters, shall be equall.

The same argument shall serve in a Triangulate.

6. If two right lines do right anglewise cut into two equall parts two sides of an assigned rectilineall, the circle of the ray from their meeting unto the angle, shall be circumscribed unto the assigned rectilineall. 5 p iiij.

As in former figures. The demonstration is the same with the former. For the three rayes, by the 2 e vij, are equall: And the meeting of them, by the 17 e x, is the center.

And thus is the common adscription of a circle: The adscription of a rectilineall followeth, and first of a Triangle.

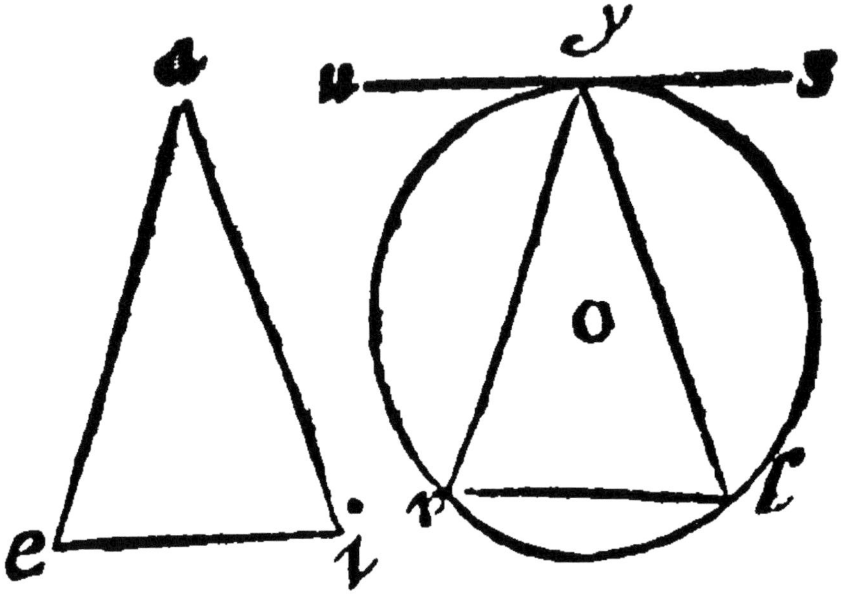

7. If two inscripts, from the touch point of a right line and a periphery, doe make two angles on each side equall to two angles of the triangle assigned be knit together, they shall inscribe a triangle into the circle given, equiangular to the triangle given è 2 p iiij.

Let the Triangle *aei* be given: And the circle, *o*, into which a Triangle equiangular to the triangle given, is to be inscribed. Therefore let the right line *uys*, touch the periphery *yrl*: And from the touch *y*, let the inscripts *yr*, and *yl*, make with the tangent two angles *uyr*, and *syl*, equall to the assigned angles *aei*, and *aie*: And let them be knit together with the right line *rl*: They shall by the 27 e xvj, make the angle of the alterne segments equall to the angles *uyr*, and *syl*. Therefore by the 4 e vij seeing that two are equall, the other must needs be equall to the remainder.

The circumscription here is also speciall.

8 If two angles in the center of a circle given, be equall at a common ray to the outter angles of a triangle given, right lines touching a periphery in the shankes of the angles, shall circumscribe a triangle about the circle given like to the triangle given. 3 p iiij.

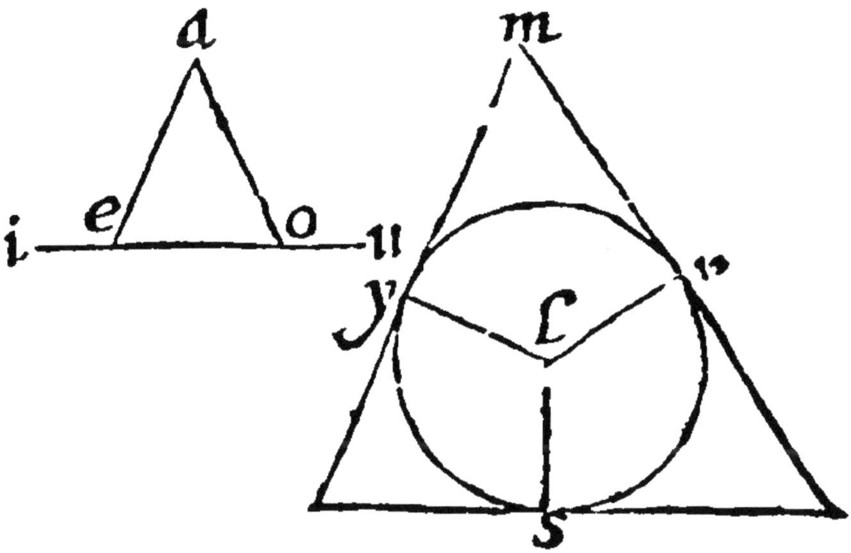

Let there be a Triangle, and in it the outter angles *aei*, and *aou*: The Circle let it be *ysr*; And in the center *l*, let the angles *ylr*, and *slr*; at the common side *lr*, bee made equall to the said outter angles *aei*, and *aou*. I say the angles of the circumscribed triangle, are equall to the angles of the triangle given. For the foure inner angles of the quadrangle *ylrm*, are equall to the foure right angles, by the 6 e x: And two of them, to wit, at *y* and *r*, are right angles, by the construction: For they are made by the secant and touch line, from the touch point by the center, by the 20 e xv. Therefore the remainders at *l* and *m*, are equall to two right angles: To which two *aei* and *aeo* are equall. But the angle at *l*, is equall to the outter: Therefore the remainder *m*, is equall to *aeo*. The same shall be sayd of the angles *aoe*, and *aou*. Therefore two being equall, the rest at *a* and *i*, shall be equall.

Therefore

9. If a triangle be a rectangle, an obtusangle, an acute angle, the center of the circumscribed triangle is in the side, out of the sides, and within the sides: And contrariwise. 5 e iiij.

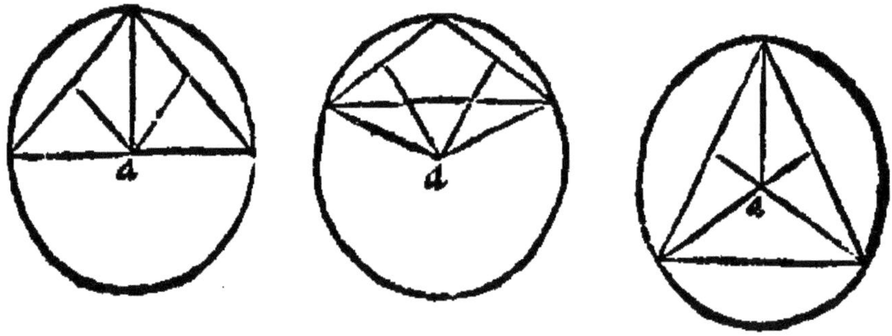

As, thou seest in these three figures, underneath, the center *a*.

Of Geometry, *the eighteenth Booke, Of the adscription of a Triangulate.*

Such is the Adscription of a triangle: The adscription of an ordinate triangulate is now to be taught. And first the common adscription, and yet out of the former adscription, after this manner.

1. If right lines doe touch a periphery in the angles of the inscript ordinate triangulate, they shall onto a circle cirumscribe a triangulate homogeneall to the inscribed triangulate.

The examples shall be laid downe according as the species or severall kindes doe come in order. The speciall inscription therefore shall first be taught, and that by one side, which reiterated, as oft as need shall require, may fill up the whole periphery. For that *Euclide* did in the quindecangle one of the kindes, we will doe it in all the rest.

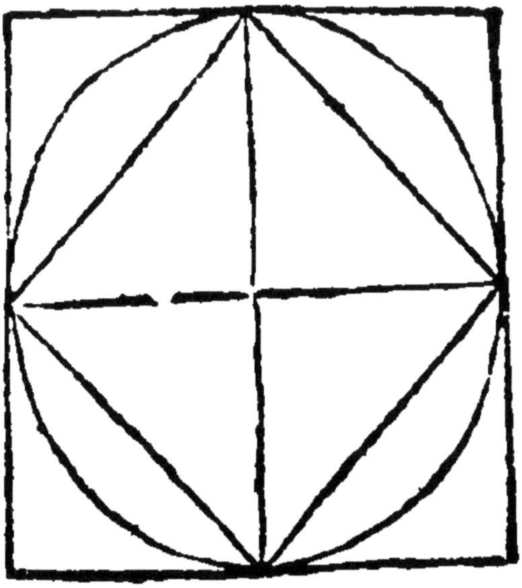

2. If the diameters doe cut one another right-angle-wise, a right line subtended or drawne against the right angle, shall be the side of the quadrate. è 6 p iiij.

As here. For the shankes of the angle are the raies whose diameters knit together shall make foure rectangled triangles, equall in shankes: And by the 2 e vij, equall in bases. Therefore they they shall inscribe a quadrate.

Therefore

3. A quadrate inscribed is the halfe of that which is circumscribed.

Because the side of the circumscribed (which here is equall to the diameter of the circle) is of power double, to the side of the inscript, by the 9 e xij.

And

4. It is greater than the halfe of the circumscribed Circle.

Because the circumscribed quadrate, which is his double, is greater than the whole circle.

For the inscribing or other multangled odde-sided figures we must needes use the helpe of a triangle, each of whose angles at the base is manifold to the other: In a Quinquangle first, that which is double unto the remainder, which is thus found.

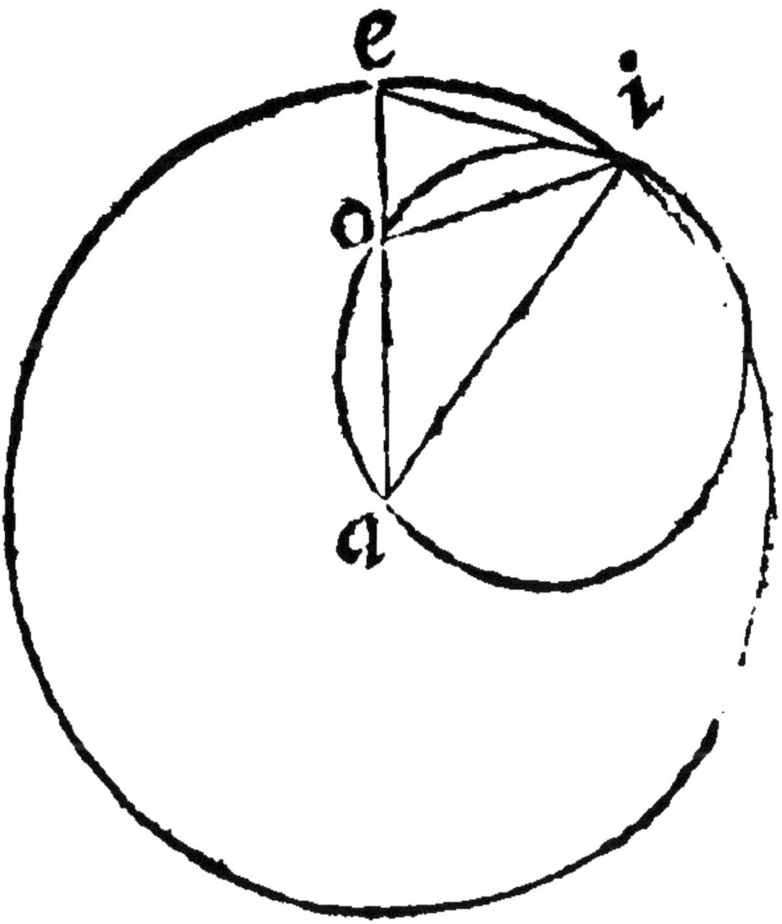

> *5. If a right line be cut proportionally, the base of that triangle whose shankes shall be equall to the whole line cut, and the base to the greater segment of the same, shall have each of the angles at base double to the remainder: And the base shall be the side of the quinquangle inscribed with the triangle into a circle. 10, and 11. p iiij.*

Here first thou shalt take for the fabricke or making of the Triangle, for the ray the right line *ae* by the 3 e xiiij, cut proportionally in *o*: A circle also shalt thou make upon the center *a*, with the ray *ae*: And then shalt thou by the 6 e xv, inscribe a right line equall to the greater segment: And shalt knit the same inscript with the whole line cut with another right line. This triangle shall be your desire. For by the 17 e vj, the angles at the base *ei* are equall, so that looke whatsoever is prooved of the one, is by and by also prooved of the other. Then let *oi* be drawne; And a Circle, by the 8 e xvij, circumscribed about the triangle *aoi*. This circle the right line *ei*, shall touch, by the 27 e xv. Because, by the grant, the right line *ae*, is cut proportionally, therefore the Oblong of the secant and outter segment, is equall to the quadrate of the greater segment, to which by the grant, the base *ei*, is equall.

Here therefore the angle *aie* is the double of the angle at *a*: because it is equall to the angles *aio*, and *oai*, which are equall betweene themselves. For by the 27 e xvj it is equall to the angle *oai* in the alterne segment. And the remainder *aio*, is equall to it selfe. Therefore also the angle *aei*, is equall to the same two angles, because it is equall to the angle *aie*. But the outter angle *eoi*, is equall to the same two, by the 15 e vj. Therefore the angles *ioe* and *oei* (because they are equall to the same) they are equall betweene themselves. Wherefore by the 17 e vj, the sides *oi* and *ei* are equall. And there also *ao* and *oi*: And the angles *oai* & *oia* are equall by the 17 e vj. Wherefore seeing that to both the angle *aie* is equall, it shall be the double of either of the equalls.

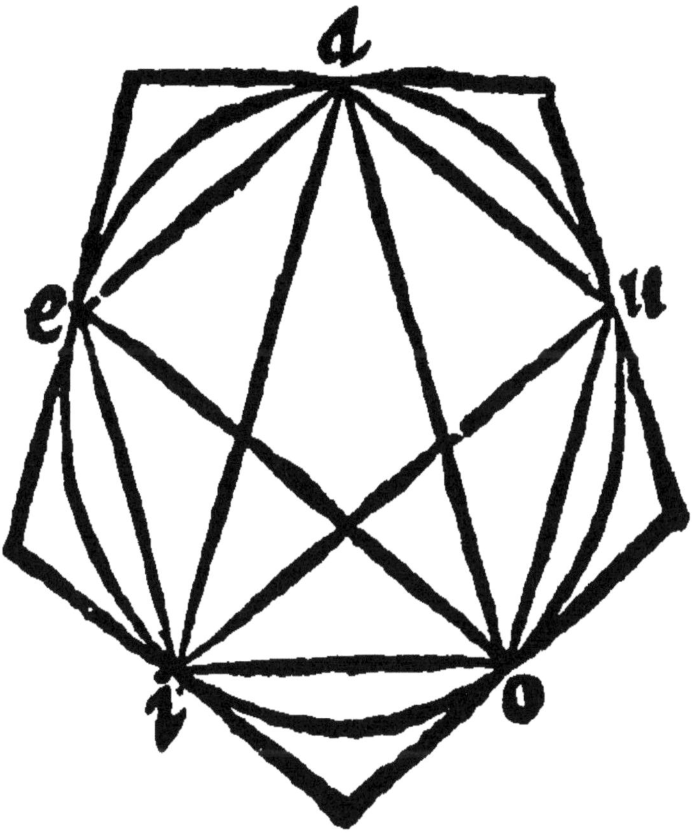

But the base *ei*, is the side of the equilater quinquangle. For if two right lines halfing both the angles of a triangle which is the double of the remainder, be knit together with a right line, both one to another, and with the angles, shall inscribe unto a circle an equilater triangle, whose one side shall be the base it selfe: As here seeing the angles *eoa*, *eoi*, *uio*, *uia*, *iao*, are equal in the periphery, the peripheries, by the 7 e, xvj. subtending them are equall: And therefore, by the 32 e, xv. the subtenses *ae*, *ei*, *io*, *ou*, *ua*, are also equall. Now of those five, one is *ae*. Therefore a right line proportionall cut, doth thus make the adscription of a quinquangle: And from thence againe is afforded a line proportionally cut.

6 If two right lines doe subtend on each side two angles of an inscript quinquangle, they are cut proportionally, and the greater segments are the sides of the said inscript è 8, p xiij.

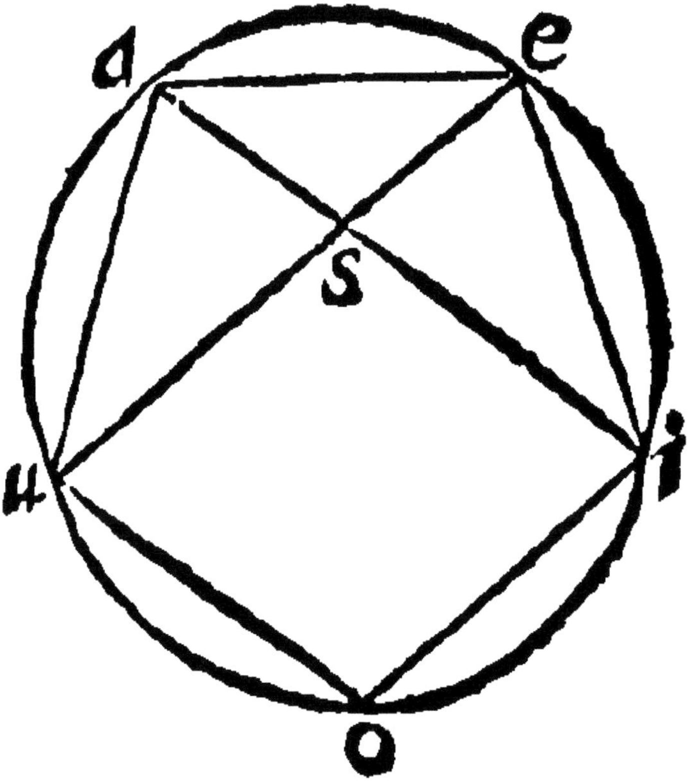

As here, Let *ai*, and *eu*, subtending the angles on each side *aei*, and *eau*: I say, That they are proportionally cut in the point *s*: And the greater segments *si*, and *su*, are equall to *ae*, the side of the quinquangle. For here two triangles are equiangles: First *aei*, and *uae*, are equall by the grant, and by the 2 e, vij. Therefore the angles *aie*, and *aes*, are equall. Then *aei*, and *ase*, are equall: Because the angle at *a*, is common to both: Therefore the other is equall to the remainder, by the 4 e, 7. Now, by the 12. e, vij. as *ia*, is unto *ae*, that is, as by and by shall appeare, unto *is*: so is *ea*, unto *as*: Therefore, by the 1 e, xiiij. *ia*, is cut proportionally in *s*. But the side *ea*, is equall to *is*: Because both of them is equall to the side *ei*, that by the grant, this by the 17. e, vj. For the angles at the base, *ise*, and *ies*, are equall, as being indeed the doubles of the same. For *ise*, by the 16. e vj. is equall to the two inner, which are equall to the angle at *u*, by the 17 e vj. and by the former conclusion. Therefore it is the double of the angles *aes*: Whose double also is the angle *uei*, by the 7 e. xvj. insisting indeede upon a double periphery.

And from hence the fabricke or construction of an ordinate quinquangle upon a right line given, is manifest.

Therefore

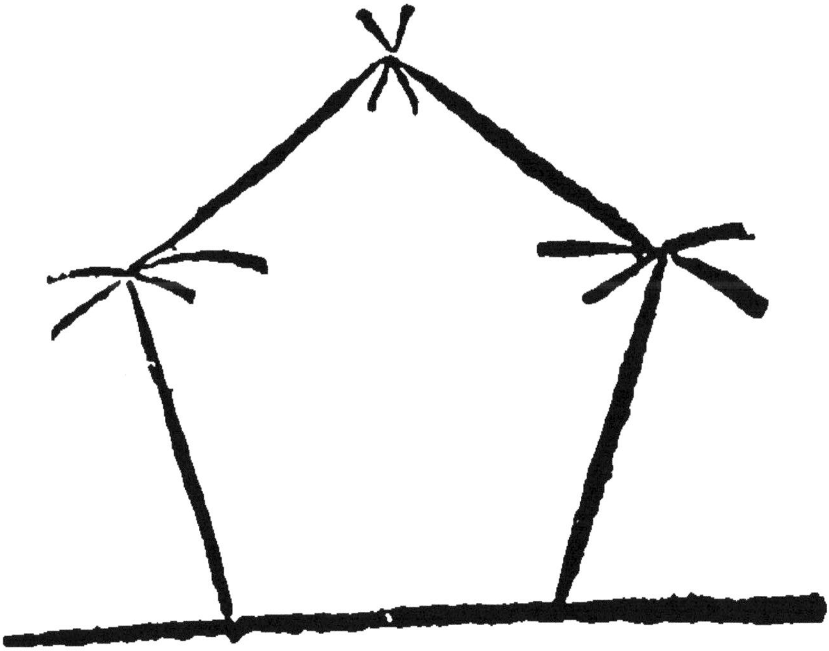

7 If a right line given, cut proportionall, be continued at each end with the greater segment, and sixe peripheries at the distance of the line given shall meete, two on each side from the ends of the line given and the continued, two others from their meetings, right lines drawne from their meetings, & the ends of the assigned shall make an ordinate quinquangle upon the assigned.

The example is thus.

8 If the diameter of a circle circumscribed about a quinquangle be rationall, it is irrationall unto the side of the inscribed quinquangle, è 11. p xiij.

So before the segments of a right line proportionally cut were irrationall.

The other triangulates hereafter multiplied from the ternary, quaternary, or quinary of the sides, may be inscribed into a circle by an inscript triangle, quadrate, or quinquangle. Therefore by a triangle there may be inscribed a triangulate of 6. 12, 24, 48, angles: By a quadrate, a triangulate of 8. 16, 32, 64, angles. By a quinquangle, a triangulate of 10, 20, 40, 80. angles, &c.

9 The ray of a circle is the side of the inscript sexangle. è 15 p iiij.

A sexangle is inscribed by an inscript equilaterall triangle, by halfing of the three angles of the said triangle: But it is done more speedily by the ray or semidiameter of the circle, sixe times continually inscribed. As in the circle given, let the diameter be ae*; And upon the center* o*, with the ray* ie*, let the periphery* uio*, be described: And from the points* o *and* u*, let the diameters be* oy*, and* us*; These knit both one with another, and also with the diameter* ae *shall inscribe an equilaterall sexangle into the circle given, whose side shal be equal* *to the ray of the same circle. As* eu, *is equal to* ui, *because they both equall to the same* ie, *by the 29 e, iiij.*

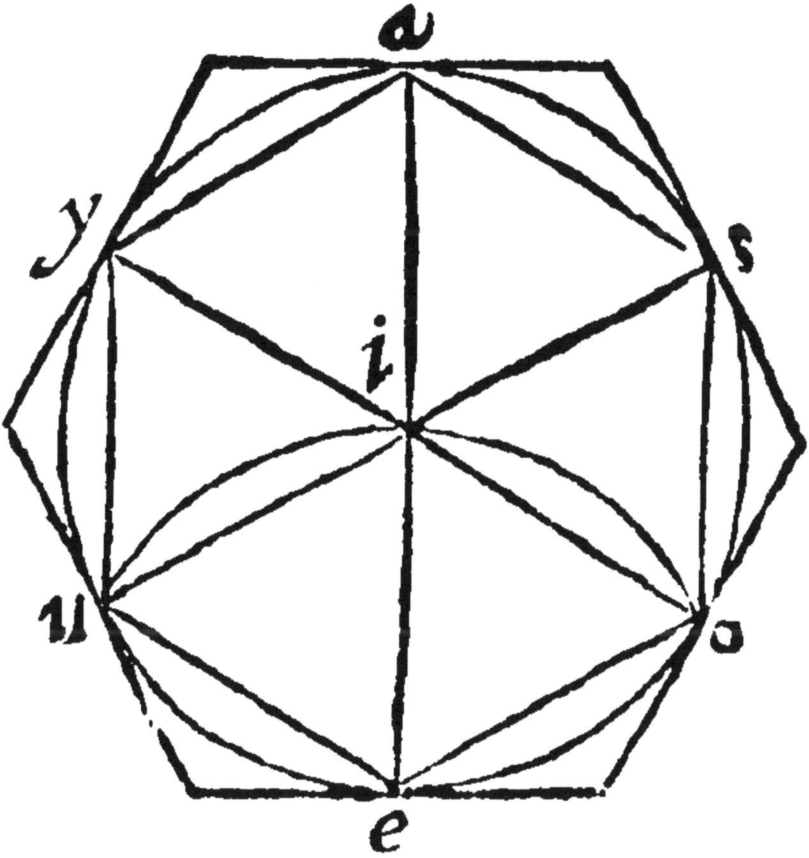

There fore *eiu*, is an equilater triangle: And likewise *eio*, is an equilater. The angles also in the center are ⅔ of one rightangle: And therefore they are equall. And by the 14. e v, the angle *sio*, is ⅓. of two rightangles: And by the 15. e v. the angles at the toppe are also equall. Wherefore sixe are equall: And therefore, by the 7 e xvj. and 32. e, xv, all the bases are equall, both betweene themselves, and as was even now made manifest, to the ray of the circle given. Therefore the sexangle inscript by the ray of a circle is an equilater; And by the 1 e xvij. equiangled.

Therefore

10 Three ordinate sexangles doe fill up a place.

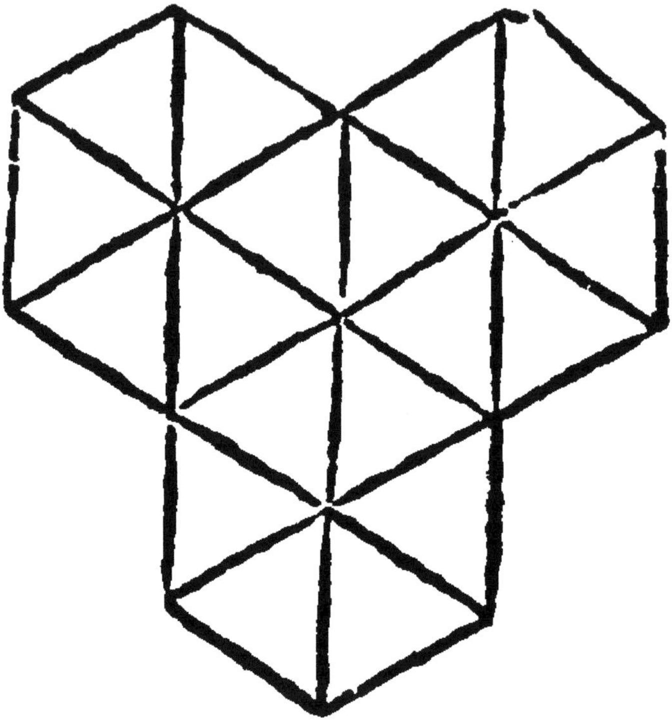

As here. For they are sixe equilater triangles, if you shal resolve the sexangles into sixe triangls: Or els because the angle of an ordinate sexangle is as much as one right angle and ⅓. of a right angle.

Furthermore also no one figure amongst the plaines doth fill up a place. A Quinquangle doth not: For three angles a quinquangle may make only 3.3/5 angles which is too little. And foure would make 4.4/5 which is as much too great. The angles of a septangle would make onely two rightangles, and 6/7 of one: Three would make 3, and 9/7, that is in the whole 4.2/7, which is too much, &c. to him that by induction shall thus make triall, it will appeare, That a plaine place may be filled up by three sorts of ordinate plaines onely.

And

11 If right lines from one angle of an inscript sexangle unto the third angle on each side be knit together, they shall inscribe an equilater triangle into the circle given.

As here; Because the sides shall be subtended to equall peripheries: Therefore by the 32 e xv. they shall be equall betweene themselves: And againe, on the contrary, by such a like triangle, by halfing the angles, a sexangle is inscribed.

12 The side of an inscribed equilater triangle hath a treble power, unto the ray of the circle 12. p xiij.

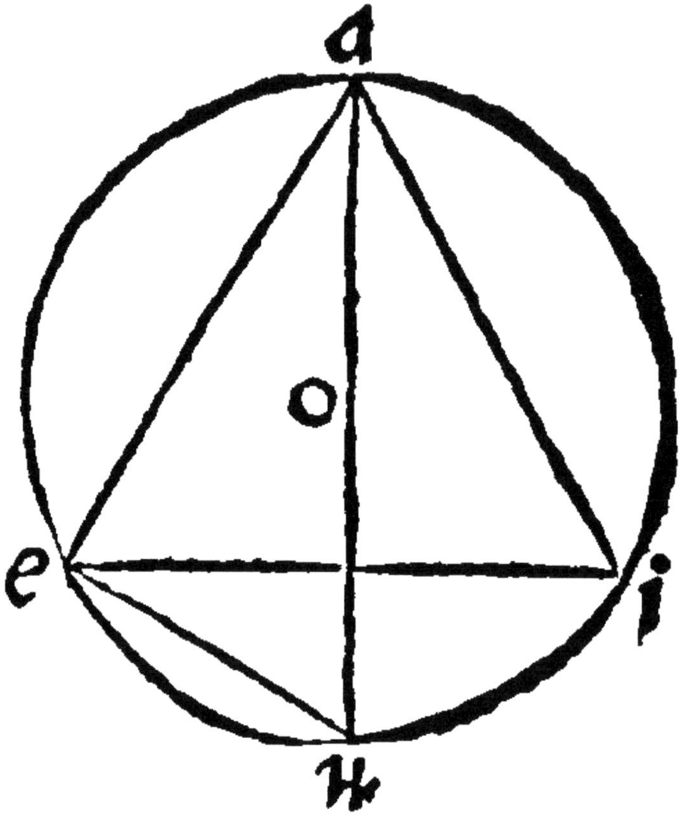

As here, with *ae*, one side of the triangle *aei*, two third parts of the halfe periphery are imployed: For with one side one third of the whole *eu*, is imployed: Therefore *eu*, is the other third part, that is, the sixth part of the whole periphery. Therefore the inscript *eu*, is the ray of the circle, by the 9 e. Now the power of the diameter *aou*, by the 14 e xij. is foure times so much as is the power of the ray, that is, of *eu*: And by 21. e xvj, and 9 e xij, *ae*, and *eu*, are of the same power; take away *eu*, and the side *ae*, shall be of treble power unto the ray.

13 If the side of a sexangle be cut proportionally, the greater segment shall be the side of the decangle.

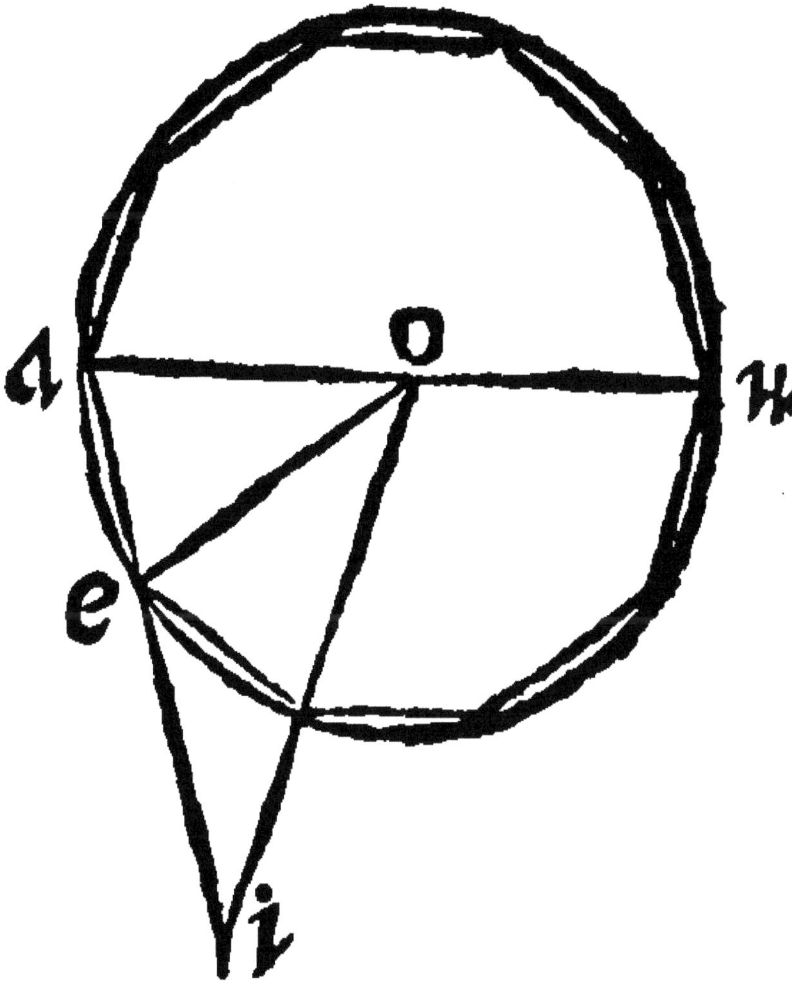

Pappus lib. 5. ca. 24. & Campanus ad 3 p xiiij. Let the ray *ao*, or side of the sexangle be cut proportionally, by the 3 e xiiij: And let *ae*, be equall to the greater segment. I say that *ae*, is the side of the decangle. For if it be moreover continued with the whole ray unto *i*, the whole *aei*, shall by the 4 e xiiij. be cut proportionally: and the greater segment *ei*, shal be the same ray. For the if the right line *iea*, be cut proportionally, it shall be as *ia*, is unto *ie*, that is to *oa*, to wit, unto the ray: so *ao*, shal be unto *ae*. Therefore, by the 15. e vij. the triangles *iao*, and *oae*, are equiangles: And the angle *aoe*, is equall to the angle *oia*. But the angle *uoe*, is foure times as great as the angle *aoe*: for it is equall to the two inner at *a*, and *e*, by the 15 e vj: which are equall between themselves, by the 10 e v. and by the 17 e vj. And therefore it is the double of *aeo*, which is the double, for the same cause, of *aio*, equall to the same *aoe*. Therefore *uoe*, is the quadruple of the said *aoe*. Therefore *ue*, is the quadruple of the periphery *ea*. Therefore the whole *uea*, is the quintuple of the same *ea*: And the whole periphery is decuple unto it. And the subtense *ae*, is the side of the decangle.

Therefore

14 If a decangle and a sexangle be inscribed in the same circle, a right line continued and made of both sides, shall be cut proportionally, and the greater segment shall be the side of a sexangle; and if the greater segment of a right line cut proportionally be the side of an hexagon, the rest shall be the side of a decagon. 9. p xiij.

The comparison of the decangle and the sexangle with the quinangle followeth.

15 If a decangle, a sexangle, and a pentangle be inscribed into the same circle the side of the pentangle shall in power countervaile the sides of the others. And if a right line inscribed do countervaile the sides of the sexangle and decangle, it is the side of the pentangle. 10. p xiiij.

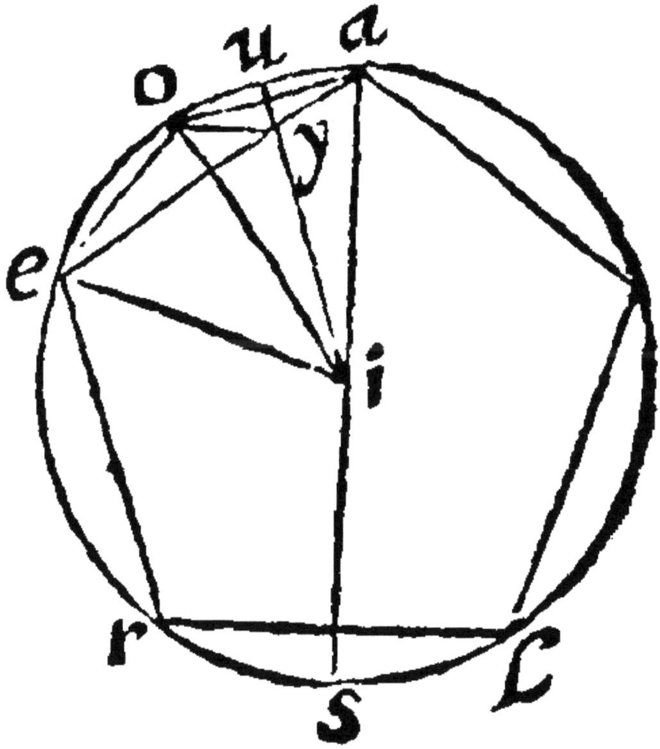

Let the side of the inscribed quinquangle be *ae*: of the sexangle, *ei*: Of the decangle *ao*. I say, the side *ae*, doth in power countervaile the rest. For let there be two perpēdiculars: The first *io*, the second *iu*, cutting the sides of the quinquangle and decangle into halves: And the meeting of the second perpendicular with the side of the quinquangle let it be *y*. The syllogisme of the demonstration is this: The oblongs of the side of the quinquangle, and the segments of the same, are equall to the quadrates of the other sides. But the quadrate of the same whole side, is equall to the oblongs of the whole, and the segments, by the 3 e, xiij. Therefore it is equall to the quadrates of the other sides.

Let the proportion of this syllogisme be demonstrated: For this part onely remaineth doubtfull. Therefore two triangles, *aei*, and *yei*, are equiangles, having one common angle at *e*: And also two equall ones *aei*, and *eiy*, the halfes, to wit, of the same *eis*: Because that is, by the 17 e, vj: one of the two equalls, unto the which *eis*, the out angle, is equall, by the 15 e. vj. And this doth insist upon a halfe periphery. For the halfe periphery *als*, is equall to the halfe periphery *ars*: and also *al*, is equall to *ar*. Therefore the remnant *ls*, is equall to the remnant *rs*: And the whole *rl*, is the double of the same *rs*: And therefore *er*, is the double of *eo*: And *rs*, the double of *ou*. For the bisegments are manifest by the 10 e, xv. and the 11 e, xvj. Therefore the periphery *ers*, is the double of the periphery *eou*: And therefore the angle *eiu*, is the halfe of the angle *eis*, by the 7 e, xvj. Therefore two angles of two triangles are equall: Wherefore the remainder, by the 4 e vij, is equall to the remainder. Wherefore by the 12 e, vij, as the side *ae*, is to *ei*: so is *ei*, to *ey*. Therefore by the 8 e xij, the oblong of the extreames is equall to the quadrate of the meane.

Now let *oy*, be knit together with a straight: Here againe the two triangles *aoe*, and *aoy*, are equiangles, having one common angle at *a*: And *aoy*, and *oea*, therefore also equall: Because both are equall to the angle at *a*: That by the 17 e, vj: This by the 2 e, vij: Because the perpendicular halfing the side of the decangle, doth make two triangles, equicrurall, and equall by the right angle of their shankes: And therefore they are equiangles. Therefore as *ea*, is to *ao*: so is *ea*, to *ay*. Wherefore by the 8 e, xij. the oblong of the two extremes is equall to the quadrate of the meane: And the proposition of the syllogisme, which was to be demonstrated. The converse from hence as manifest *Euclide* doth use at the 16 p xiij.

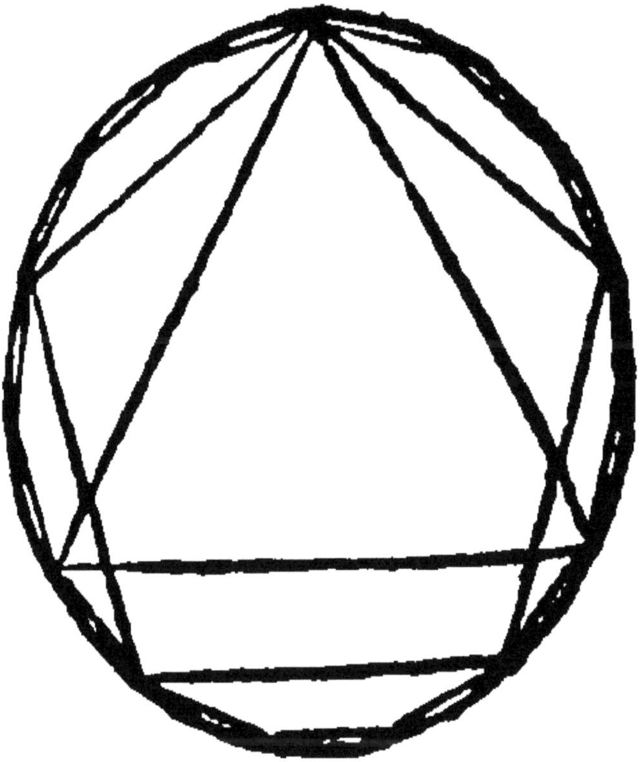

16. If a triangle and a quinquangle be inscribed into the same Circle at the same point, the right line inscribed betweene the bases of the both opposite to the said point, shall be the side of the inscribed quindecangle. 16. p. iiij.

For the side of the equilaterall triangle doth subtend 1/3 of the whole pheriphery. And two sides of the ordinate quinquangle doe subtend 2/5 of the same. Now 2/5 - 1/3 is 1/15: Therefore the space betweene the triangle, and the quinquangle shall be the 1/15 of the whole periphery.

Therefore

> *17. If a quinquangle and a sexangle be inscribed into the same circle at the same point, the periphery intercepted beweene both their sides, shall be the thirtieth part of the whole periphery.*

As here. Therefore the inscription of ordinate triangulates, of a Quadrate, Quinquangle, Sexangle, Decangle, Quindecangle is easie to bee performed by one side given or found, which reiterated as oft as need shall require, shal subtend the whole periphery. *Jun. 4.* A. C. CIƆIƆCXXII *Campana pulsante pro*. H. W.

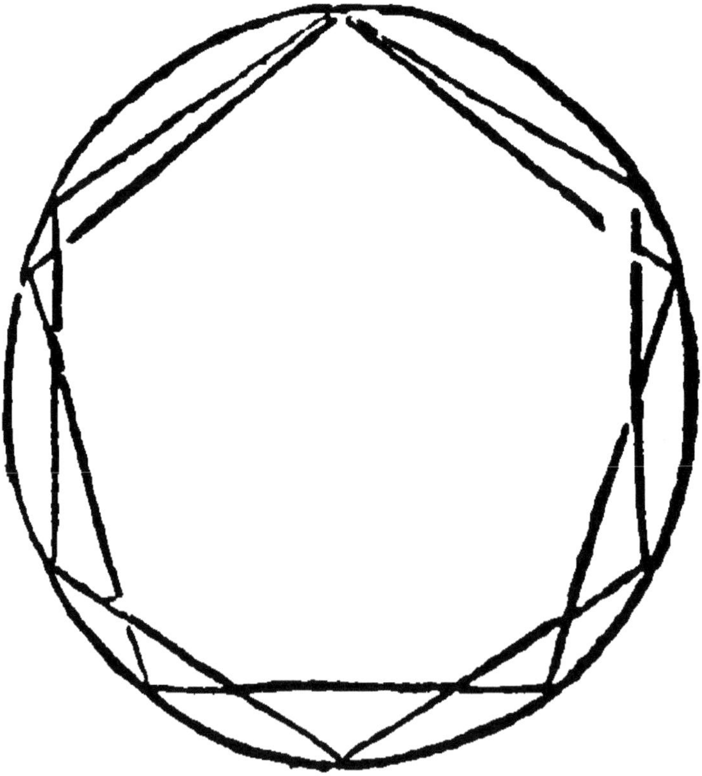

Of Geometry *the ninteenth Booke; Of the Measuring of ordinate Multangle and of a* Circle.

Out of the Adscription of a Circle and a Rectilineall is drawne the Geodesy of ordinate Multangles, and first of the Circle it selfe. For the meeting of two right lines equally, dividing two angles is the center of the circumscribed Circle: From the center unto the angle is the ray: And then if the quadrate of halfe the side be taken out of the quadrate of the ray, the side of the remainder shall be the perpendicular, by the 9 e xij. Therefore a speciall theoreme is here thus made:

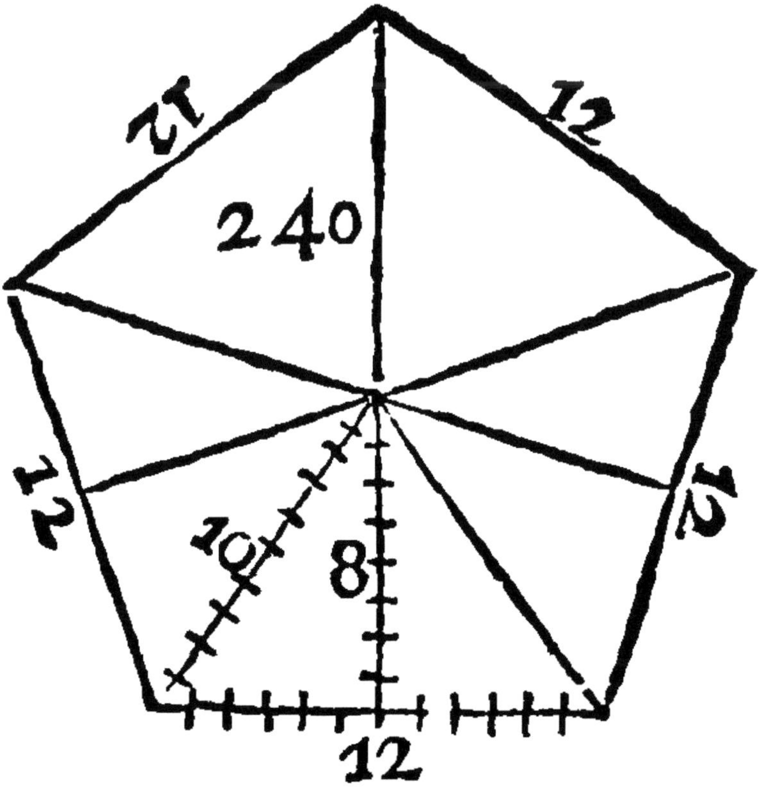

1. A plaine made of the perpendicular from the center unto the side, and of halfe the perimeter, is the content of an ordinate multangle.

As here; The quadrate of 10, the ray is 100. The quadrate of 6, the halfe of the side 12, is 36: And 100. 36 is 64, the quadrate of the Perpendicular, whose side 8, is the Perpendicular it selfe. Now the whole periphery of the Quinquangle, is 60. The halfe thereof therefore is 30. And the product of 30, by 8, is 240, for the content of the sayd quinquangle.

The Demonstration here also is of the certaine antecedent cause thereof. For of five triangles in a quinquangle, the plaine of the perpendicular, and of halfe the base is one of them, as in the former hath beene taught: Therefore five such doe make the whole quinquangle. But that multiplication, is a multiplication of the Perpendicular by the Perimeter or bout-line.

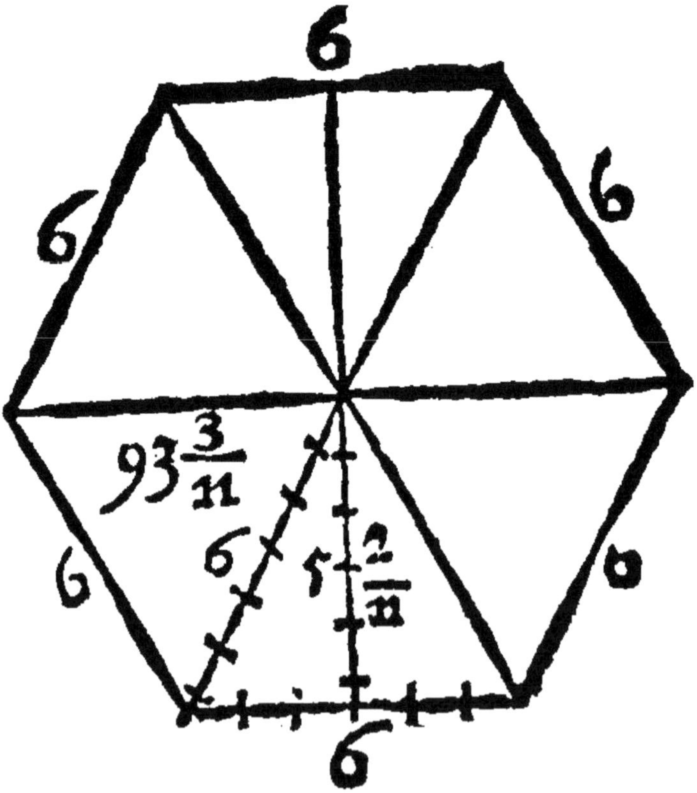

In an ordinate Sexangle also the ray, by the 9 e xviij, is knowne by the side of the sexangle. As here, the quadrate of 6, the ray is 36. The quadrate of 3, the halfe of the side, is 9: And 36 - 9. are 27, for the quadrate of the Perpendicular, whose side 5.2/11 is the perpendicular it selfe. Now the whole perimeter, as you see, is 36. Therefore the halfe is 18. And the product of 18 by 5.2/11 is 93.3/11 for the content of the sexangle given.

Lastly in all ordinate Multangles this theoreme shall satisfie thee.

2 The periphery is the triple of the diameter and almost one seaventh part of it.

Or the Periphery conteineth the diameter three times and almost one seventh of the same diameter. That it is triple of it, sixe raies, (that is three diameters) about which the periphery, the 9 e xviij, is circumscribed doth plainely shew: And therefore the continent is the greater: But the excesse is not altogether so much as one seventh part. For there doth want an unity of one seventh: And yet is the same excesse farre greater than one eighth part. Therefore because the difference was neerer to one seventh, than it was to one eighth, therefore one seventh was taken, as neerest unto the truth, for the truth it selfe.

Therefore

3. The plaine of the ray, and of halfe the periphery is the content of the circle.

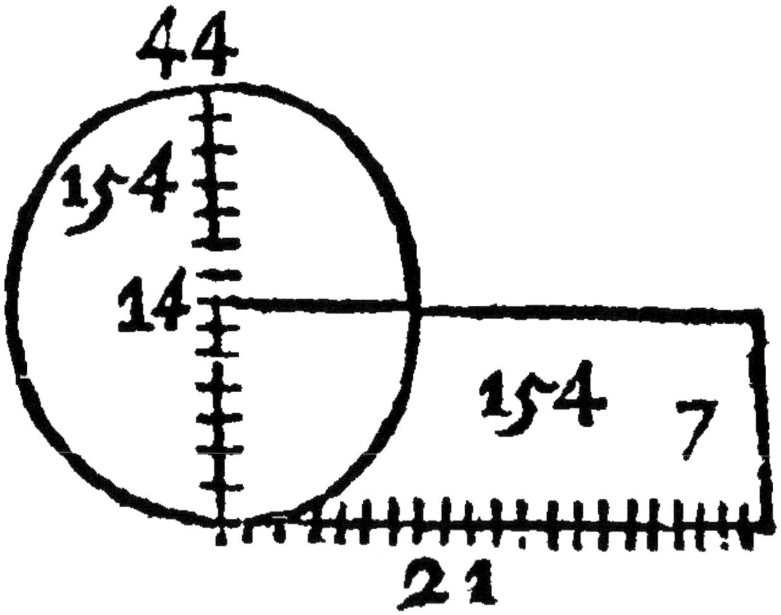

For here 7, the ray, of halfe the diameter 14, Multiplying 22, the halfe of the periphery 44, maketh the oblong 154, for the content of the circle. In the diameter two opposite sides, and likewise in the perimeter the two other opposite sides of the rectangle are conteined. Therefore the halfes of those two are taken, of the which the rectangle is comprehended.

And

4. As 14 is unto 11, so is the quadrate of the diameter unto the Circle.

For here 3 bounds of the proportion are given in *potentia*: The fourth is found by the multiplication of the third by the second, and by the Division of the product by the first: As here the Quadrate of the diameter 14, is 196. The product of 196 by 11 is 2156. Lastly 2156 divided by 14, the first bound, giveth in the Quotient 154, for the content of the circle sought. This ariseth by an analysis out of the quadrate and Circle measured. For the reason of 196, unto a 154; is the reason of 14 unto 11, as will appeare by the reduction of the bounds.

This is the second manner of squaring of a circle taught by *Euclide* as *Hero* telleth us, but otherwise layd downe, namely after this manner. *If from the quadrate of the diameter you shall take away 3/14 parts of the same, the remainder shall be the content of the Circle.* As if 196, the quadrate be divided by 14, the quotient likewise shall be 14. Now thrise 14, are 42: And 196 - 42, are 154, the quadrate equall to the circle.

Out of that same reason or rate of the pheriphery and diameter ariseth the manner of measuring of the Parts of a circle, as of a Semicircle, a Sector, a Section, both greater and lesser.

And

5. The plaine of the ray and one quarter of the periphery, is the content of the semicircle.

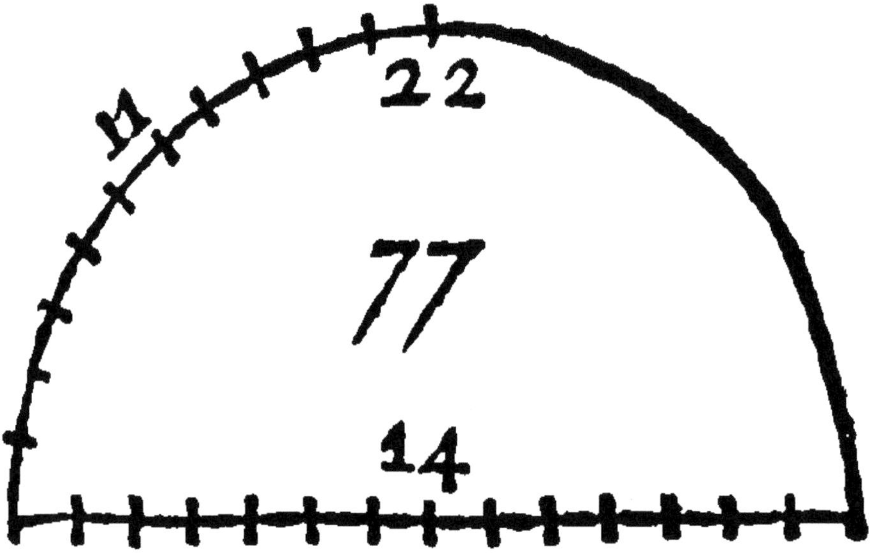

As here thou seest: For the product of 7, the halfe of the diameter, multiplyed by 11, the quarter of the periphery, doth make 77, for the content of the semicircle.

This may also be done by taking of the halfe of the circle now measured.

And

6. The plaine made of the ray and halfe the base, is the content of the Sector.

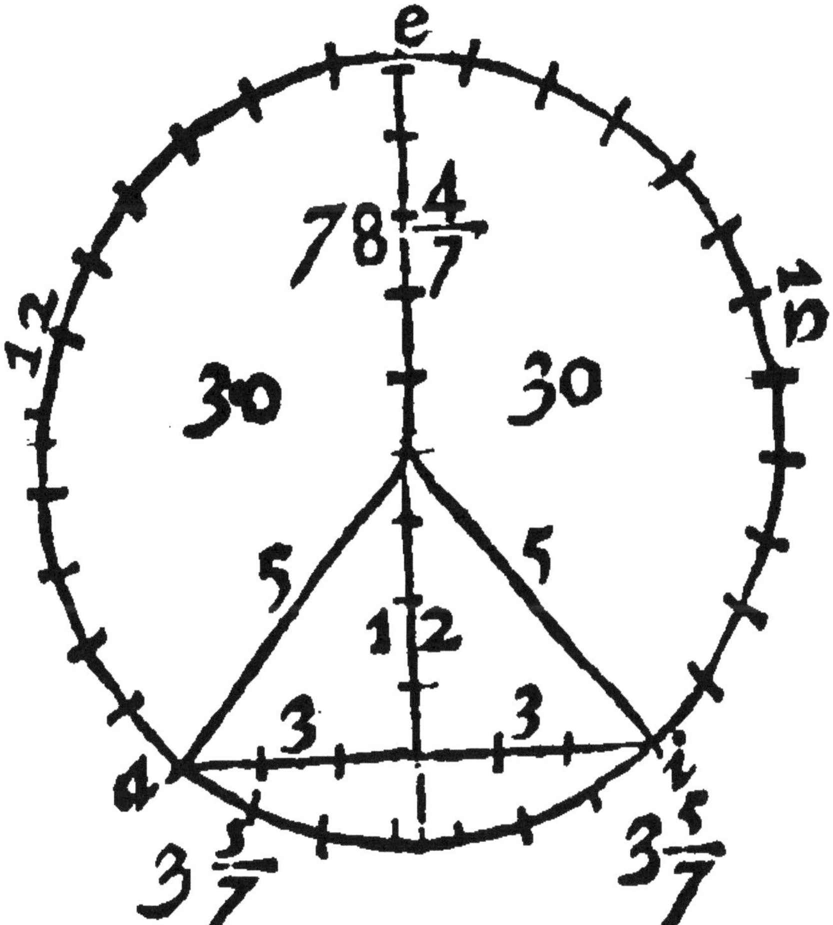

Here are three sectours, *ae* the base of 12 foote: And *ei* in like manner of 12 foote. The other or remainder *ia* of 7 *f.* and 3/7 of one foote. The diameter is 10 foote. Multiply therefore 5, halfe of the diameter, by 6 halfe of the base, and the product 30, shall be the content of the first sector. The same shall also be for the second sectour. Againe multiply the same ray or semidiameters 5, by 3.5/7, the halfe of 7.3/7, the product of 18.4/7 shall be the content of the third sector. Lastly, 30 + 30 + 18.4/7 are 78.4/7, the content of the whole circle.

And

7. If a triangle, made of two raies and the base of the greater section, be added unto the two sectors in it, the whole shall be the content of the greater section: If the same be taken from his owne sector, the remainder shall be the content of the lesser.

In the former figure the greater section is *aei*: The lesser is *ai*. The base of them both is as you see, 6. The perpendicular from the toppe of the triangle, or his heighth is 4. Therefore the content of the triangle is 12. Wherefore 30 + 30 + 12, that is 72, is the content of the greater section *aei*. And the lesser sectour, as in the former was taught, is 18.4/7. Therefore 18.4/7 - 12, that is, 6.4/7, is the content of *ai*, the lesser section.

And

8. A circle of unequall isoperimetrall plaines is the greatest.

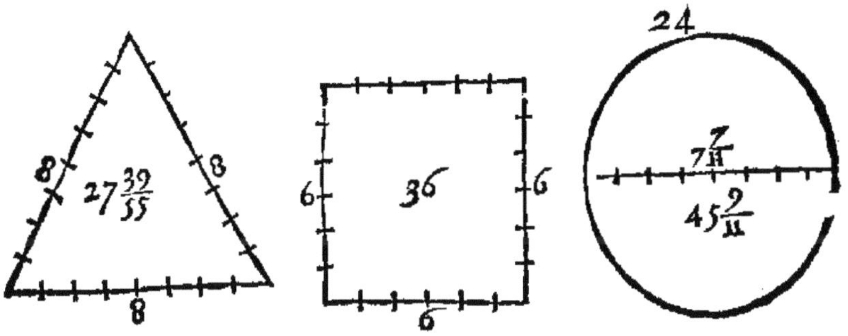

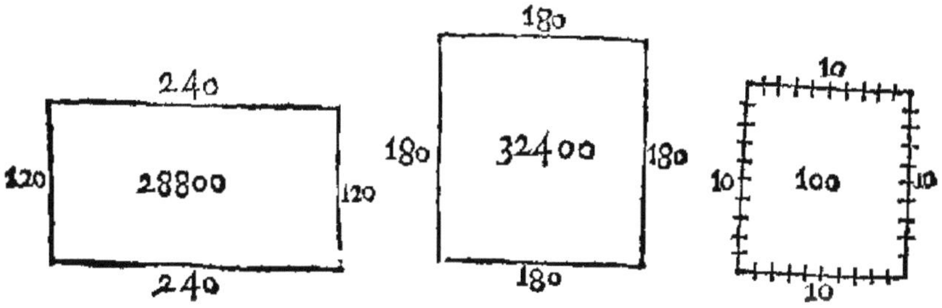

The reason is because it is the most ordinate, and comprehended of most bounds; see the 7, and 15 e iiij. As the Circle *a*, of 24 perimeter, is greater then any rectilineall figure, of equall perimeter to it, as the Quadrate *e*, or the Triangle *i*.

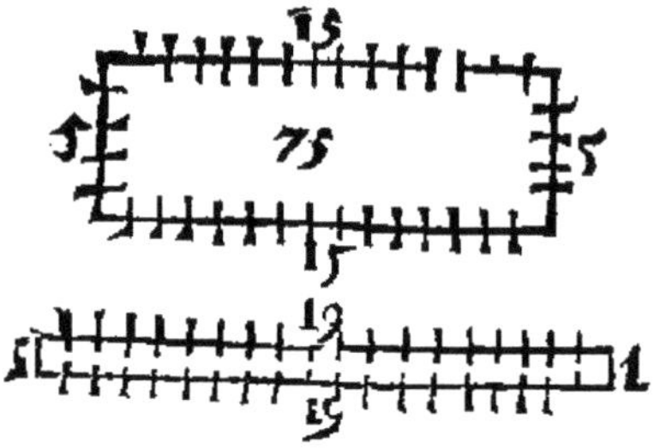
15
5
75
5
15
19
19

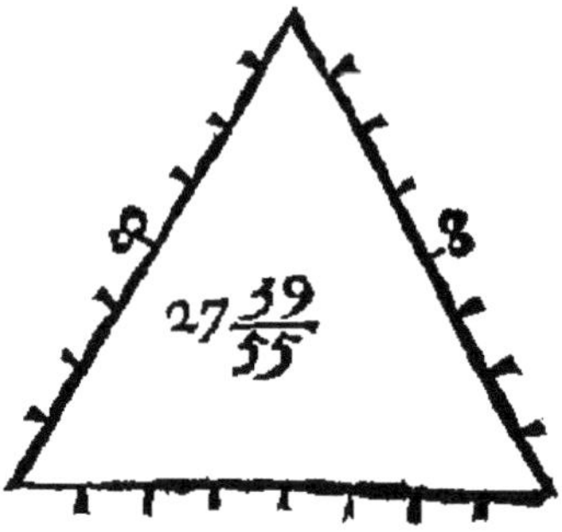
8
8
27 39/55

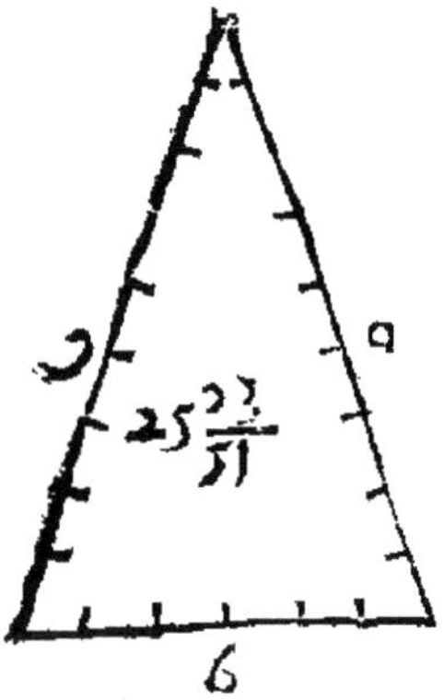
9
9
25 23/51
6

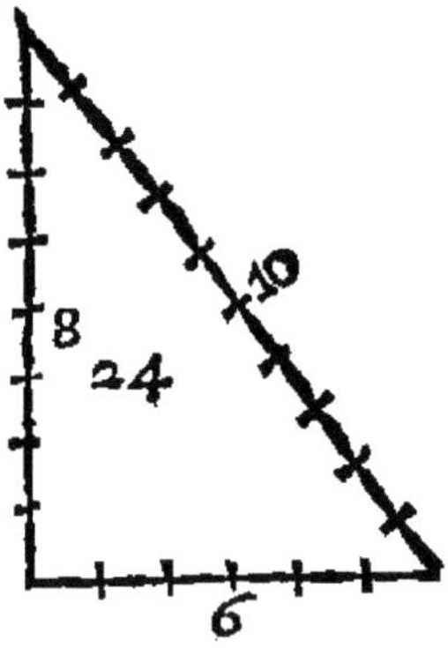
10
8
24
6

Of Geometry *the twentieth Booke, Of a Bossed surface.*

1. A bossed surface is a surface which lyeth unequally betweene his bounds.

It is contrary unto a Plaine surface, as wee heard at the 4 e v.

2. A bossed surface is either a sphericall, or varium.
3. A sphericall surface is a bossed surface equally distant from the center of the space inclosed.

Therefore

4. It is made by the turning about of an halfe circumference the diameter standeth still. è 14 d xj.

As here if thou shalt conceive the space betweene the periphery and the diameter to be empty.

5. The greatest periphery in a sphericall surface is that which cutteth it into two equall parts.

Those things which were before spoken of a circle, the same almost are hither to bee referred. The greatest periphery of a sphericall doth answere unto the Diameter of a Circle.

Therefore

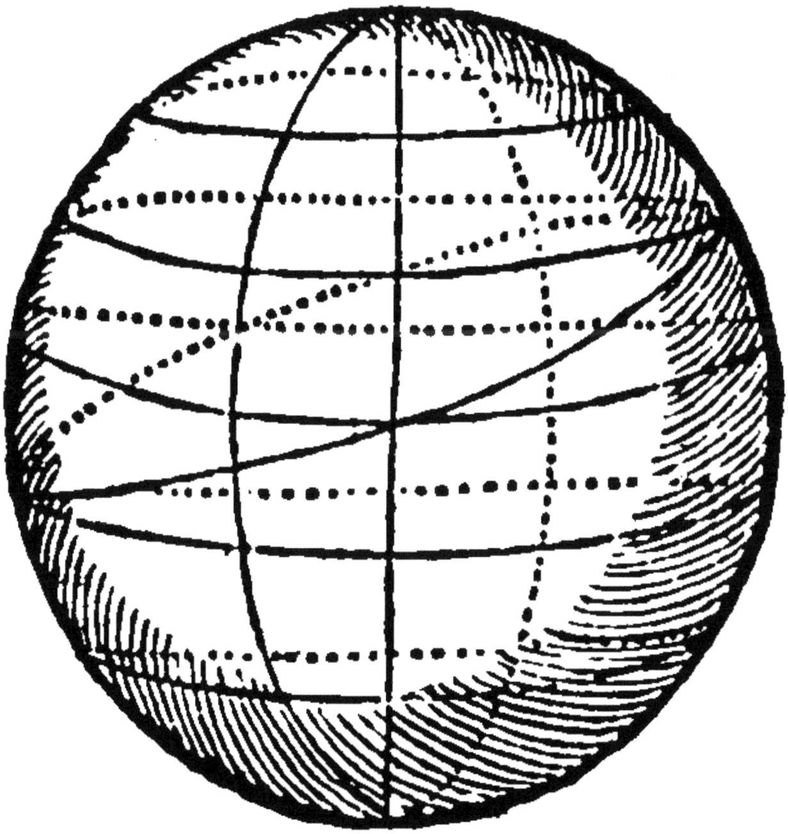

6. That periphery that is neerer to the greatest, is greater than that which is farther off: And on each side those two which are equally distant from the greatest, are equall.

The very like unto those which are taught at the 15, 16, 17, 18. e. xv. may here againe be repeated: As here.

7 The plaine made of the greatest periphery and his diameter is the sphericall.

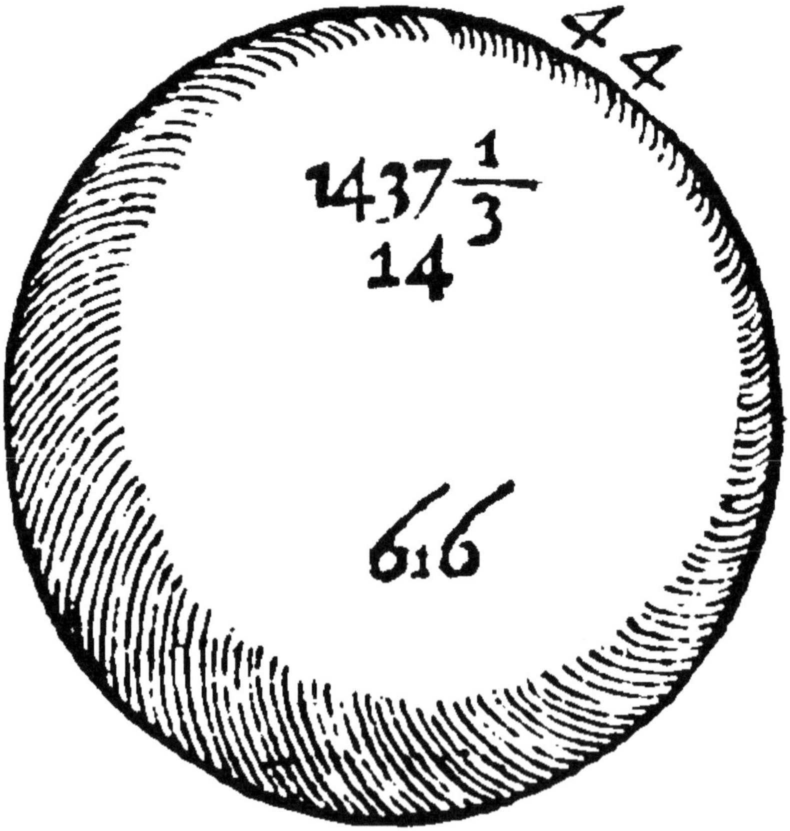

So the plaine made of the diameter 14. and of 44. the greatest periphery, which is 616. is the sphericall surface. So before the content of a circle was measured by a rectangle both of the halfe diameter, and periphery. But here, by the whole periphery and whole diameter, there is made a rectangle for the measure of the sphericall, foure times so great as was that other: Because by the 1 e vj. like plaines (such as here are conceived to be made of both halfe the diameter, and halfe the periphery, and both of the whole diameter and whole periphery) are in a doubled reason of their homologall sides.

Therefore

8 A plaine of the greatest circle and 4, is the sphericall.

This consectarium is manifest out of the former element.

And

9 As 7 is to 22. so is the quadrate of the diameter unto the sphericall.

For 7, and 22, are the two least bounds in the reason of the diameter unto the periphery: But in a circle, as 14, is to 11, so is the quadrate of the diameter unto the circle. The analogie doth answer fitly: Because here thou multipliest by the double, and dividest by the halfe: There contrariwise thou multipliest by the halfe, and dividest by the double. Therefore there one single circle is made, here the quadruple of that. This is, therefore the analogy of a circle and sphericall; from whence ariseth the hemispherical, the greater and the lesser section.

And

10 The plaine of the greatest periphery and the ray, is the hemisphericall.

As here, the greatest periphery is 44. the ray 7. The product therefore of 44. by 7. that is, 308. is the hemisphericall.

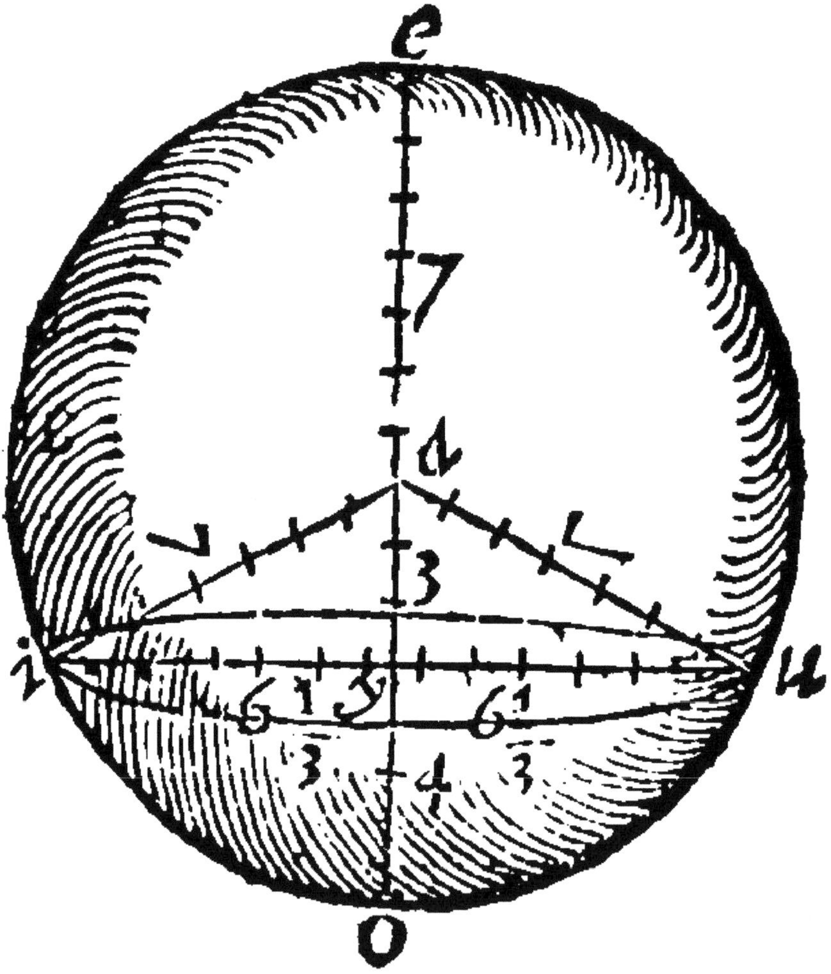

11 If looke what the part be of the ray perpendicular from the center unto the base of the greater section, so much the hemisphericall be increased, the whole shall be the greater section of the sphericall: But if it be so much decreased, the remainder shall be the lesser.

As in the example, the part of the third ray, that is, of 3/7, is from the center: such like part of the hemispherical 308, is 132. (For the 7, part of 308. is 44. And three times 44. is 132.) Therefore 132. added to 308. do make 440. for the greater section of the sphericall. And 132. taken from 308. doe leave 176. for the lesser section of the same.

12 The varium is a bossed surface, whose base is a periphery, the side a right line from the bound of the toppe, unto the bound of the base.

13 A varium is a conicall or a cylinderlike forme.

14 A conicall surface is that which from the periphery beneath doth equally waxe lesse and lesse unto the very toppe.

Therefore

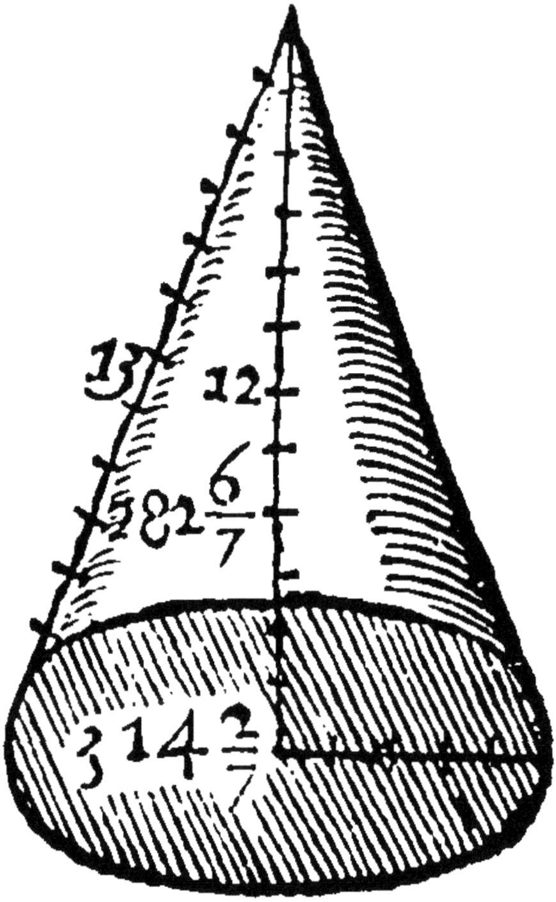

15. It is made by turning about of the side about the periphery beneath.
16 The plaine of the side and halfe the base is the conicall surface.

As in the example next aforegoing, the side is 13. The halfe periphery is 15.5/7: And the product of 15.5/7 by 13. is 204.2/7. for the conicall surface. To which if you shall adde the circle underneath, you shall have the whole surface.

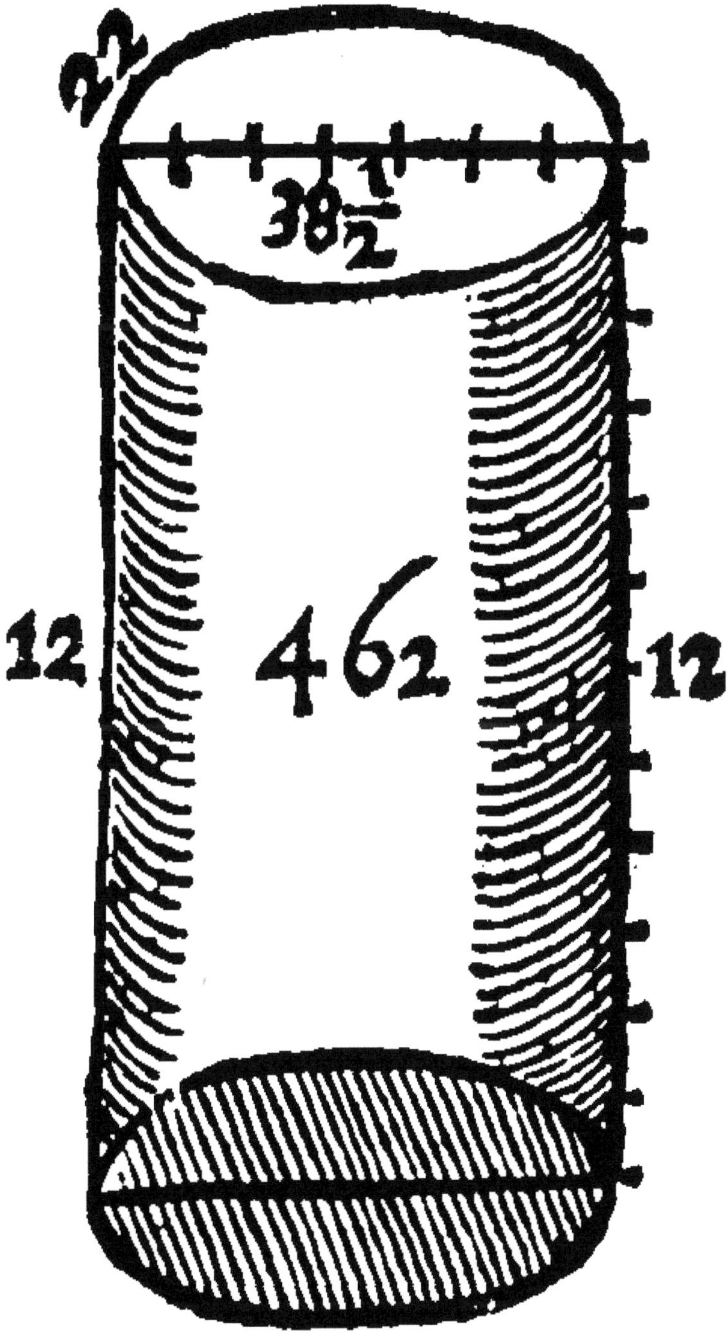
22
$38\frac{1}{2}$
12
462
12

> *17 A cylinderlike forme is that which from the periphery underneath unto the the upper one, equall and parallell unto it, is equally raised.*

Therefore

> *18 It is made by the turning of the side about two equall and parallell peripheries.*

> *19 The plaine of his side and heighth is the cylinderlike surface.*

As here the periphery is 22. as is gathered by the Diameter, which is 7. The heighth is 12. The base therefore is 38.1/2. And 38.1/2 by 12. are 462. for the cylinderlike surface. To which if you shall adde both the bases on each side, to wit, 38.1/2. twise, or 77. once, the whole surface shall be 539.

Geometry, *the one and twentieth Book, Of Lines and Surfaces in solids.*

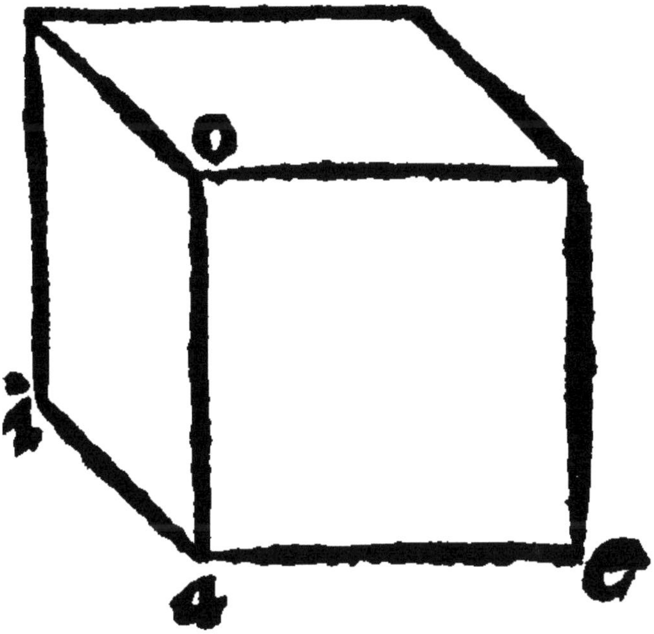

1 A body or solid is a lineate broad and high 1 d xj.

For length onely is proper to a line: Length and breadth, to a surface: Length breath, and heighth joyntly, belong unto a body: This threefold perfection of a magnitude, is proper to a body: Whereby wee doe understand that are in a body, not onely lines of length, and surfaces of breadth, (for so a body should consist of lines and surfaces.) But we do conceive a solidity in length, breadth and heighth. For every part of a body is also a body. And therefore a solid we doe understand the body it selfe. As in the body *aeio*, the length is *ae*; the breadth, *ai*, And the heighth, *ao*.

2 The bound of a solid is a surface 2 d xj.

The bound of a line is a point: and yet neither is a point a line, or any part of a line. The bound of a surface is a line: And yet a line is not a surface, or any part of a surface. So now the bound of a body is a surface: And yet a surface is not a body, or any part of a body. A magnitude is one thing; a bound of a magnitude is another thing, as appeared at the 5 e j.

As they were called plaine lines, which are conceived to be in a plaine, so those are named solid both lines and surfaces which are considered in a solid; And their perpendicle and parallelisme are hither to be recalled from simple lines.

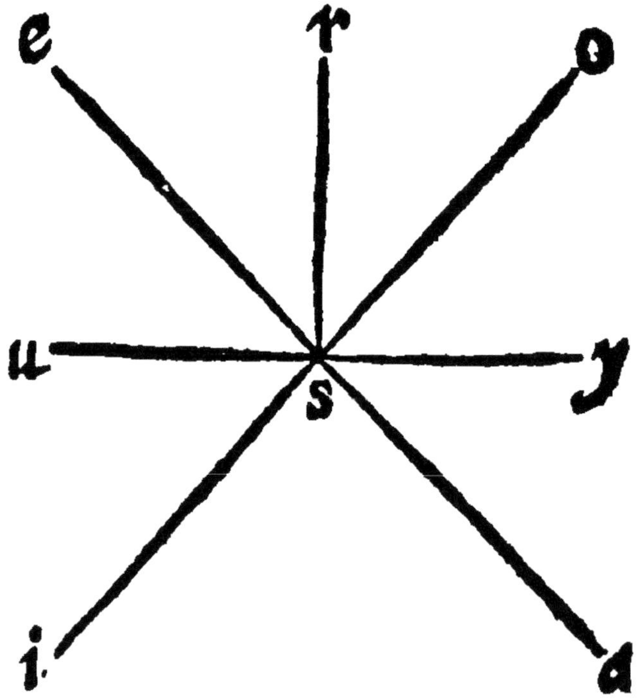

3 If a right line be unto right lines cut in a plaine underneath, perpendicular in the common intersection, it is perpendicular to the plaine beneath: And if it be perpendicular, it is unto right lines, cut in the same plaine, perpendicular in the common intersection è 3 d and 4 p xj.

Perpendicularity was in the former attributed to lines considered in a surface. Therefore from thence is repeated this consectary of the perpendicle of a line with the surface it selfe.

If thou shalt conceive the right lines, *ae*, *io*, *uy*, to cut one another in the plaine beneath, in the common intersections: And the line *rs*, falling from above, to be to every one of them perpendicular in the common point *s*, thou hast an example of this consectary.

4 If three right lines cutting one another, be unto the same right line perpendicular in the common section, they are in the same plaine 5. p xj.

For by the perpendicle and common section is understood an equall state on all parts, and therefore the same plaine: as in the former example, *as*, *ys*, *os*, suppose them to be to *sr*, the same loftie line, perpendicular, they shall be in the same nearer plaine *aiueoy*.

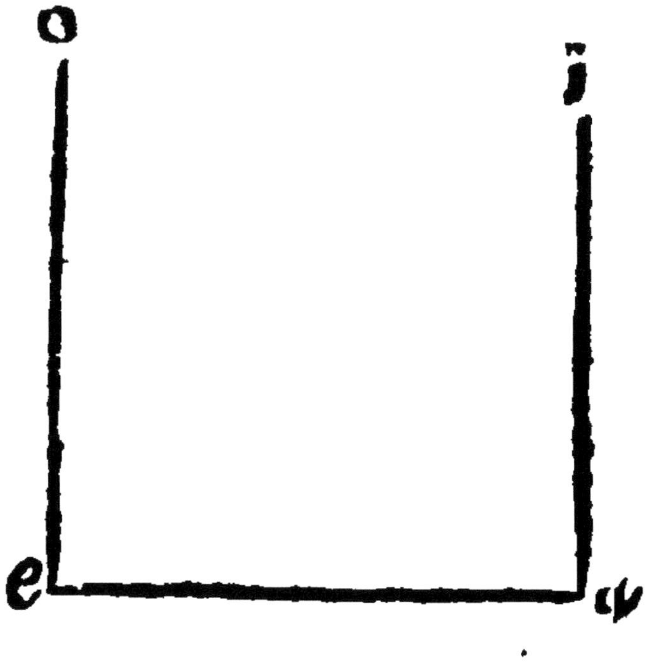

5 If two right lines be perpendicular to the under-plaine, they are parallells: And if the one two parallells be perpendicular to the under plaine, the other is also perpendicular to the same. 6. 8 p xj.

The cause is out of the first law or rule parallells. For if two right lines be perpendicular to the same under plaine, being joyned together by a right line, they shall make their inner corners equall to two right angles: And therefore they shall be parallells, by the 21. e v. And if in two parallells knit together with a right line, one of the inner angles, be a right angle: the other also shall be a right angle. Because they are divided by a common perpendicular; As in the example. If the angles at *a*, and *e*, be right angles, *ai*, and *eo*, are parallells, and contrariwise, if *ai*, and *eo* be parallells, and the angle at *a*, be a right angle, the angle at *e*, also shall be a right angle.

6 If right lines in diverse plaines be unto the same right line parallel, they are also parallell betweene themselves. 9 p xj.

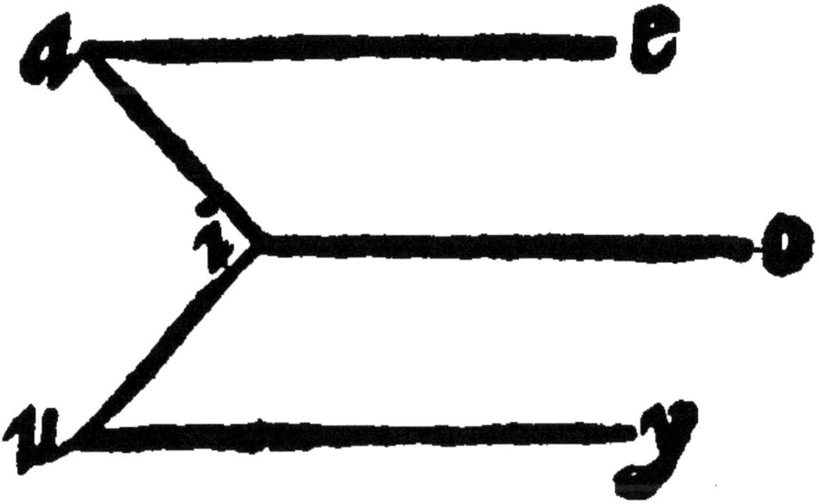

As here *ae*, and *uy*, right lines in diverse plaines suppose them to be parallell to *io*: I say, they are parallell one to another. For from the point *i*, let *ia*, and *iu*, be erected at right angles to *io* to cut the parallells, by the 17. e v. Therefore, by the 3 e, *oi*, seeing that it is perpedicular to *ia*, and *iu*, two lines cutting one another, it is perpendicular to the plaine beneath. Therefore by the the 6 e, *yu*, and *ea*, are perpendicular to the same plaine: And therefore, by the same, they are parallell.

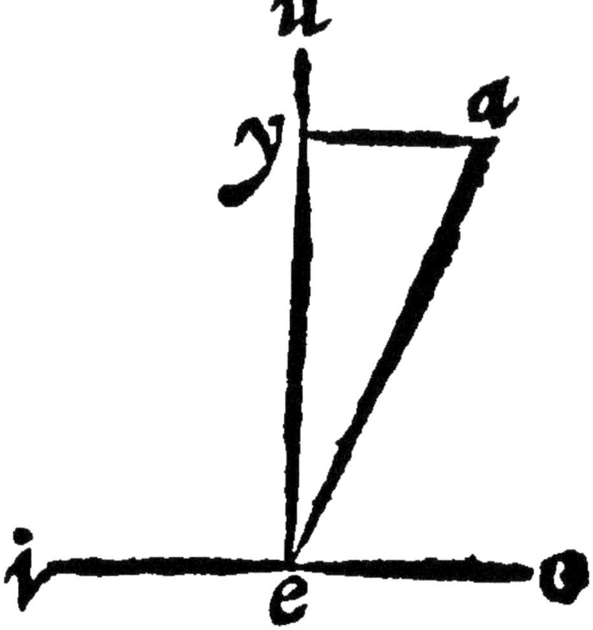

7 If two right lines be perpendiculars, the first from a point above, unto a right line underneath, the second from the common section in the plaine underneath, a third, from the sayd point perpendicular to the second, shall be perpendicular to the plaine beneath. è 11 p xj.

It is a consectary out of the 3 e. As for example, if from a lofty point *a*, *ae*, be by the 18 e v, perpendicular to *e*, a point of the right line *io* underneath: And from *e* the common section, by the 17 e v, there be *eu*, another perpendicular: Lastly *ay*, a lofty right line, be by the 18 e v, perpendicular unto *eu*, at the point *y*, *ay* shall be perpendicular unto the plaine underneath. For that *ae* is perpendicular to *io*, the same *ae* declineth neither to the right hand, nor to the left, by the 13 e ij. And in that againe *ay* is perpendicular to *eu*, it leaneth neither forward nor backeward. Therefore it lyeth equally or indifferently, betweene the foure quarters of the world.

If the right line *io*, doe with equall angles agree to *r*, the third element.

8. If a right line from a point assigned of a plaine underneath, be parallell to a right line perpendicular to the same plaine, it shall also be perpendicular to the plaine underneath. ex 12 p xj.

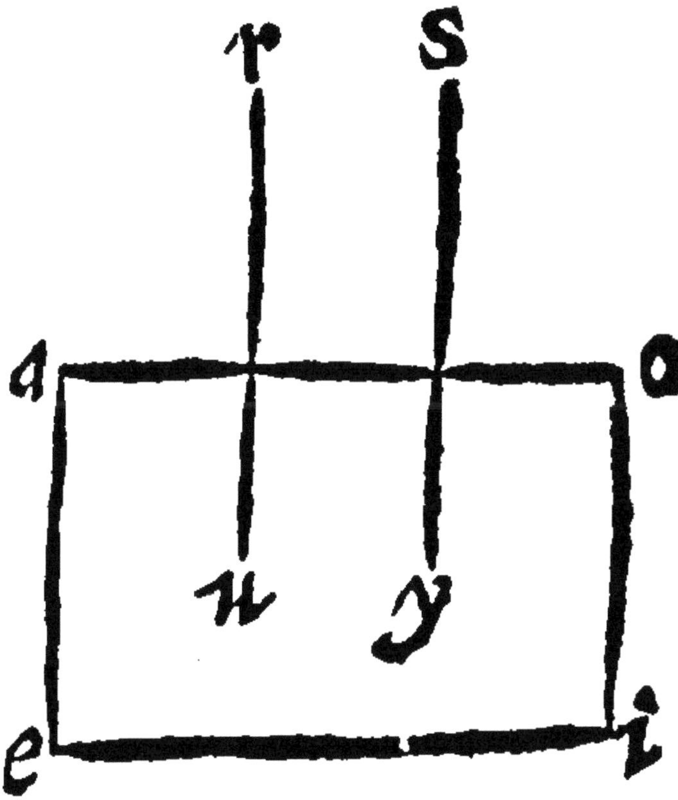

As for example let the plaine be *aeio*: And the assigned point in it *u*: From this point a lofty perpendicular is to be erected. Let there be made from the point *y*, the perpendicular *ys*, unto the plaine underneath, by the 7 e. And to it let *ur*, be made parallell by the 24 e v. Now *ur*, seeing it is parallell to a perpendicular upon the plaine underneath, it shall be perpendicular to the same, by the 5 e.

9. If a right line in one of the plaines cut, perpendicular to the common section, be perpendicular to the other, the plaines are perpendicular: And if the plaines be perpendicular, a right line in the one perpendicular to the common section is perpendicular to the other è 4 d, and 38 p xj.

The perpendicularity of plaines, is drawne out of the former condition of the perpendicle: And the state of plaines on each side equall betweene themselves, is fetch'd from a perpendicularity of a right line falling upon a plaine. Because from hence it is understood that the plaine it selfe doth lye indifferently betweene all parts signified by right lines: Which in a Booke with the pages each way opened, is perceived by the verses or lines of the pages, both to the section and plaine underneath, perpendicular as here thou seest.

10. If a right line be perpendicular to a plaine, all plaines by it, are perpendicular to the same: And if two plaines be unto any other plaine perpendiculars, the common section is perpendicular to the same. e 15, and 19 p. xj.

The first is a consectary drawne out of the 9 e. And the latter is from hence manifest, because that same common section is a right line, in any manner of lofty plaines intersected, perpendicular both to the common section and plaine underneath. For if the common section, were not perpendicular to the plaine underneath, neither should the plaines cutting one another be perpendicular to the plaine underneath, but some one should be oblique, against the grant, as here thou seest.

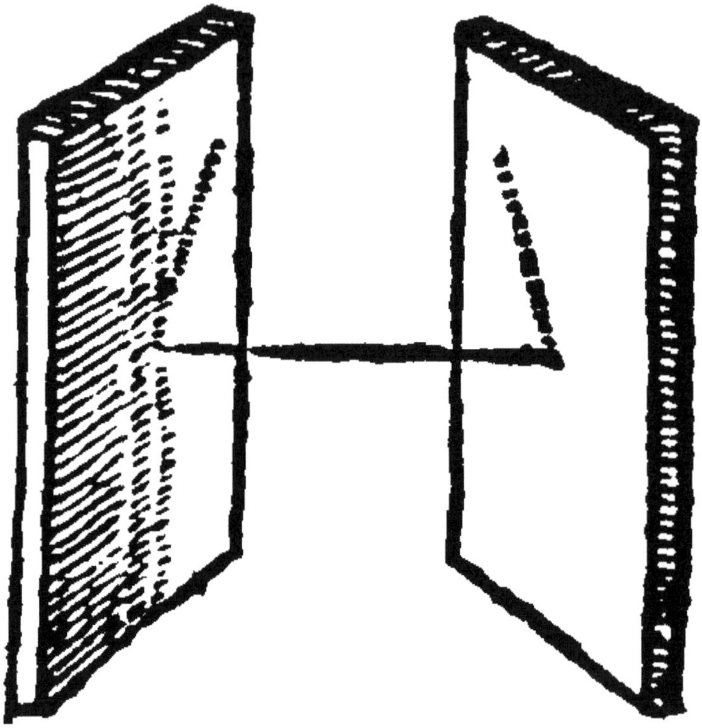

11. Plaines are parallell which doe leane no way. 8 d xj.

And

12. Those which divided by a common perpendicle. 14 p xj.

It is a consectary out of the 3, and 6 e. For if the middle right line be perpendicular to both the plaines, it is also to the right lines on either side cut, perpendicular in the common intersection: And the inner angles on each side, being right angles, will evince them to be parallels.

It is also out of the definition of parallels, at the 15 e ij.

And

13. If two paires of right in them be joyntly bounded, they are parallell. 15 p xj.

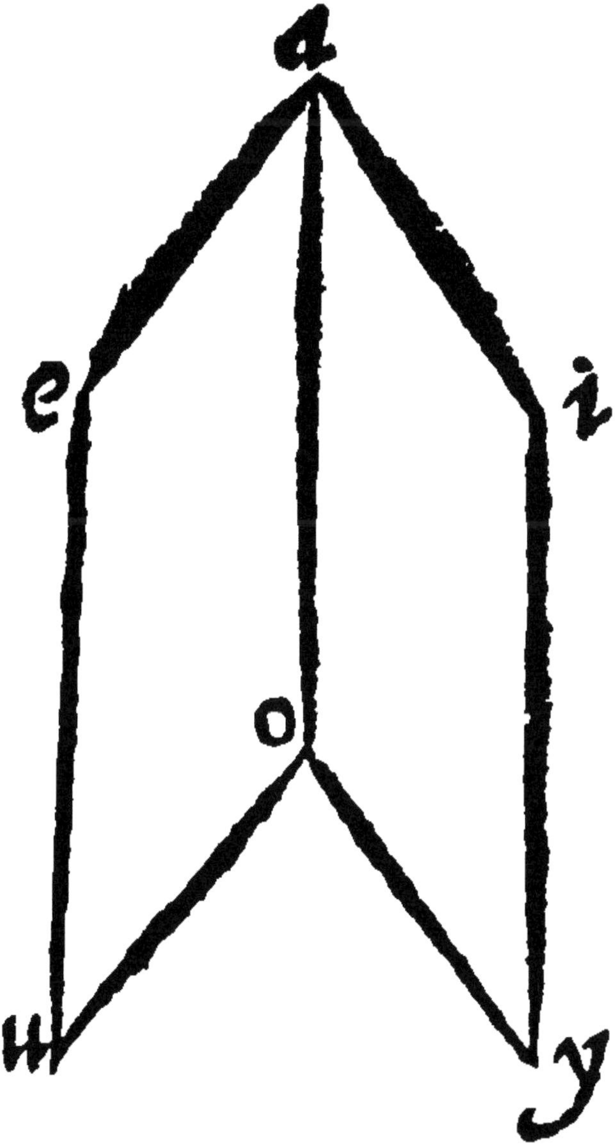

Such are the opposite walls in the toppe or ridge of houses. As let *aei*, and *uoy*, be plaine which have two payres of right lines, *ea*, and *ia*: Item *uo*, and *yo*, joyntly bounded in *a*, and *o*: And parallels, to wit *ea*, against *uo*: and *ia*, against *yo*. I say that the plaines themselves are parallels: For the right lines *ue*, and *oa*: item *yi*, and *oa*, doe knit together equall parallels, they shal by the 27 e v, be equall and parallels: And so they shall prove the equidistancie.

The same will fall out if thou shalt imagine the joyntly bounded to infinitely drawn out; for the plaines also infinitely extended shall be parallell.

14. If two parallell plaines are cut with another plaine, the common sections are parallels, 16 p xj.

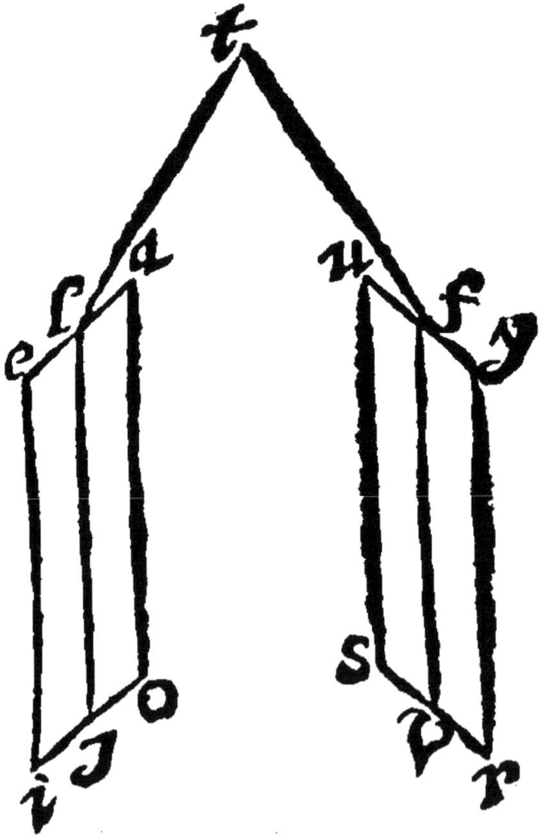

As here thou seest the parallell plaines *aeio*, and *uysr*, cut by the plaine *ljvf*, the common sections *lj*, and *fv*, shall also be parallell: Otherwise they themselves, and therefore also the plaines in which they are, shall meete, as in the point *t*, which is against the grant.

The twenty second Booke, of P. Ramus Geometry, Of a Pyramis.

1. The axis of a solid is the diameter about which it is turned, e 15, 19, 22 d xj.

The Axis or Axeltree is commonly thought to be proper to the sphere or globe, as here *ae*: But it is attributed to other kindes of solids, as well as to that.

2. A right solid is that whose axis is perpendicular to the center of the base.

Thus *Serenus* and *Apllonius* doe define a Cone and a Cylinder: And these onely *Euclide* considered: Yea and indeed stereometry entertaineth no other kinde of solid but that which is right or perpendicular.

3. If solids be comprehended of homogeneall surfaces, equall in multitude and magnitude, they are equall. 10 d xj.

Equality of lines and surfaces was not informed by any peculiar rule; farther than out of reason and common sense, and in most places congruency and application was enough and did satisfie to the full: But here the congruency of Bodies is judged by their surfaces. Two cubes are equall, whose sixe sides or plaine surfaces, are equall, &c.

4. If solids be comprehended of surfaces in multitude equall and like, they are equall, 9 d xj.

This is a consectary drawne out of the general difinition of like figures, at the 19 e. iiij. For there like figures were defined to be equiangled and proportionall in the shankes of the equall angles: But in like plaine solids the angles are esteemed to be equall out of the similitude of their like plaines: And the equall shankes are the same plaine surfaces, and therefore they are proportionall, equall and alike.

5 Like solids have a treble reason of their homologall sides, and two meane proportionalls. 33. p xj. 8 p xij.

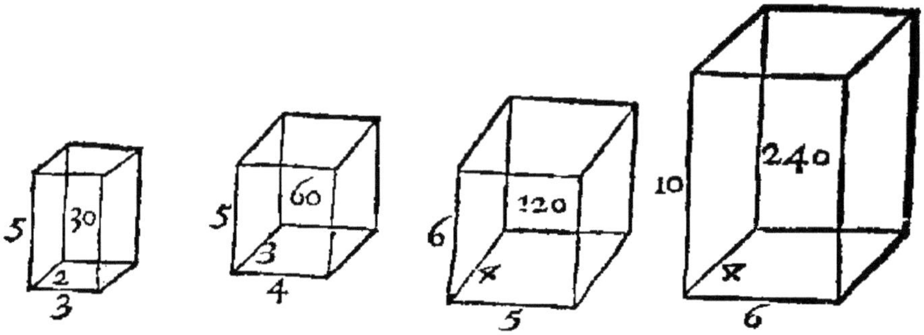

It is a consectary drawne out of the 24 e. iiij. as the example from thence repeated shall make manifest.

6 A solid is plaine or embosed.

7 A plaine solid is that which is comprehended of plaine surfaces.

8 The plaine angles comprehending a solid angle, are lesse than foure right angles. 21. p xj.

For if they should be equall to foure right angles, they would fill up a place by the 27 e, iiij. neither would they at all make an angle, much lesse therefore would they doe it if they were greater.

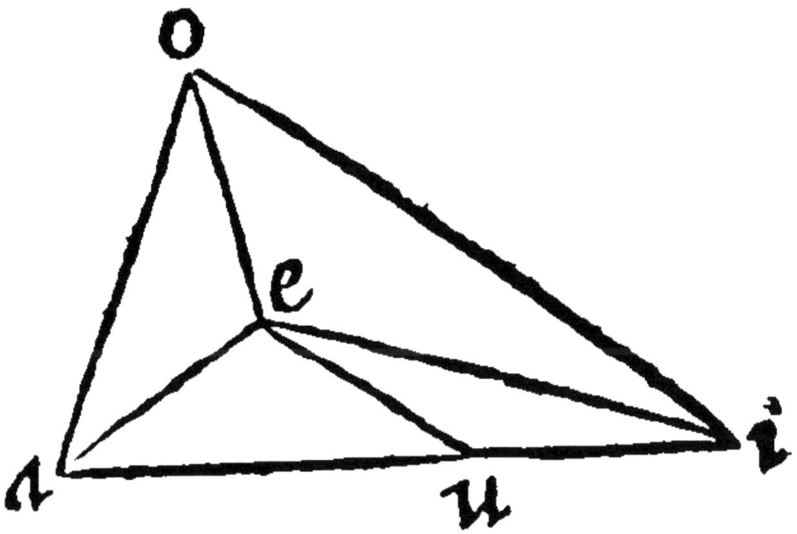

9 If three plaine angles lesse than foure right angles, do comprehend a solid angle, any two of them are greater than the other: And if any two of them be greater than the other, then may comprehend a solid angle, 21. and 23. p xj.

It is an analogy unto the 10 e vj. and the cause is in a readinesse. For if two plaine angles be equall to the remainder, they shall with that third include no space betweene them: But if thou shalt conceit to fit the plaine to the shankes, with the congruity they should of two make one: but much lesse if they be lesser.

The converse from hence also is manifest.

Euclide doth thus demonstrate it: First if three angles are equall, then by and by two are conceived to be greater than the remainder. But if they be unequall, let the angle *aei*, be greater than the angle *aeo*: And let *aeu*, equall to *aeo*, be cut off from the greater *aei*: And let *eu*, be equall to *eo*. Now by the 2 e, vij. two triangles *aeu*, and *aeo*, are equall in their bases *au*, and *ao*. Item *ao*, and *ei*, are greater than *ai*, and *ao*: And *ao*, is equall to *au*. Therefore *oi*, is greater than *iu*. Here two triangles, *uei*, and *ieo*, equall in two shankes; and the base *oi*, greater than the base *iu*. Therefore, by the 5 e vij. the angle *oei*, is greater than the angle *ieu*. Therefore two angles *aeo*, and *oei*, are greater than *aei*.

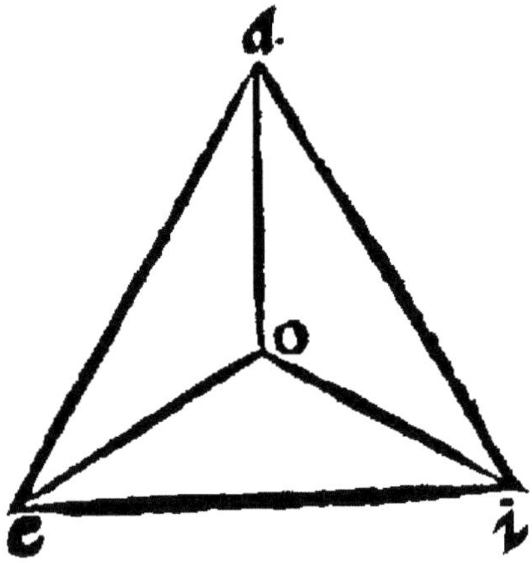

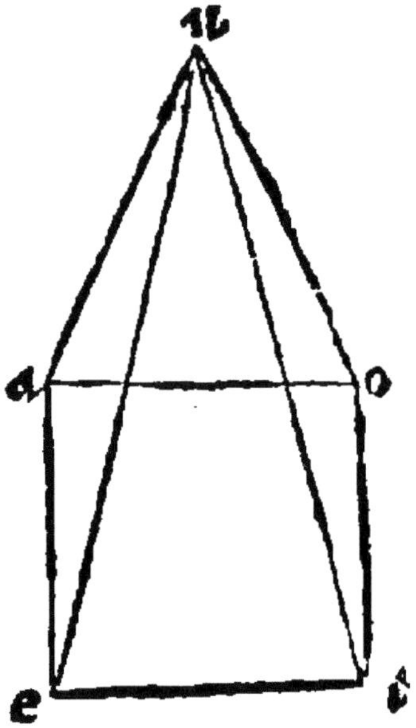

10 A plaine solid is a Pyramis or a Pyramidate.

11 A Pyramis is a plaine solid from a rectilineall base equally decreasing.

As here thou conceivest from the triangular base *aei*, unto the toppe *o*, the triangles *aoe*, *aoi*, and *eoi*, to be reared up.

In the pyramis *aeiou*, thou seest from the quadrangular base *aeio*, unto the toppe *u*, foure triangles in like manner to be raised.

Therefore

12 The sides of a pyramis are one more than are the base.

The sides are here named *Hedræ*.

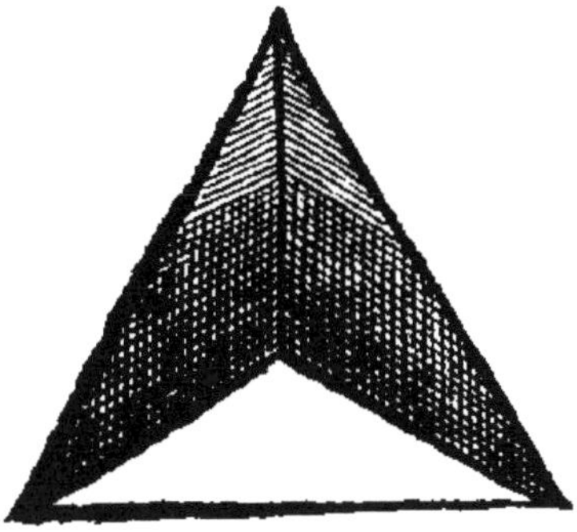

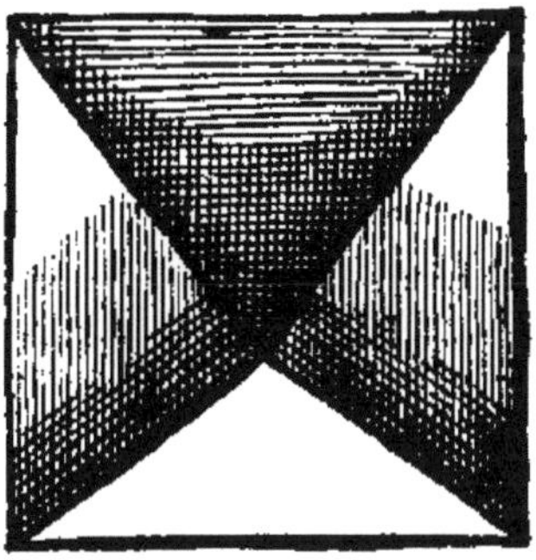

And

13 A pyramis is the first figure of solids.

For a pyramis in solids, is as a triangle is in plaines. For a pyramis may be resolved into other solid figures, but it cannot be resolved into any one more simple than it selfe, and which consists of fewer sides than it doth.

Therefore

14 Pyramides of equall heighth, are as their bases are 5 e, and 6. p xij.

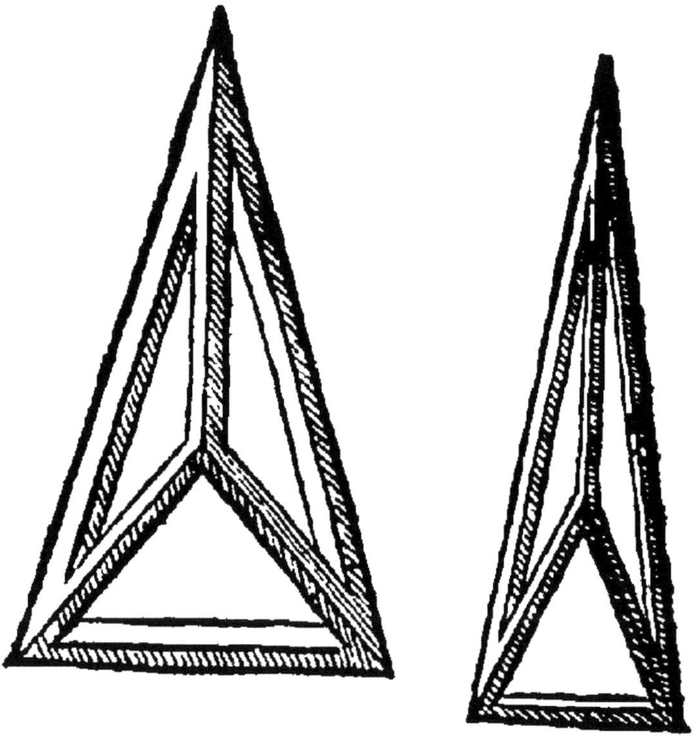

And

15 Those which are reciprocall in base and heighth are equall 9 p xij.

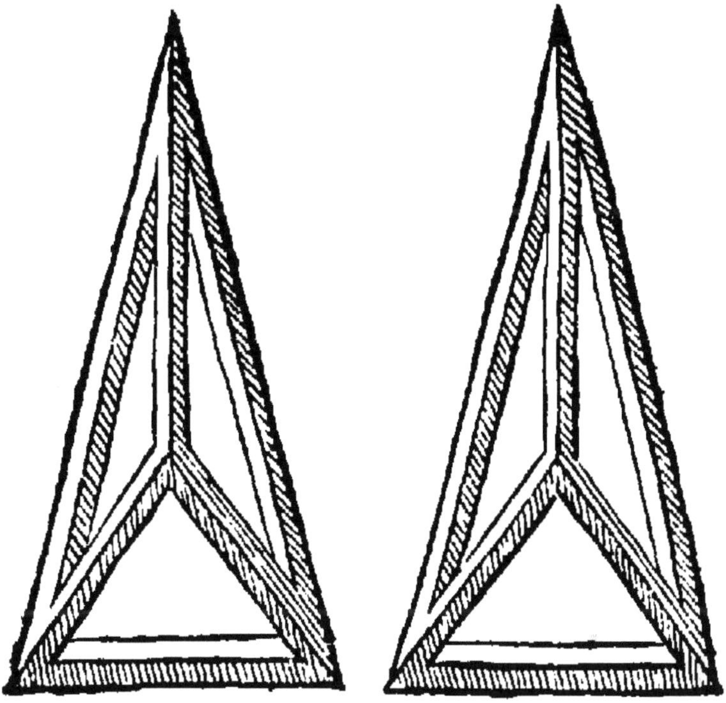

These consectaries are drawne out of the 16, 18 e. iiij.

16 A tetraedrum is an ordinate pyramis comprehended of foure triangles 26. d xj.

As here thou seest. In rectilineall plaines we have in the former signified, in every kinde there is but one ordinate figure: Amongst the triangles the equilater: Amongst the quadrangles, the Quadrate: so now of all kinde of Pyramides, there is one kinde ordinate onely, and that is the Tetraedrum. And yet not every Tetraedrum is such, but that only which is comprehended of triangles, not onely severally ordinate, but equall one to another altogether alike.

Therefore

17 The edges of a tetraedrum are sixe, the plaine angles twelve, the solide angles foure.

For a Tetraedrum is comprehended of foure triangles, each of them having three sides, and three corners a peece: And every side is twise taken: Therefore the number of edges is but halfe so many.

And

18 Twelve tetraedra's doe fill up a solid place.

Because 8. solid right angles filling a place, and 12. angles of the tetraedrum are equall betweene themselves, seeing that both of them are comprehended of 24 plaine right-angles. For a solid right angle is comprehended of three plaine right angles: And therefore 8. are comprehended of 24. In like manner the angle of a Tetraedrum is comprehended of three plaine equilaters, that is of sixe third of one right angle: and therefore of two right angles: Therefore 12 are comprehended of 24.

And

19. If foure ordinate and equall triangles be joyned together in solid angles, they shall comprehend a tetraedrum.

This fabricke or construction is very easie, as you may see in these examples: For if thou shalt joyne or fold together these triangles here thus expressed, thou shalt make a tetraedrum.

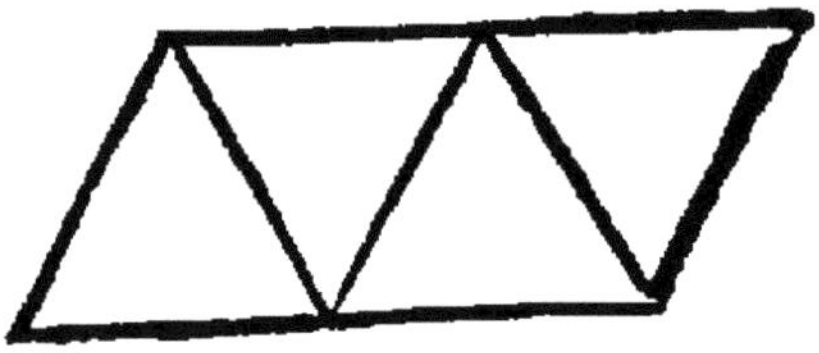

20. If a right line whose power is sesquialter unto the side of an equilater triangle, be cut after a double reason, the double segment perpendicular to the center of the triangle, knit together with the angles thereof shall comprehend a tetraedrum. 13 p xiij.

For a solid to be comprehended of right lines understand plaines comprehended of right lines, as in other places following.

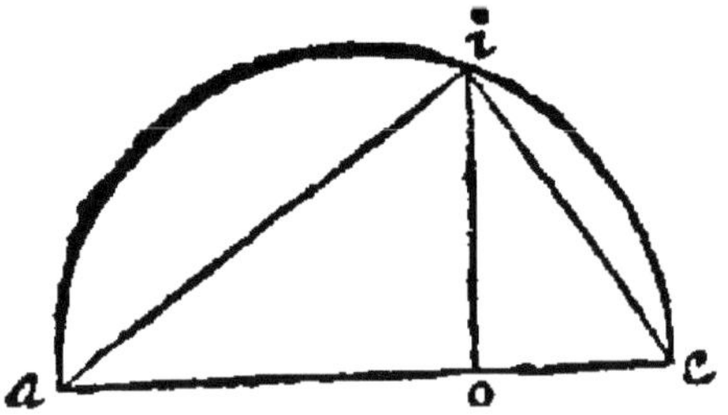

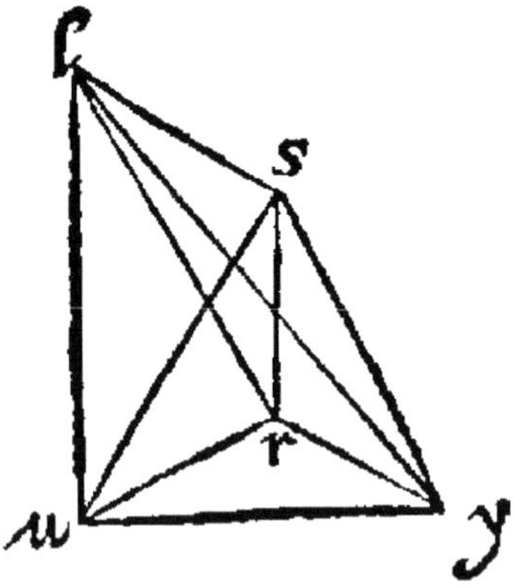

As here, Let first *ae* be the right line whose power is sesquialter unto *ai* the side of the equilater triangle, as in the forme was manifest at the 13 e xij. And let it be by the 29 e v, be cut in a double reason in *o*: And let the double segment *ao*, be perpendicular to the equilater triangle *uys*, unto the center *r*, by the 7 e xxj. And let *lr* be knit with the angles, by *lu*, *ls*, *ly*. I say that the triangles *uys, usl, uyl,* are equilater and equall, because all the sides are equall. First the three lower ones are equall by the grant: And the three higher ones are equall by the 9 e xij. And every one of the higher ones are equall to the under one. For if a Circle bee supposed to bee circumscribed about the triangle, the side shall be of treble power to the ray *ur,* by the 12 e xviij. But the higher one also is of treble power to the same ray, as is manifest in the first figure of the ray *oi,* which is for the ray of the second figure *ur*. For as *ao,* is to *oi,* so by the 9 e viij, is *oi,* unto *oe*: And by the 25 e iiij, as the first rect line *ao,* is unto the third *oe*: so is the quadrate *ao,* unto the quadrate *oi*. And by compounding *ao* with *oe*; As *ae* is to *oe*; so are the quadrates *ao*; and *oi,* that is, by the 9 e xij, the quadrate *ai,* unto the quadrate *oi,* But *ae* is the triple of *oe*. Therefore the quadrate *ai,* is the triple of the quadrate *oi*. Wherefore the higher side equall to *ai,* is of treble power to the ray: And therefore also all the sides are equall: And therefore againe the triangles themselves are equall.

The twenty third Booke of Geometry, *of a* Prisma.

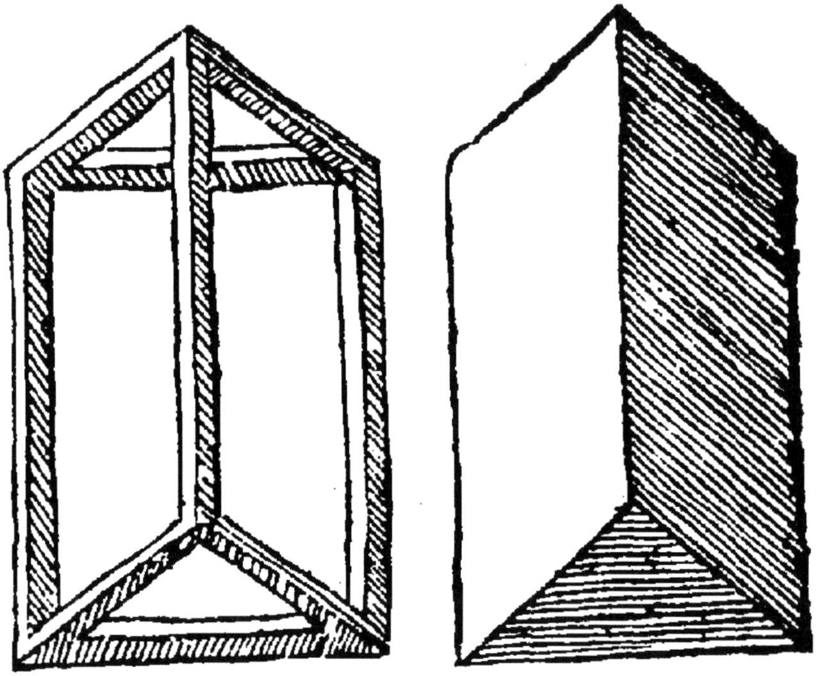

1 *A Pyramidate is a plaine solid comprehended of pyramides.*

2. *A pyramidate is a Prisma, or a mingled polyedrum.*

3. *A prisma is a pyramidate whose opposite plaines are equall, alike, and parallell, the rest parallelogramme. 13 d xj.*

As here thou seest. The base of a pyramis was but one: Of a Prisma, they are two, and they opposite one against another, First equall; Then like: Next parallell. The other are parallelogramme.

Therefore

4. *The flattes of a prisma are two more than are the angles in the base.*

And indeed as the augmentation of a Pyramis from a quaternary is infinite: so is it of a Prisma from a quinary: As if it be from a triangular, quadrangular, or quinquangular base; you shal have a Pentraedrum, Hexaedrum, Heptaedrum, and so in infinite.

5. *The plaine of the base and heighth is the solidity of a right prisma*.

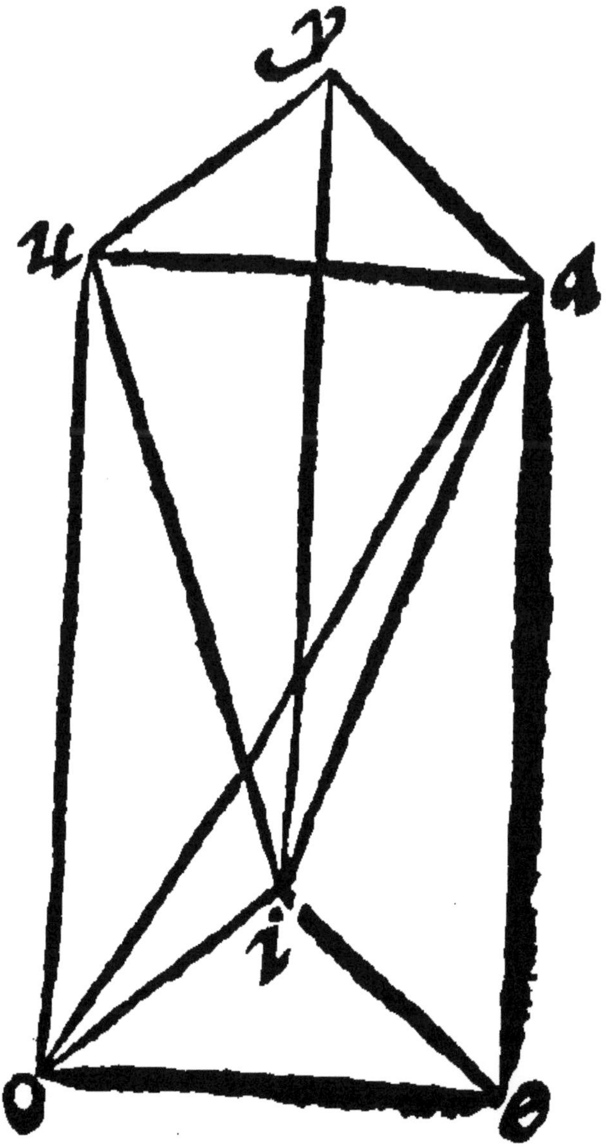

6. *A prisma is the triple of a pyramis of equall base and heighth. è 7 p. xij.*

As in the example a prisma pentaedrum is cut into three equall pyramides. For the first consisting of the plaines *aei, aeo, aoi, eio*; is equall to the second consisting of the plaines *aoi, aou, aiu, iou*, by the 10 e vij. Because it is equall to it both in common base and heighth. Therefore the first and second are equall. And the same second is equall to it selfe, seeing the base is *iou*, and the toppe *a*. Then also it is equall to the third consisting of the plaines *aiu, aiy, uiy, auy*. Therefore three are equall.

If the base be triangular, the Prisma may be resolved into prisma's of triangular bases, and the theoreme shall be concluded as afore.

Therefore

7. *The plaine made of the base and the third part of the heighth is the solidity of a pyramis of equall base and heighth.*

The heighth of a pyramis shall be found, if you shall take the square of the ray of the base out of the quadrate of the side: for the side of the remainder, by the 9 e xij, shall be the altitude or heighth, as in the example following.

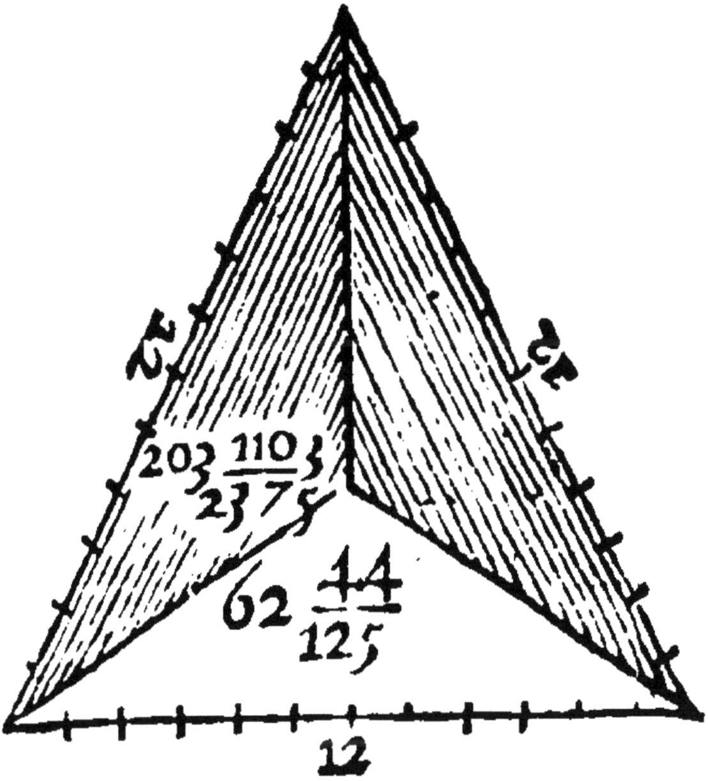

Here the content of the triangle by the 18 e xij, is found to be 62.44/125 for the base of the pyramis. The altitude is 9.15/19: Because by the 12 e xviij, the side is of treble power to the ray. But if from 144, the quadrate of 12 the side, you take the subtriple *i.* 48, the remainder 96, by the 9 e xij, shall be the square of the heighth. And the side of the quadrate shall be 9.15/19. Now the third part of 9.15/19 is 3.5/19. And the plaine of 62.44/125 and 3.5/19, shall be 203.1103/2375 for the solidity of the pyramis.

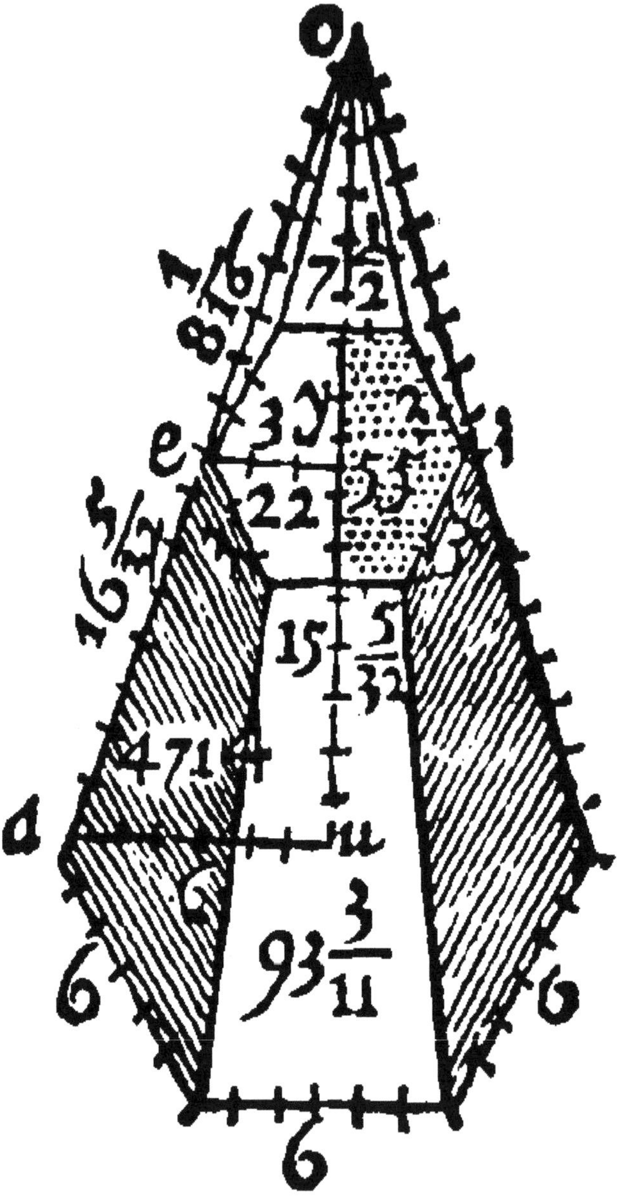
o
7 1/2
e
22
15 5/32
a
u
93 3/11
6
6
6

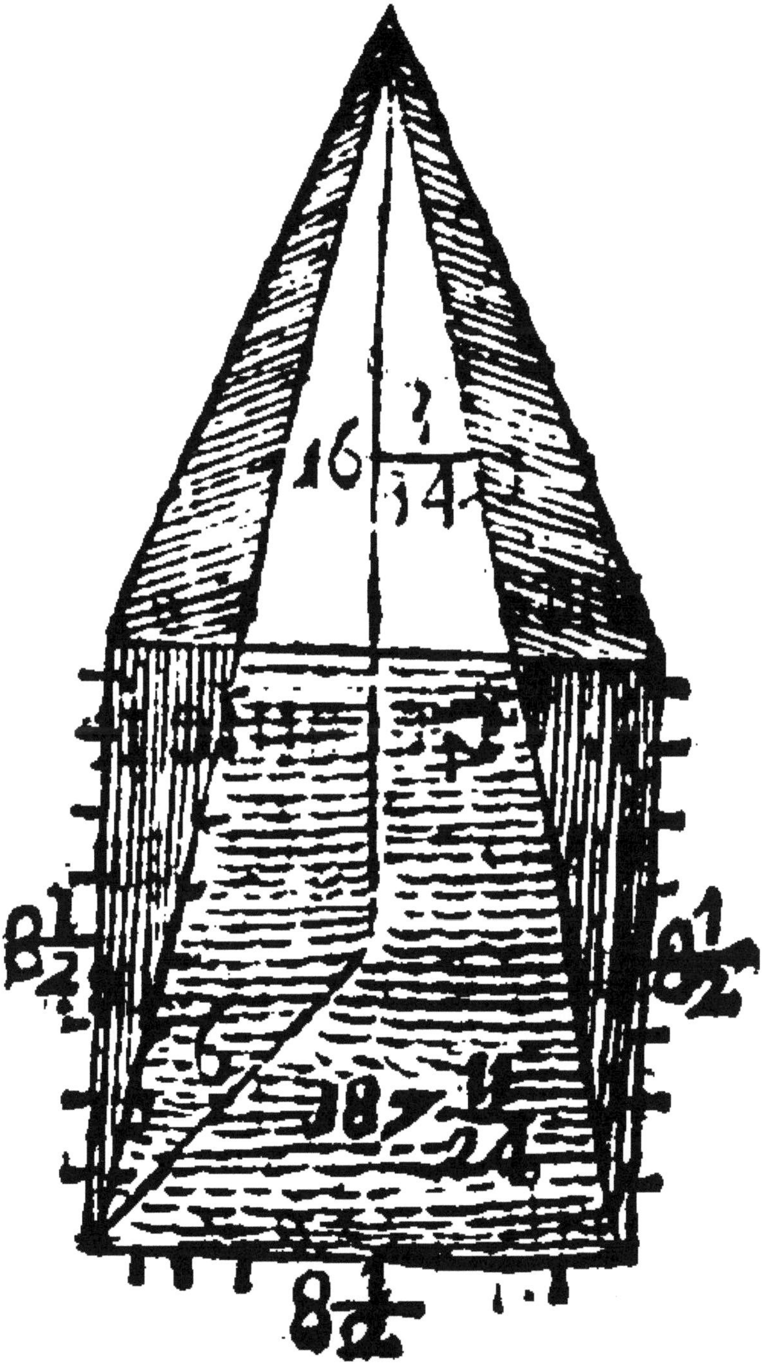
16

So in the example following, Let 36, the quadrate of 6 the ray, be taken out of 292.9/1156 the quadrate of the side 17.3/34 the side 16.3/34 of 256.9/1156 the remainder shall be the height, whose third part is 5.37/102; the plaine of which by the base 72.1/4 shall be 387.11/24 for the solidity of the pyramis given.

If the pyramis be unperfit, first measure the whole, and then that part which is wanting: Lastly from the whole subtract that which was wanting, and the remaine shall be the solidity of the unperfect pyramis given: As here, let *ao*, the side of the whole be 16.5/12, *eo* the side of the particular be 8.1/16. Therefore the perpendicular of the whole *ou*, shall be 15.5/32: Whose third part is 5.5/96: Of which, and the base 93.3/11 the plaine shall be 471.134/1056 for the whole pyramis. But in the lesser pyramis, 9 the square of the ray 3, taken out of 65.1/256 the quadrate of the side 8.1/16 the remaine shall be 56.1/256; whose side is almost 7-1/2 for the heighth. The third part of which is 2-1/2. The base likewise is almost 22. The plaine of which two is 55, for the solidity of the lesser pyramis: And 471 - 55 is 416, for the imperfect pyramis.

After this manner you may measure an imperfect Prisma.

8. *Homogeneall Prisma's of equall heighth are one to another as their bases are one to another, 29, 30, 31, 32 p xj.*

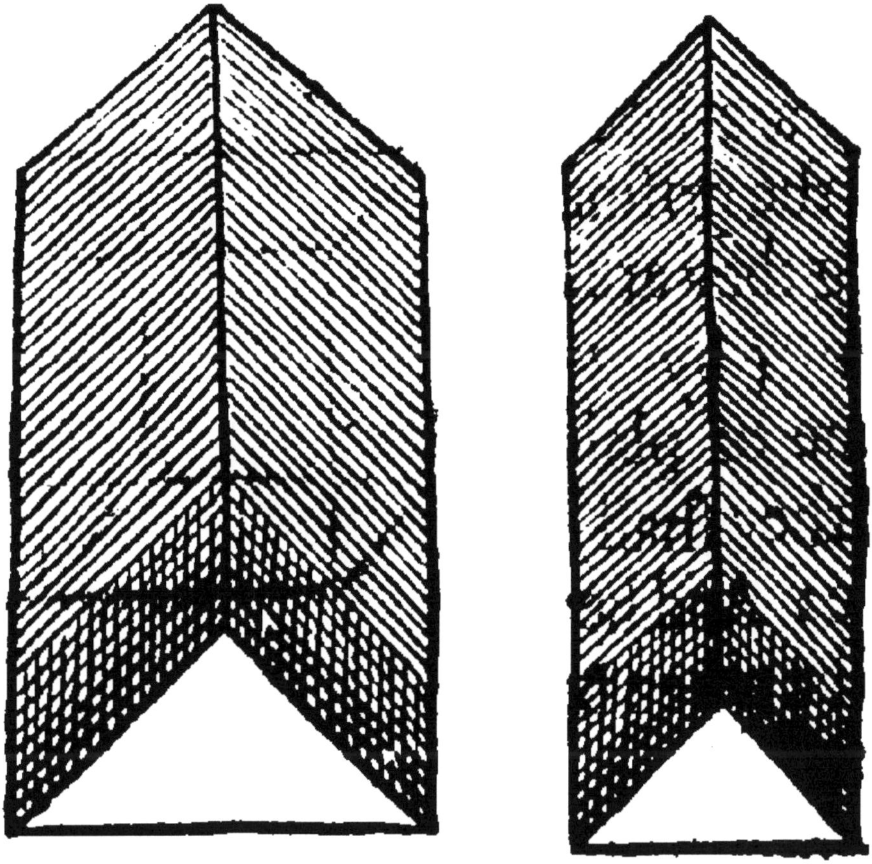

The reason is, because they consist equally of like number of pyramides. Now it is required that they be homogeneall or of like kindes; Because a Pentaedrum with an Hexaedrum will not so agree.

This element is a consectary out of the 16 e iiij.

And

9. *If they be reciprocall in base and heighth, they are equall.*

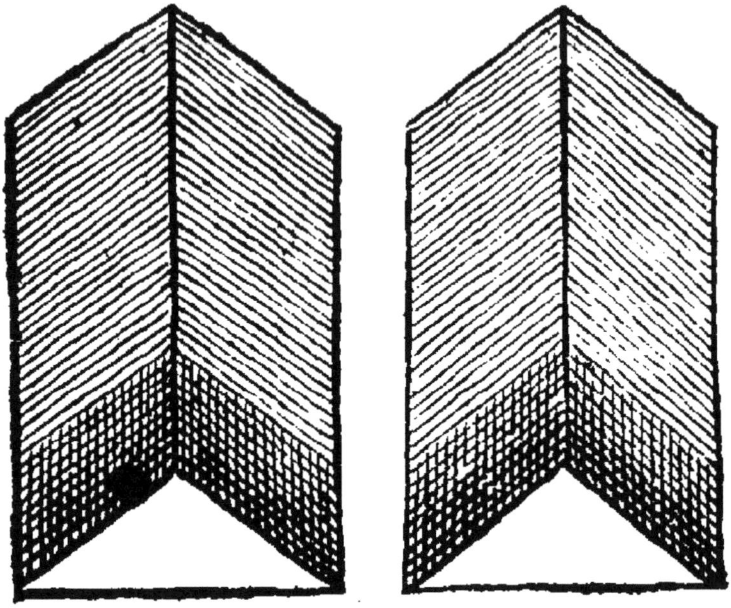

This is a Consectary out the 18 e iiij.

And

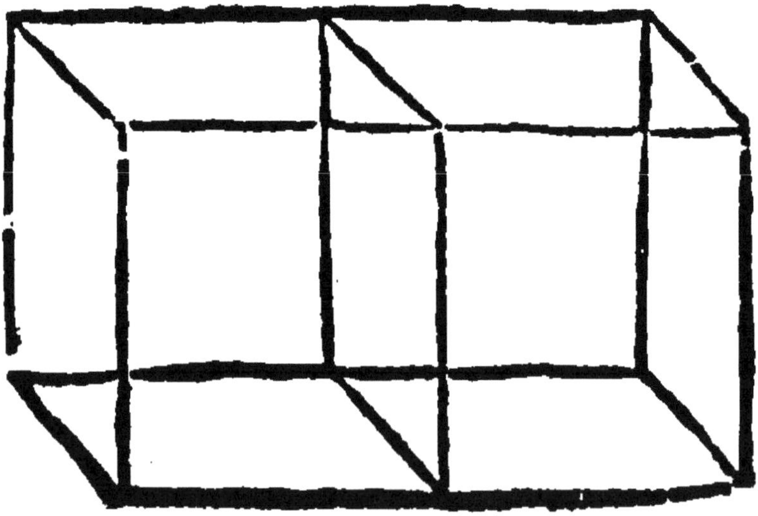

10. *If a Prisma be cut by a plaine parallell to his opposite flattes, the segments are as the bases are. 25 p. xj.*

The segments are homogeneall because the prismas. Therefore seeing they are of equall heighth (by the heighth I meane of plaine dividing them) they shall be as their bases are: And here the bases are to be taken opposite to the heighth.

11. *A Prisma is either a Pentaedrum, or Compounded of pentaedra's.*

Here the resolution sheweth the composition.

12 *If of two pentaedra's, the one of a triangular base, the other of a parallelogramme base, double unto the triangular, be of equall heighth, they are equall 40. p xj.*

The cause is manifest and briefe: Because they be the halfes of the same prisma: As here thou maist perceive in a prisma cut into two halfes by the diagoni's of the opposite sides.

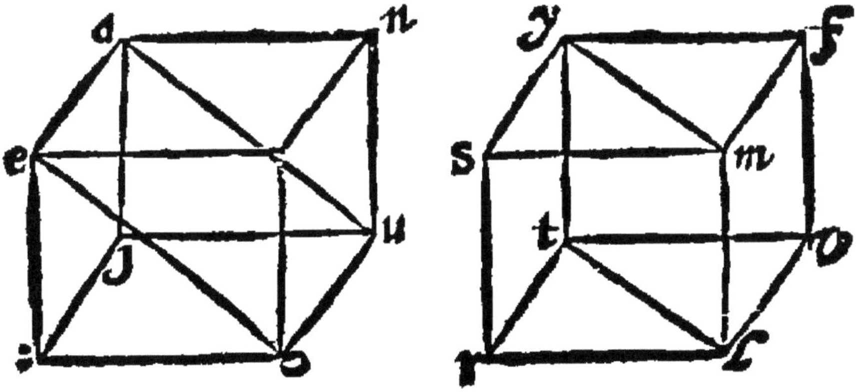

Euclide doth demonstrate it thus: Let the Pentaedra's *aeiou*, and *ysrlm,* be of equall heighth: the first of a triangular base *eio*: The second of a parallelogramme base *sl*, double unto the triangular. Now let both of them be double and made up, so that first be *aeioun.* The second *ysrlvf*. Now againe, by the grant, the base *sl*, is the double of the base *eio*,: whose double is the base *eo*, by the 12 e x. Therefore the bases *sl*, and *eo*, are equall: And therefore seeing the prisma's, by the grant, here are of equall heighth, as the bases by the conclusion are equall, the prisma's are equall; And therefore also their halfes *aeiou*, and *ysmlr*, are equall.

The measuring of a pentaedrall prisma was even now generally taught: The matter in speciall may be conceived in these two examples following.

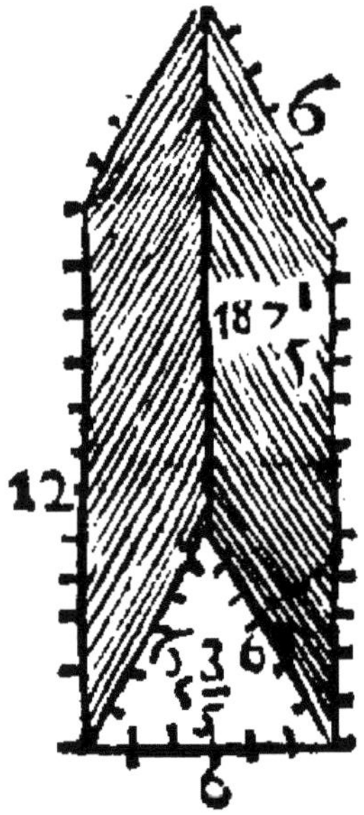

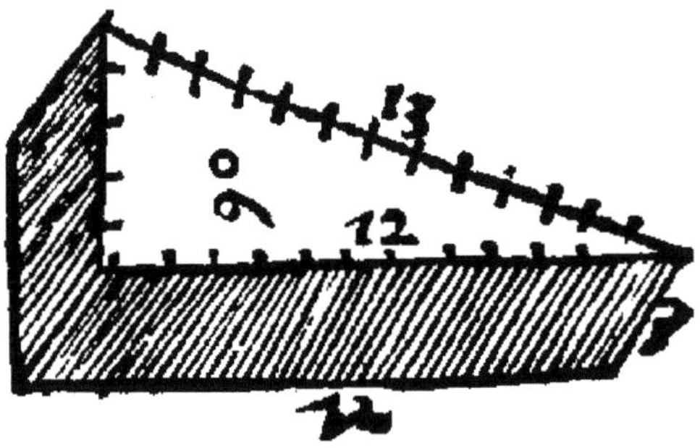

The plaine of 18. the perimeter of the triangular base, and 12, the heighth is 216. This added to the triangular base, 15.18/31. or 15.3/5, almost twise taken, that is, 31.1/5, doth make 247.1/5, for the summe of the whole surface. But the plaine of the same base 15.3/5, and the heighth 12. is 187.1/5, for the whole solidity.

So in the pentaedrum, the second prisma, which is called *Cuneus*, (a wedge) of the sharpnesse, and which also more properly of cutting is called a prisma, the whole surface is 150, and the solidity 90.

13 *A prisma compounded of pentaedra's, is either an Hexaedrum or Polyedrum: And the Hexaedrum is either a Parallelepipedum or a Trapezium.*

14 *A parallelepipedum is that whose opposite plaines are parallelogrammes ê 24. p xj.*

Therefore a Parallelepipedum in solids, answereth to a Parallelogramme in plaines. For here the opposite *Hedræ* or flattes are parallell: There the opposite sides are parallell.

Therefore

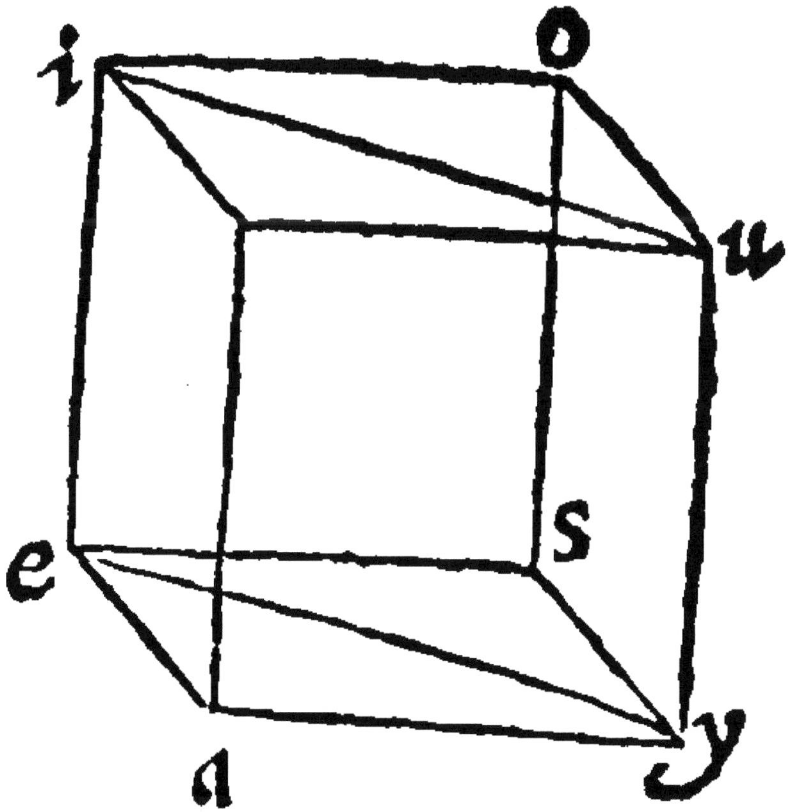

15 *It is cut into two halfes with a plaine by the diagonies of the opposite sides. 28 p xj. It answereth to the 34. p j.*

Let the Prisma be of sixe bases *ai, yo, ye, ui, si, au*. The diagonies doe cut into halfes, by the 10. e x. the opposite bases: And the other opposite bases or the two prisma's cut, are equall by the 3 e. Wherefore two prisma's are comprehended of bases, equall both in multitude and magnitude: therfore they are equall.

And

16 *If it be halfed by two plaines halfing the opposite sides, the common bisection and diagony doe halfe one another 39. p xj.*

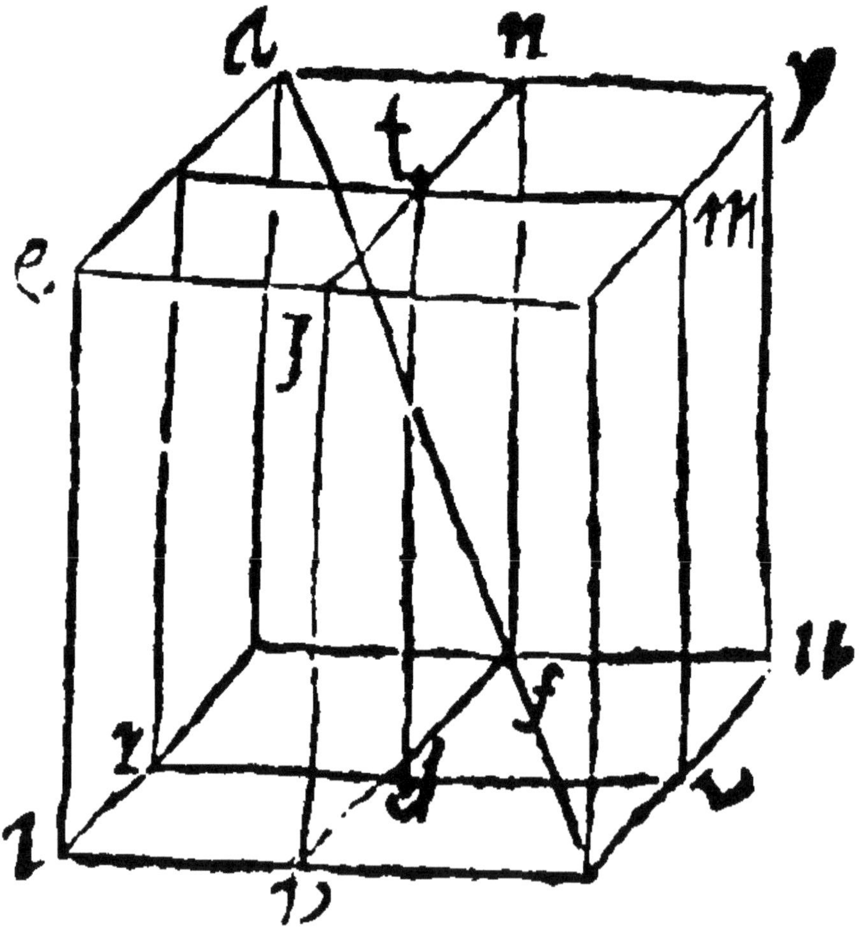

Because here the diameters (such as is that bisection) are halfed betweene themselves [or doe halfe one another.] Let the parallelepipedum *aeiouy*, be cut in to *y* the halfs by two plains, fro *srlm*, *uivf*, halfing the opposite sides: Here the common section *ts*, and the diagony *ao*, doe cut one another.

17 *If three lines be proportionall, the parallelepipedum of meane shall be equall to the equiangled parallelepipedum of all them. è 36. p xj.*

It is a consectary out of the 8 e.

18 *Eight rectangled parallelepiped's doe fill a solid place.*

19 *The Figurate of a rectangled parallelepipedum is called a solid, made of three numbers 17. d vij.*

As if thou shalt multiply 1, 2, 3. continually, thou shalt make the solid 6. Item if thou shalt in like manner multiply 2, 3, 4. thou shalt make the solid 24. And the sides of that solid 6 solid shall be 1, 2, 3. Of 24, they shall be 2, 3, 4.

Therefore

20 *If two solids be alike, they have their sides proportionalls, and two meane proportionalls 21 d vij, 19. 21. p viij.*

It is a consectary out of the 5 e xxij. But the meane proportionalls are made of the sides of the like solids, to wit, of the second, third, and fourth: Item of the third, fourth, and fifth, as here thou seest.

Of Geometry *the twentie fourth Book.* *Of a Cube.*

1 *A Rightangled parallelepipedum is either a Cube, or an Oblong.*

2 *A Cube is a right angled parallelepipedum of equall flattes, 25. d. xj.*

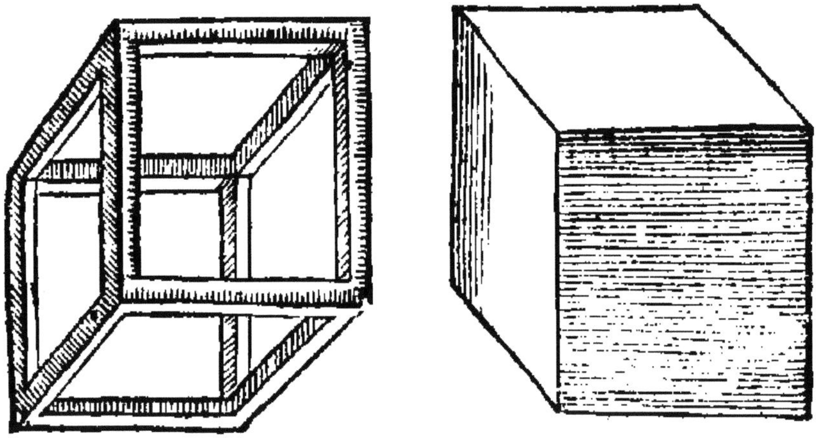

As here thou seest in these two figures.

Therefore

3 *The sides of a cube are 12. the plaine angles 24. the solid 8.*

Therefore

4 *If sixe equall quadrates be joyned with solid angles, they shall comprehend a cube.*

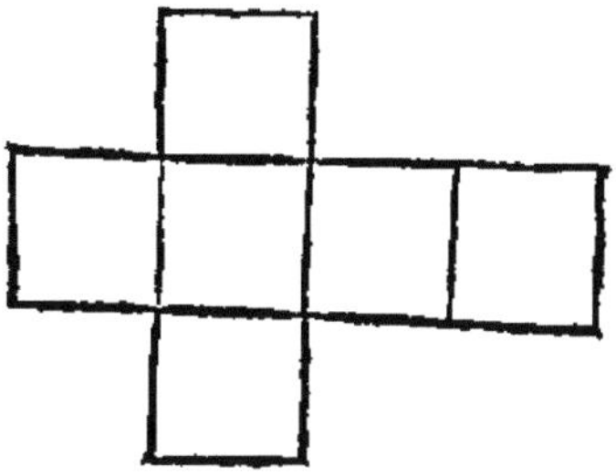

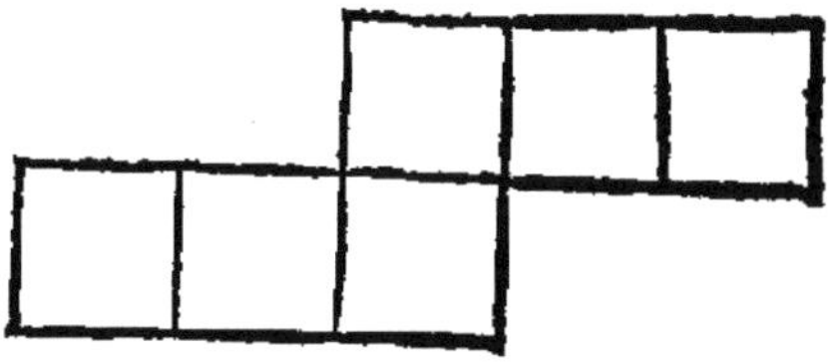

As here in these two examples.

Therefore

5 *If from the angles of a quadrate, perpendiculars equall to the sides be tied together aloft, they shall comprehend a Cube. è 15 p xj.*

It is a consectary following upon the former consectary: For then shall sixe equall quadrates be knit together:

6 *The diagony of a Cube is of treble power unto the side.*

For the Diagony of a quadrate is of double power to the side, by the 12 e, xij. And the Diagony of a Cube is of as much power as the side the diagony of the quadrate, by the same *e*. Therefore it is of treble power to the side.

7 *If of foure right lines continually, proportionally the first be the halfe of the fourth, the cube of the first shall be the halfe of the Cube of the second è 33 p xj.*

It is a consectary out of the 25 e, iiij. From hence *Hippocrates* first found how to answer *Apollo's* Probleme.

8 *The solid plaine of a cube is called a Cube, to wit, a solid of equall sides. 19, d vij.*

Therefore

9 *It is made of a number multiplied into his owne quadrate.*

So is a Cube made by multiplying a number by it selfe, and the product againe by the first. Such are these nine first cubes made of the nine first Arithmeticall figures.

1	2	3	4	5	6	7	8	9	Latera.
1	4	9	16	25	36	49	64	81	Quadrates.
1	8	27	64	125	216	343	512	729	Cubes.

This is the generall invention of a Cube, both Geometricall and Arithmeticall.

10 *If a right line be cut into two segments, the Cube of the whole shall be equall to the Cubes of the segments, and a double solid thrice comprehended of the quadrate of his owne segment and the other segment.*

As for example, the side 12, let it be cut into two segments 10 and 2. The cube of 12. the whole, which is 1728, shall be equall to two cubes 1000, and 8 made of the segments 10. and 2. And a double solid; of which the first 600. is thrise comprehended of 100. the quadrate of his segment 10. and of 2. the other segment: The second 120. is thrice comprehended of 4, the quadrate of his owne segment, and of 10. the other segment. Now 1000 + 600 + 120. + 8, is equall to 1728: And therefore a right. &c.

But the genesis of the whole cube will make all this whole matter more apparant, to wit, how the extreme and meane solids are made. Let therefore a cube be made of three equall sides, 12, 12, and 12: And first of all let the second side be multiplied by the first, after this manner: And not adding the severall figures of the same degree, as was taught in multiplication, but multiply againe every one of them by the other side; and lastly, add the figures of the same degrees severally, thus:

12
12
———
24
12
12
———
48
24
24
12
———
1,6,12,8

Or thus,

12
12
———
4
20
20
100
———
12
———
8
40
40
40
200
200
200
1000
———
1728

Therefore

11. *The side of the first severall cube is the other side of the second solide: And the quadrate of the same side is the other side of the first solide, whose other side is the side of the second cube; and the quadrate of the same other side is the other side of the second solid.*

In that equation therefore of foure solids with one solid, thou shalt consider a peculiar making and composition: First that the last cube be made of the last segment 2: Then that the second solid of 4, the quadrate of his owne segment, and of 10, the other segment be thrise comprehended: Lastly that the first solid of 100, the square of his owne segment 10 and the other segment 2, be also thrice comprehended: Lastly, that the Cube 1000, be made of the greater segment 10. Out of this making &c.

And thus much of the Cube: Of other sorts of parallelepipedes, as of the Oblong, the Rhombe, the Rhomboides, and of the Trapezium, and many flatted pentaedra's there is no peculiar stereometry. The measuring of a Prisma hath in the former beene generally declared, and is now onely farther be made more plaine by speciall examples; as here:

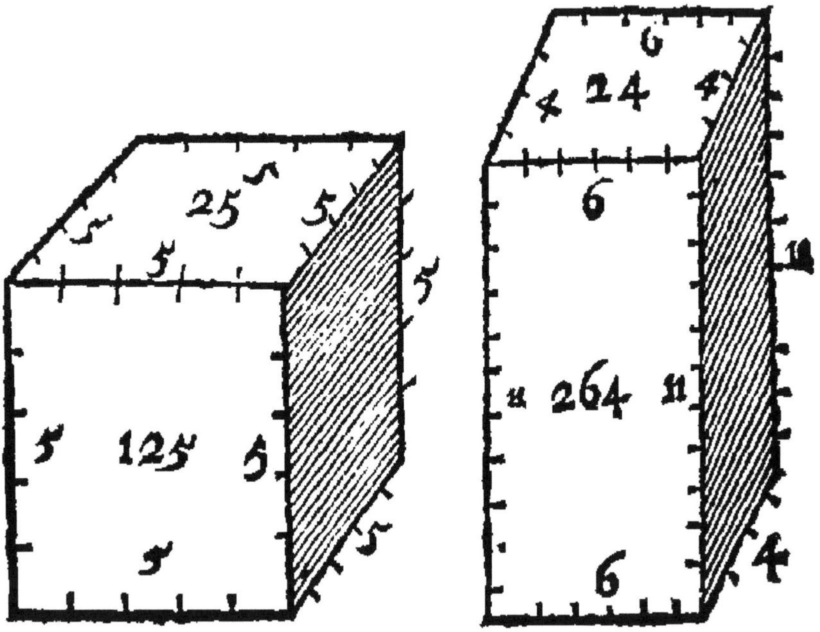

The plaine of the perimeter of the base 20, and the altitude 5 is 100. This added to 25 and 25, both the bases that is to 50, maketh 150, for the whole surface. Now the plaine of 25 the base, and the heighth 5 is 125, for the whole solidity.

So in the Oblong, the plaine of the base's perimeter 20, and the heighth 11, is 220, which added to the bases 24 and 24, that is 48, maketh 268, for the whole surface. But the plaine of the base 24, and the height 11, is 264, for the solidity.

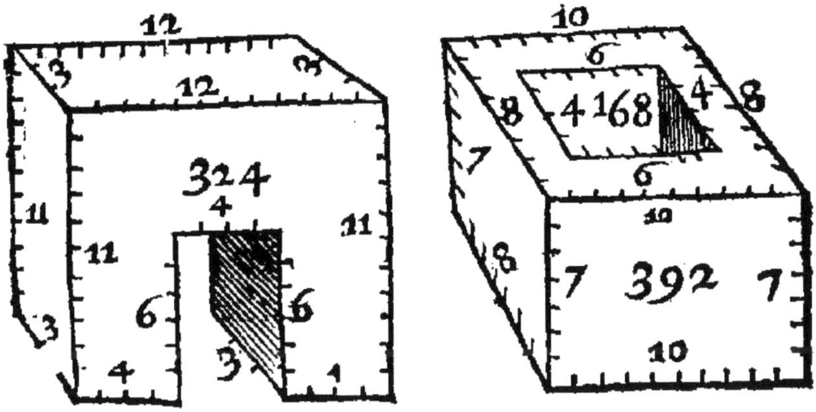

The same also Geodesie or manner of measuring is used in the measuring of rectangled walls or gates and doores, which have either any window, or any hollow or voyde space cut out of them, if those voyde places be taken out of them; as here thou seest in the next following example. The thickenesse is 3 foote; the breadth 12, the heighth 11. Therefore the whole solidity is 396. Now the Gate way is of thickenesse 3 foote, of breadth 4: of heighth 6. And therefore the whole solidity of the Gate is 72 foote. But 396 - 72 are 314. Therefore the solidity of the rest of the wall remaining is 324.

In the second example, the length is 10. The breadth 8, the heighth 7. Therefore the whole body if it were found, were 560 foote. But there is an hollow in it, whose length is 6, breadth 5, heighth 7. Therefore the cavity or hollow place is 168. Now 560 - 168 is 392, for the solidity of the rest of the sound body.

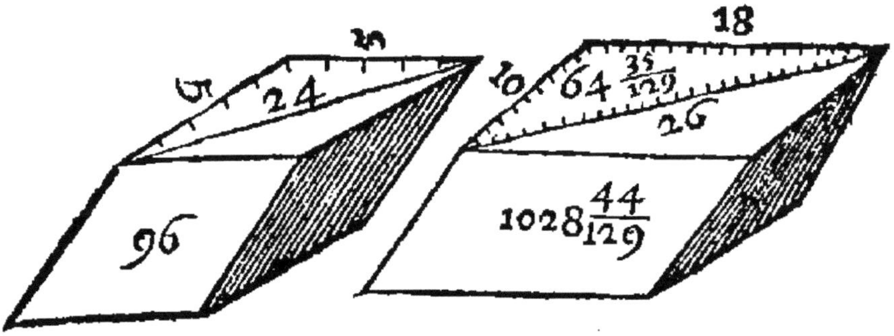

Thus are such kinde of walls whether of mudde, bricke, or stone, of most large houses to bee measured. The same manner of Geodesy is also to be used in the measuring of a Rhombe, Rhomboides, Trapezium or mensall, and any kinde of multangled body. The base is first to be measured, as in the former: Then out of that and the heighth the solidity shall be manifested: As in the Rhombe the base is 24, the heighth 4. Therefore the solidity is 96.

In the Rhomboides, the base is 64.35/129: The heigh 16. Therefore the solidity is 1028.44/129.

The same is the geodesy of a trapezium, as in these examples: The surface of the first is 198: The solidity 192.1/2.

The surface of the second is 158.3/49: The solidity is 91.29/49.

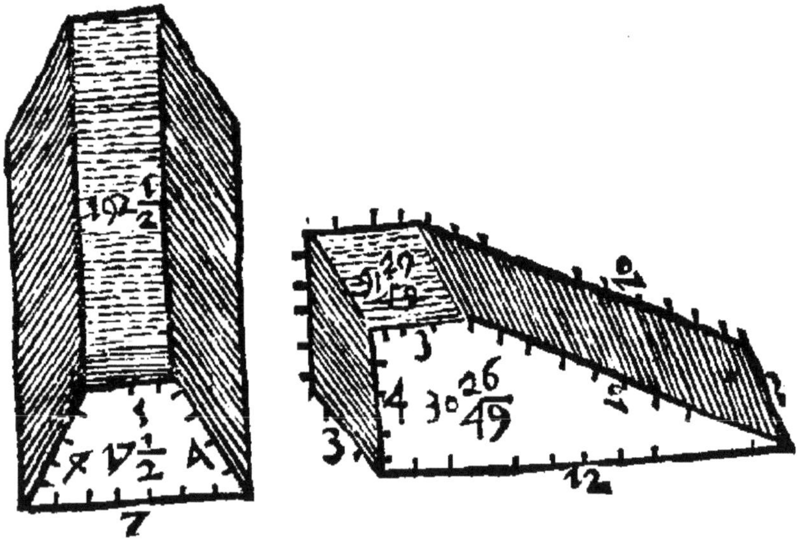

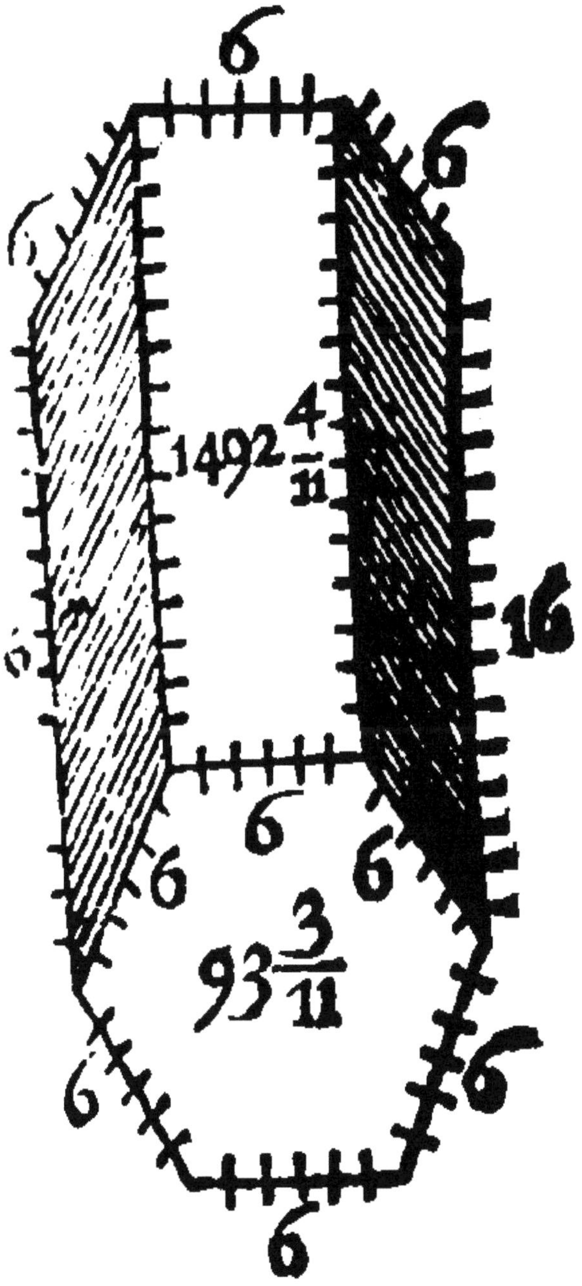

The same shall be also the geodesy of a many flatted Prisma: As here thou seest in an Octoedrum of a sexangular base: The surface shall bee 762.6/11: The solidity 1492.4/11.

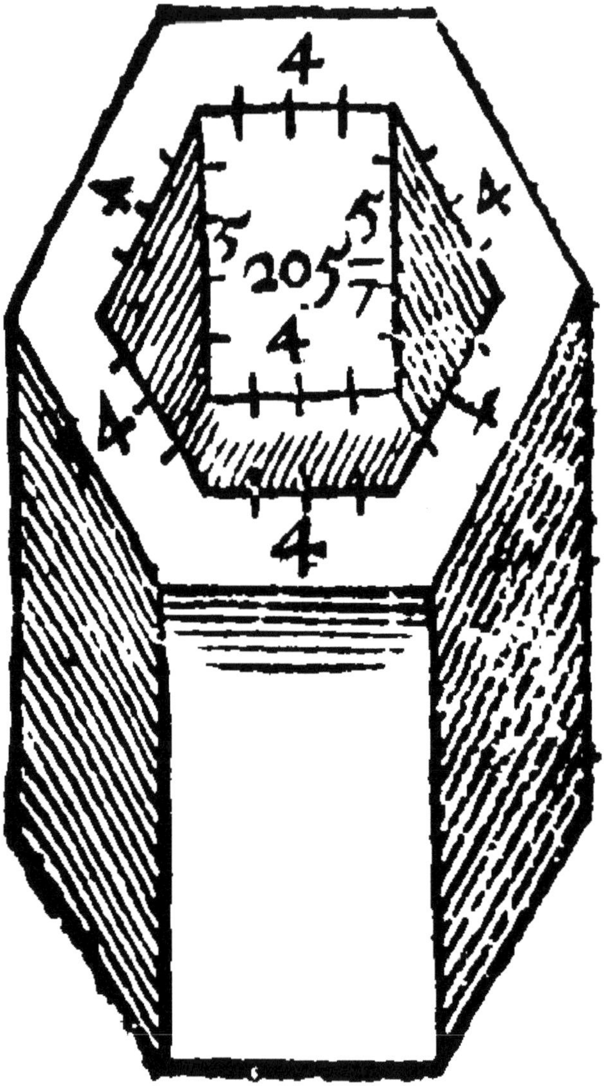

And from hence also may the capacity or content of vessels or measures, made after any manner of plaine solid bee esteemed and judged of as here thou seest. For here the plaine of the sexangular base is 41.1/7; (For the ray, by the 9 e xviij, is the side:) and the heighth 5, shall be 205.5/7. Therefore if a cubicall foote doe conteine 4 quarters, as we commonly call them, then shall the vessell conteine 822.6/7 quartes, that is almost 823 quartes.

Of Geometry *the twenty fifth Booke; Of mingled ordinate* Polyedra's.

1. *A mingled ordinate polyedrum is a pyramidate, compounded of pyramides with their toppes meeting in the center, and their bases onely outwardly appearing.*

Seeing therefore a Mingled ordinate pyramidate is thus made or compounded of pyramides the geodesy of it shall be had from the Geodesy of the pyramides compounding it: And one Base multiplyed by the number of all the bases shall make the surface of the body. And one Pyramis by the number of all the pyramides; shall make the solidity.

2 *The heighth of the compounding pyramis is found by the ray of the circle circumscribed about the base, and by the semidiagony of the polyedrum.*

The base of the pyramis appeareth to the eye: The heighth lieth hidde within, but it is discovered by a right angle triangle, whose base is the semidiagony or halfe diagony, the shankes the ray of the circle, and the perpendicular of the heighth. Therefore subtracting the quadrate of the ray, from the quadrate of the halfe diagony the side of the remainder, by the 9 e xij. shall be the heighth. But the ray of the circle shall have a speciall invention, according to the kindes of the base, first of a triangular, and then next of a quinquangular.

3 *A mingled ordinate polyedrum hath either a triangular, or a quinquangular base.*

The division of a Polyhedron ariseth from the bases upon which it standeth.

4 *If a quadrate of a triangular base be divided into three parts, the side of the third part shall be the ray of the circle circumscribed about the base.*

As is manifest by the 12 e. xviij. And this is the invention or way to finde out the circular ray for an octoedrum, and an icosoedrum.

5 *A mingled ordinate polyedrum of a triangular base, is either an Octoedrum, or an Icosoedrum.*

This division also ariseth from the bases of the figures.

6 *An octoedrum is a mingled ordinate polyedrum, which is comprehended of eight triangles. 27 d xj.*

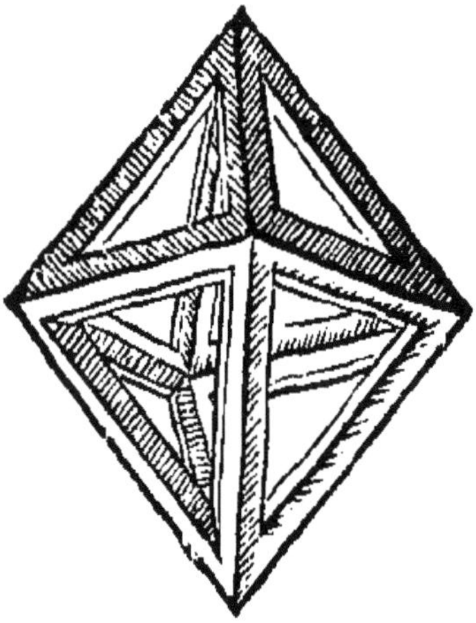

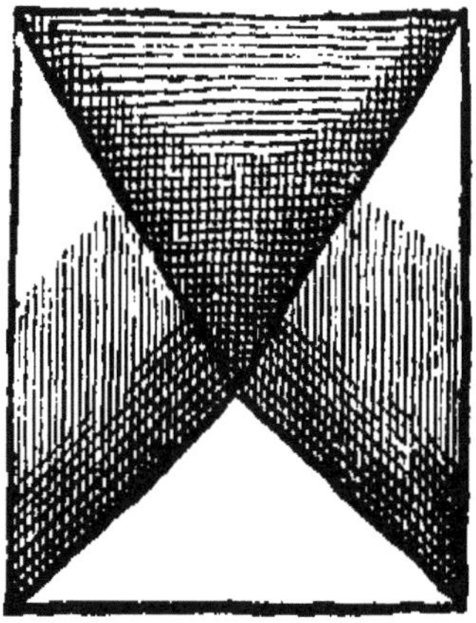

As here thou seest, in this Monogrammum and solidum, that is lines and solid octahedrum.

Therefore

7 *The sides of an octoedrum are 12. the plaine angles 24, and the solid 6.*

And

8 *Nine octoedra's doe fill a solid place.*

For foure angles of a Tetraedrum are equall to three angles of the Octoedrum: And therefore 12. are equall to nine. Therefore nine angles of an octaedrum doe countervaile eight solid right angles.

And

9 *If eight triangles, equilaters and equall be joyned together by their edges; they shall comprehend an octaedreum.*

This construction is easie, as it is manifest in the example following: Where thou seest as it were two equilater and equall triangles of a double pentaedrum to cut one another.

10 *If a right line of each side perpendicular to the center of a quadrate and equall to the halfe diagony be tied together with the angles, it shall comprehend an octaedrum, 14. d xiij.*

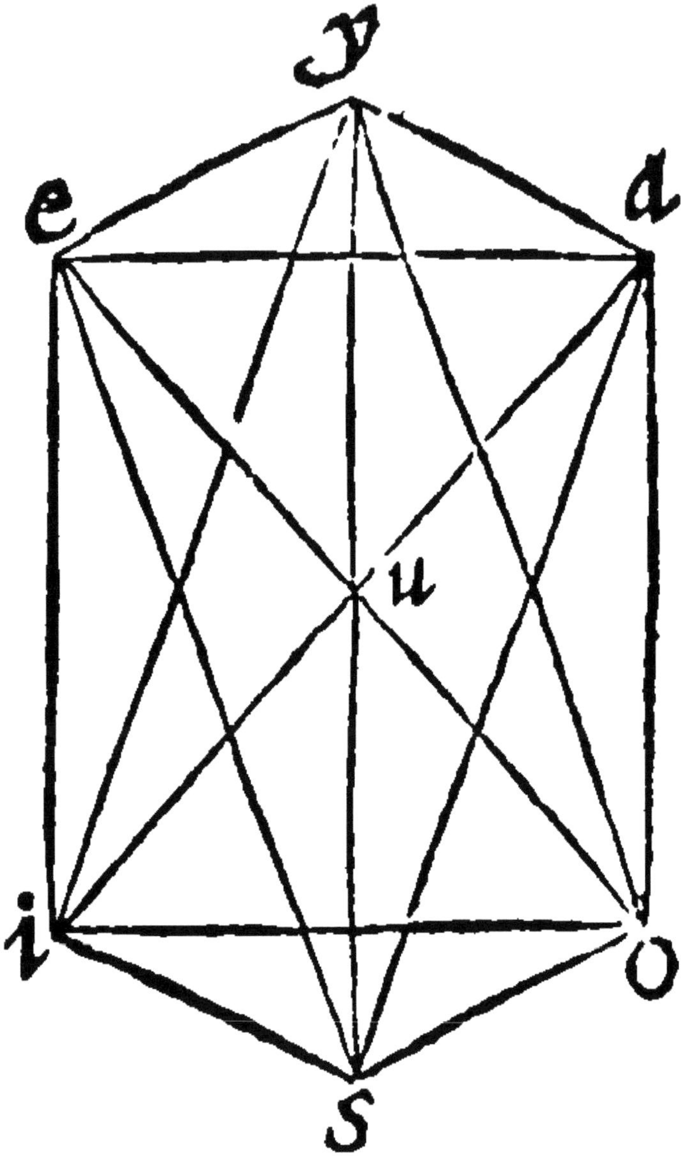

For the perpendicular *yu*, and *su*, with the semidiagoni's, *ua, uo, ui, ue*, shall be made equall by the 2 e vij, the eight sides *ya, ye, yo, yi, se, si, sa, so*; And also eight triangles.

Therefore;

11 *The Diagony of an octaedrum is of double power to the side.*

As is manifest by the 9 e xij.

And

12 *If the quadrate of the side of an octaedrum, be doubled, the side of the double shall be the diagony.*

As in the figure following, the side is 6. The quadrate is 36. the double is 72. whose side 8.8/17, is the diagony.

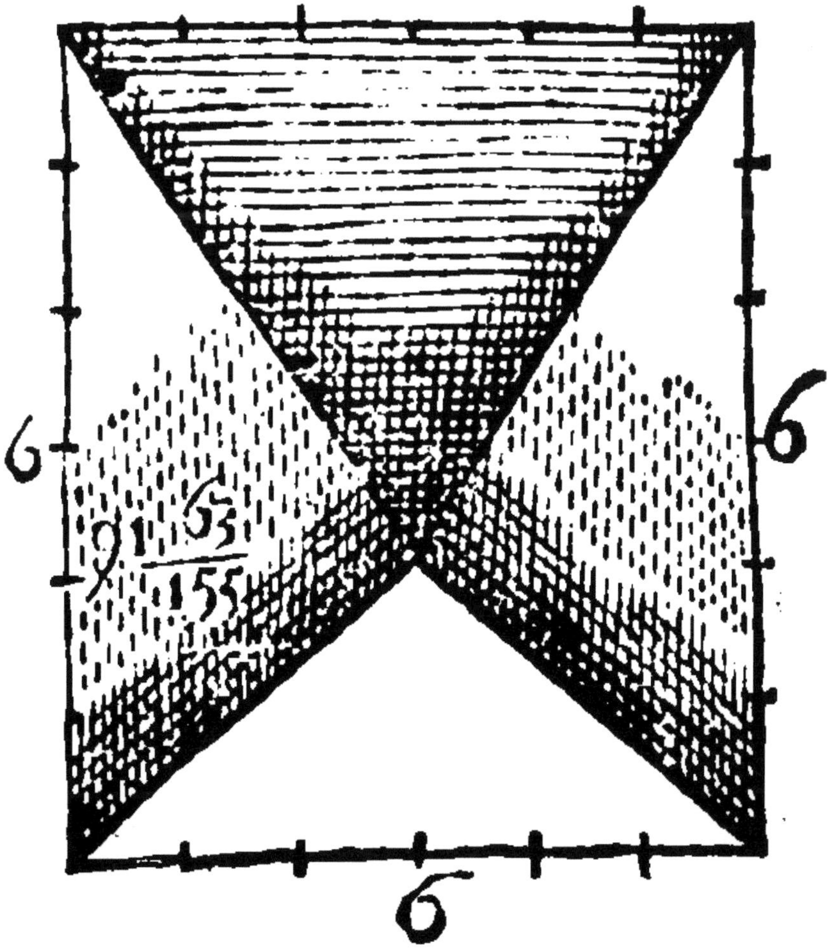

And from hence doth arise the geodesy of the octaedrum. For the semidiagony is 4.4/17. whose quadrate is 17.171/289. And the quadrate of 6, the side of the equilater triangle, being of treble power to the ray, by the 12 e, xviij. is 36. And the side of 12. the third part 3.3/7 is the ray of the circle. Wherefore 8.8/17. that is 5.21/289. is the quadrate of the perpendicular, whose side 2.1/5 is the height of the same perpendicular: whose third part againe 11/25. multiplied by 15.18/31. the triangular base doe make 11.66/155 for one of the eight pyramides: Therefore the same 11.66/155 multiplied by eight, shall make 91.63/155 for the whole octoedrum.

13 *An Icosaedrum is an ordinate polyedrum comprehended of 20 triangles 29 d xj.*

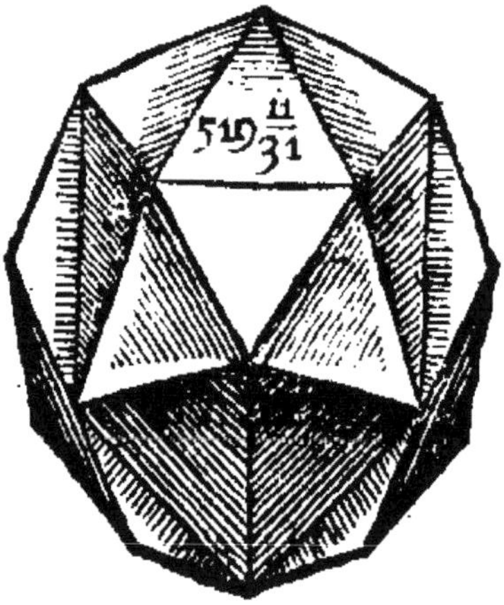

Therefore

14 *The sides of an Icosaedrum are 30. plaine angles 60. the solid 12.*

And

15 *If twentie ordinate and equall triangles be joyned with solid angles, they shall comprehend an Icosaedrum.*

This fabricke is ready end easie, as is to be seene in this example following.

16. *If ordinate figures, to wit, a double quinquangle, and one decangle be so inscribed into the same circle, that the side of both the quinquangle doe subtend two sides of the decangle, sixe right lines perpendicular to the circle and equall to his ray, five from the angles of one of the quinquangles, knit together both betweene themselves, and with the angles of the other quinquangle; the sixth from the center on each side continued with the side of the decangle, and knit therewith the five perpendiculars, here with the angles of the second quinquangle, they shall comprehend an icosaedrum. è 15 p xiij.*

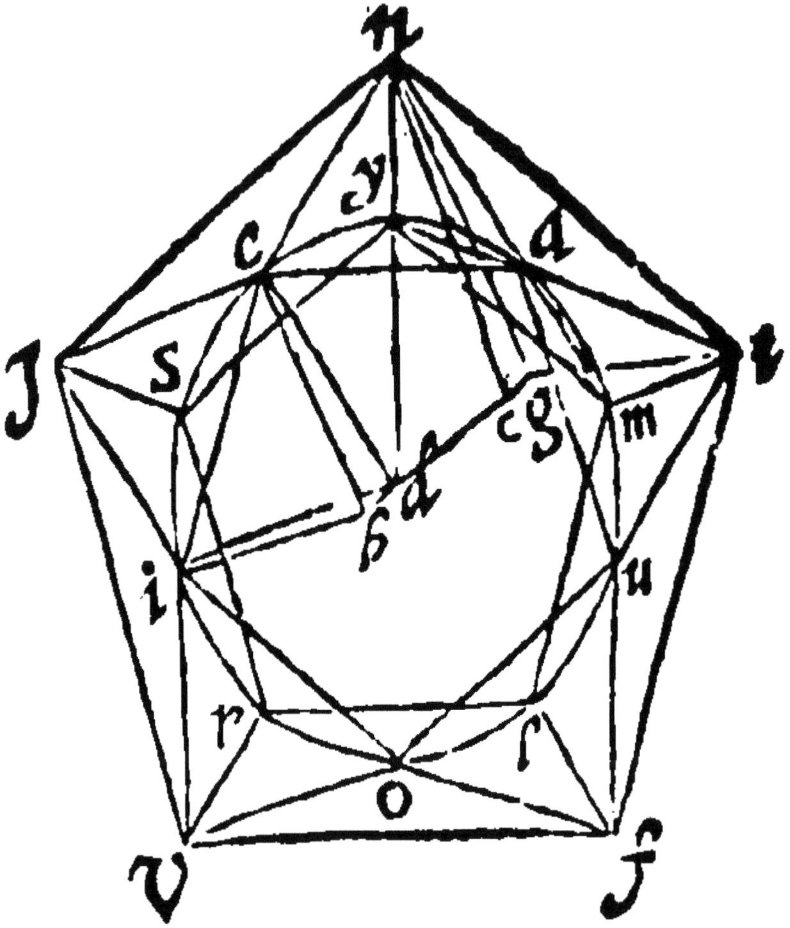

For there shall be made 20 triangles, both equilaters and equall. Let there be therefore two ordinate quinquangles, the first *aeiou*; The second *ysrlm*; each of whose sides let them subtend two sides of a decangle; to wit, *utym*, let it subtend *ya*, and *am*. Then let there be five perpendiculars from the angles of the second quinquangle *yj, sy, rv, lf, mt*. And let them be knit first one with another, by the lines *nj, jv, vf, ft, tn*. Secondarily, with the angles of the first quinquangle, by the lines *ne, ej, ji, iv, of, fu, ut, ta, an*. The sixth perpendicular from the center *d*, let it be *bg*, the ray *dc*, continued at each end with the side of the decangle, *cg*, and *db*, tied together about with the perpendiculars, as by the lines *ng, tg*: Beneath with the angles of the first quinquangle, as by the lines *be, bi*, and in other places in like manner, and let all the plaines be made up. This say I, is an Icosaedrum; And is comprehended of 20. triangles, both equilaters and equall. First, the tenne middle triangles, leaving out the perpendiculars, that they are equilaters and equall, one shall demonstrate, as *nat*. For *mt* and *yu*, because they are perpendiculars, they are also, by the 6 e xxj. parallells: And by the grant, equall. Therefore by the 27 e, v, *nt*, is equall to *ym*, the side of the quinquangle. Item *na*, by the 6 e xij. is of as great power, as both the shankes *ny*, and *ya*, that is, by the construction, as the sides of the sexangle and decangle: And, by the converse of the 15. e xviij. it is the side of the quinquangle. The same shall fall out of *ot*. Wherefore *nat*, is an equilater triangle. The same shall fall out of the other nine middle triangles, *nae*, *nej*, *eji*, *jiv*, *ivo*, *vof*, *fou*, *fut*, *uta*, *tan*.

In like manner also shall it be proved of the five upper triangles, by drawing the right lines *dy* and *cn* which as afore (because they knit together equall parallells, to wit, *dc*, and *yn*) they shall be equall. But *dy*, is the side of a sexangle: Therefore *cn*, shall be also the side of a sexangle: And *cg*, is the side of a decangle: Therefore *an*, whose power is equall to both theirs by the 9 e xij. shall by the converse of the 15 e xviij, be the side of a quinquangle: And in like manner *gt*, shall be concluded to be the side of a quinquangle. Wherefore *ngt*, is an equilater: And the foure other shall likewise be equilaters.

The other five triangles beneath shall after the like manner be concluded to be equilaters. Therefore one shall be for all, to wit, *ibe*, by drawing the raies *di*, and *de*. For *ib*, whose power, as afore, is as much as the sides of the sexangle, and decangle, shall be the side of the quinquangle: And in like sort *be*, being of equall power with *de*, and *do*, the sides of the sexangle and decangle, shall be the side of the quinquangle. Wherefore the triangle *ebi*, is an equilater: And the foure other in like manner may be shewed to be equilaters. Therefore all the side of the twenty triangles, seeing they are equall, they shall be equilater triangles: And by the 8 e, vij. equall.

17 *The diagony of an icosaedrū is irrational unto the side*.

This is the fourth example of irrationality, or incommensurability. The first was of the Diagony and side of a square or quadrate. The second was of the segments of a line proportionally cut. The third of the Diameter of a circle and the side of a quinquangle.

And

18 *The power of the diagony of an icosaedrum is five times as much as the ray of the circle.*

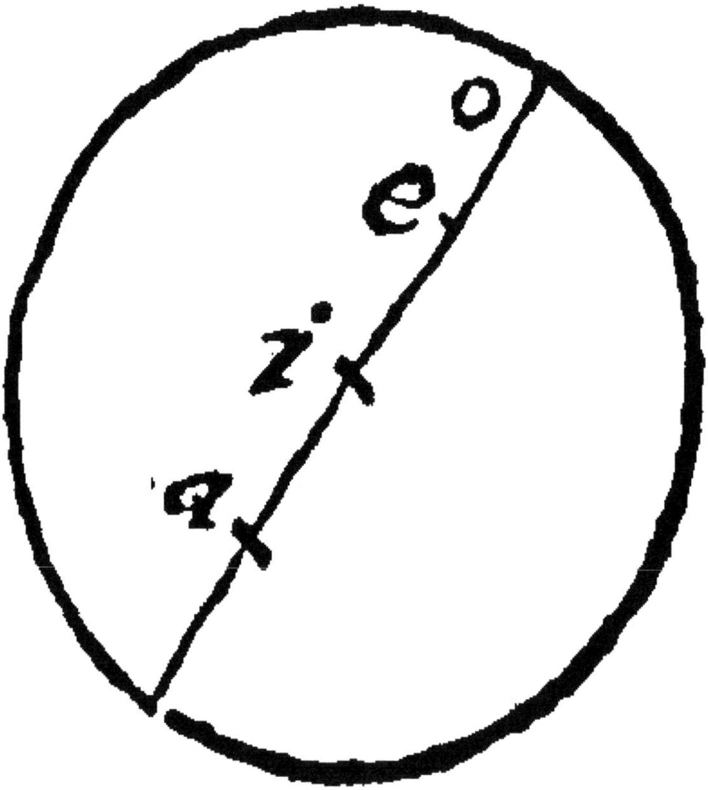

For by the 13 e, xviij. the line continually made of the side of the sexangle and decangle is cut proportionally, and the greater segment is the side of the sexangle: As here. Let the perpendicular *ae*, be cut into two equall parts in *i*. Then *eo*, that is the lesser segment continued with the halfe of the greater, that is, with *ie*. it shall by the 6 e xiiij, be of power five times so great as is the power of the same halfe. Therefore seeing that *io*, the halfe of the diagony is of power fivefold to the halfe: the whole diagony shall be of power fivefold to the whole cut.

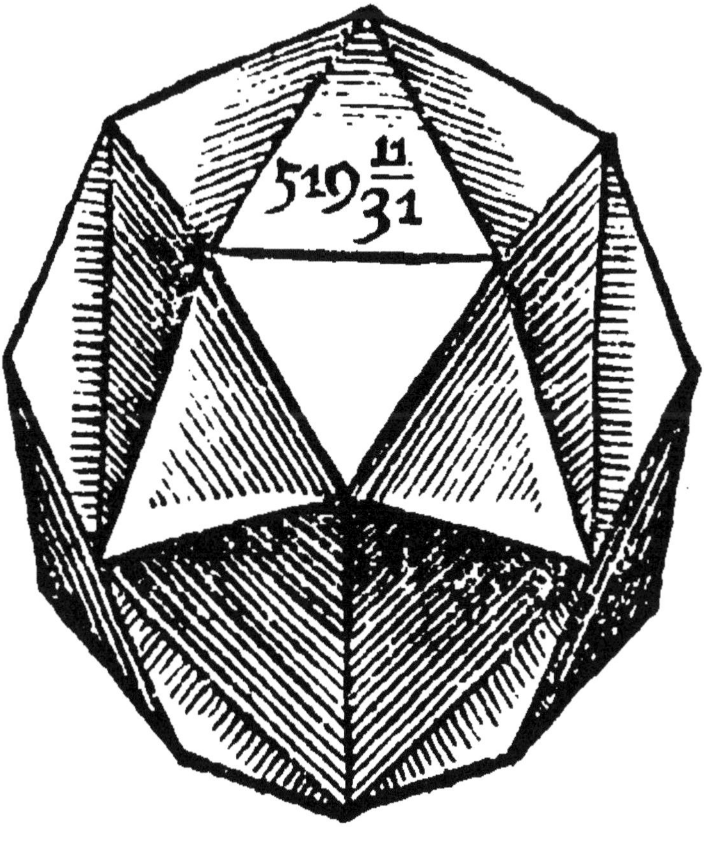

And from hence also shall be the geodesy of the Icosaedrum. For the finding out of the heighth of the pyramis, there is the semidiagony of the side of the decangle and the halfe ray of the circle: But the side of the decangle is a right line subtending the halfe periphery of the side of the quinquangle, or else the greater segment of the ray proportionally cut. For so it may be taken Geometrically, and reckoned for his measure. Therefore if the quadrate of the side of the decangle, be taken out of the quadrate of the side of the quinquangle, there shall by the 15 e xviij, remaine the quadrate of the sexangle, that is of the ray. The side of the decangle (because the side of the quinquangle here is 6) shall be 3.3/35 to wit a right line subtending the halfe periphery. Now the halfe ray shall thus be had. The quadrates of the quinquangle and decangle are 36, and 9.639/1225. And this being subducted fro that, the remaine 26.386/1225 by the 15 e xviij, shall be the quadrate or square of the sexangle: And the side of it, 5, and almost 5/7 shall be the ray: The halfe ray therefore shall be 2.6/7. To the side of the decangle 3.3/35 adde 2.6/7: the whole shal be 5.33/35 for the semi-diagony of the Icosaedrum. The ray of the circle circumscribed about the triangle, is by the 12 e xviij, the same which was before 3.3/7 to wit of the quadrate 12. Therefore if the quadrate of the circular ray, be taken out of the quadrate of the halfe diagony, there shall remaine the quadrate of the heighth and perpendicular: the quadrate of the halfe-diagony is 35.389/1225: the quadrate of the circular ray is 12. This taken out of that beneath 23.639/1225: whose side is almost 5, for the perpendicular and heighth proposed: From whence now the Pyramis is esteemed. The case of a triangular pyramis is 15.18/31. The Plaine of this base and the third part of the heighth is 25.30/31 for the solidity of one Pyramis. This multiplyed by 20 maketh 519.11/31 for the summe or whole solidity of the Icosaedum. And this is the geodesy or manner of measuring of an Icosaedrum.

19. *A mingled ordinate polyedrum of a quinquangular base is that which is comprehended of 12 quinquangles, and it is called a Dodecaedrum.*

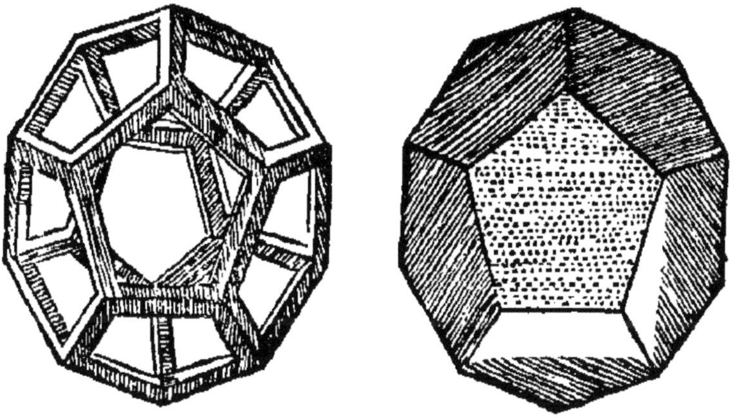

Therefore

20. *The sides of a Dodecaedrum are 30, the plaine angles 60. the solid 20.*

And

21. *If 12 ordinate equall quinquangles be joyned with solid angles, they shall comprehend a Dodecaedrum.*

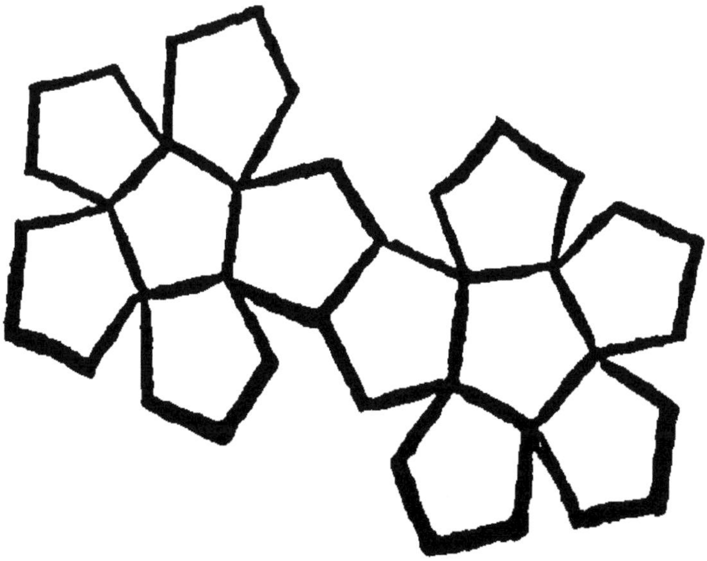

As here thou seest.

22. *If the sides of a cube be with right lines cut into two equall parts, and three bisegments of the bisecants in the abbuting plaines, neither meeting one the other, nor parallell one unto another, two of one, the third of that next unto the remainder, be so proportionally cut that the lesser segments doe bound the bisecant: three lines without the cube perpendicular unto the sayd plaines from the points of the proportionall sections, equall to the greater segment knit together, two of the same bisecant, betweene themselves and with the next angles of cube; the third with the same angles, they shall comprehend a dodecaedrum. 17 p xiij.*

Let there be two plaines for a cube for all, that one quinquangle for twelve may be described, and they abutting one upon another, aeio, *and* euyi, *having their sides halfed by the bisecantes,* sr, lm, rn, jv: *And the three bisegments or portions of the bisegments* lm, *and* rn, *neither concurring or meeting, nor parallell one to another; two of the said* lm, *to wit,* fl, *and* fm: *The third next unto the remainder, that is* lr. *And let each bisegment be cut proportionally in the points* d, c, g; *so that the lesser segments doe bound the bisecant, to wit,* dl, cm, *and* gr. *Lastly let there be three perpendiculars from the points* db, cg, *to the said* d, cp, gz: *And the two first knit one to another, by* bp: *And againe with the angles of the cube, by* be, *and* pi: *The third knit with the same angles, by* ze, *and* zi: *And let all the plaines be made up. I say first, that the five sides* bp, pi, iz, ze, *and* eb *are equall; Because, every one of them severally are the doubles of the same greater segment. For in drawing the right lines* de *and* eg, ig, *it shall be plaine of two of them; And after the same manner of the rest. First therefore* cd, *and* bp, *are equall by the 6 e xxj, and by the 27 e v. Therefore* bp, *is the double of the greater segment.*

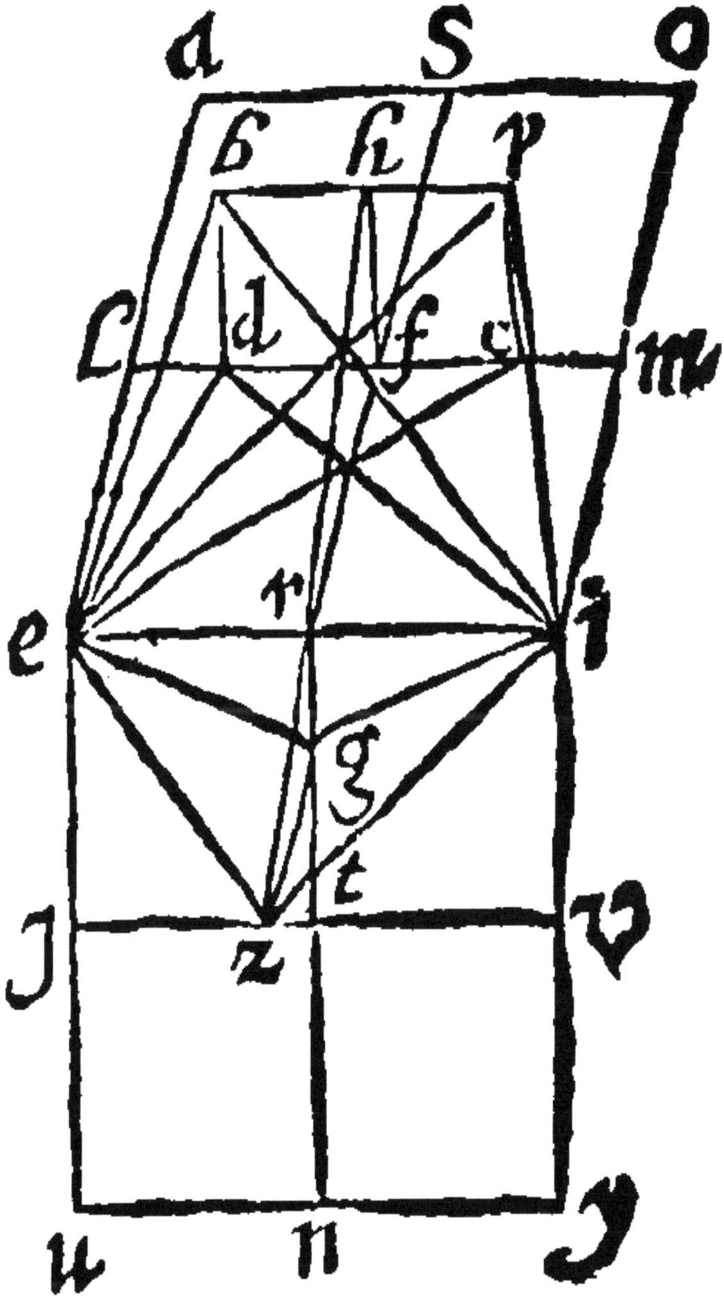

Then the whole *fl*, cut proportionally, and the lesser segment *dl*, they are by the 7 e xiiij, of treble power to the greater *fd*, that is, by the fabricke *db*. Therefore *le* wich is equall to *lf*, the line cut, and *ld*, are of treble power to the same *db*: But by the 9 e xij, *de* is of as much power as *le*, and *ld* too. Therefore *de* is of treble power to *db*. Therefore both *ed*, and *db*, are of quadruple power to *db*. But *be*, by the 9 e xij, is of as much power as *ed*, and *db*. And therefore *be*, is of quadruple value to *db*: And by the 14 e xij, it is the double of the said *db*. Therefore the two sides *eb*, and *bp*, are equall: And by the same argument *pi*, *iz*, and *ze*, are equall. Therefore the quinquangle is equilater.

I say also that it is a Plaine quinquangle: For it may be said to be an oblique quinquangle; and to be seated in two plaines. Let therefore *fh* be parallell to *db*, and *cp*: and be equall unto them. And let *hz*, be drawne: This *hz* shall be cut one line, by the 14 e vij. For as the whole *tr*, that is *rf*, is unto the greater segment that is to *fh*: so *fh*, that is *zg*, is unto *gr*. And two paire of shankes *fh*, *gr*, *fc*, *gz*, by the 6 e xxj, are alternely or crosse-wise parallell. Therefore their bases are continuall.

Hitherto it hath beene prooved that the quinquangle made is an equilater and plaine: It remaineth that it bee prooved to be Equiangled. Let therefore the right lines *ep*, and *ec*, be drawne: I say that the angles, *pbe*, and *ezi*, are equall: Because they have by the construction, the bases of equall shankes equall, being to wit in value the quadruple of *le*. For the right line *lf*, cut proportionally, and increased with the greater segment *df*, that is *fc*, is cut also proportionally, by the 4 e xiiij, and by the 7 e xiiij, the whole line proportionally cut, and the lesser segment, that is *cp*, are of treble value to the greater *fl*, that is of the sayd *le*. Therefore *el*, and *lc*, that is *ec*, and *cp*, that is *ep*, is of quadruple power to *el*: And therefore by the 14 e xij, it is the double of it: And *ei*, it selfe in like manner, by the fabricke or construction, is the double of the same. Therefore the bases are equall. And after the same manner, by drawing the right lines *id*, and *ib*, the third angle *bpi*, shall be concluded to be equal to the angle *ezi*. Therefore by the 13 e xiiij, five angles are equall.

23. *The Diagony is irrationall unto the side of the dodecahedrum.*

This is the fifth example of irrationality and incommensurability. The first was of the diagony and side of a quadrate or square. The second was of a line proportionally cut and his segments: The third is of the diameter of a Circle and the side of an inscribed quinquangle. The fourth was of the diagony and side of an icosahedrum. The fifth now is of the diagony and side of a dodecahedrum.

24 *If the side of a cube be cut proportionally, the greater segment shall be the side of a dodecahedrum.*

For that hath beene told you even now.

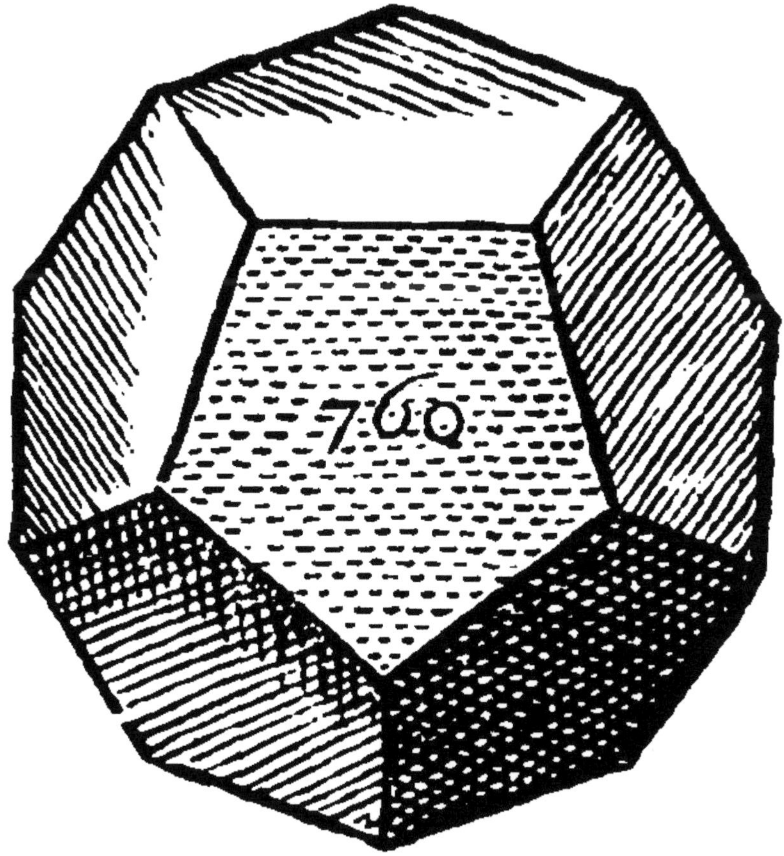

But from hence also doth arise the geodesy or mãner of measuring of a dodecahedrum. For if the quadrate of the line subtending the angle of a quinquangle be trebled, the half of the treble shall be the side of the semidiagony of the dodecahedrum: Because by the 6 e xxiiij, the diagony of the cube, that is of the dodecahedrum is of treble power to the side of the cube. But if the quadrate of the side of the decangle be taken out of the quadrate of the side of the quinquangle; The side of the remainder shall be the ray of the circle circumscribed about a quinquangle. Lastly if the quadrate of the ray, be taken of the quadrate of the half-diagony; the side of the remainder shall be the heighth of perpendicular. As if the side of the decangle be 7.3/5: The quadrate of that shall be 57.19/25: the treble of which is 173.7/25 whose side is about 13.107/131 for the side of the Dodecahedrum, therefore 6.119/131 the halfe shall be the semidiagony of the dodecahedrum. The ray of the Circle shall now thus be found. If the quadrate of the side of the decangle be taken out of the quadrate of the side of the sexangle; the side of the remainder, shall be the Ray of the Circle, by the 15 and 9 e xviij. As here the side of the Quinquangle is 4.2/3. The side of the Decangle 2.2/5: And the quadrates therefore are 21.7/9, and 5.19/25. This subducted from that leaveth 16.4/225 whose side is 4.2/15 for the Ray of the Circle.

The semidiagony and ray of the circle thus found, the altitude remaineth. Take out therefore the quadrate of the ray of the circle, 16.4/225 out of the quadrate of the semidiagony 47.12458/17161, the side of the remainder 31.2714406/3861225 is for the altitude or heighth: whose 1/3 is 5/3. The quinquangled base is almost 38. Which multiplied by 5/3 doth make 63.1/3 for the solidity of one Pyramis; which multiplied by 12, doth make 760. for the soliditie of the whole dodecaedrum.

25 *There are but five ordinate solid plaines.*

This appeareth plainely out of the nature of a solid angle, by the kindes of plaine figures. Of two plaine angles a solid angle cannot be comprehended. Of three angles of an ordinate triangle is the angle of a Tetrahedrum comprehended: Of foure, an Octahedrum: Of five, an Icosahedrum: Of sixe none can be comprehended: For sixe such like plaine angles, are equall to 12 thirds of one right angle, that is to foure right angles. But plaine angles making a solid angle, are lesser than foure right angles, by the 8 e xxij. Of seven therefore, and of more it is, much lesse possible. Of three quadrate angles the angle of a cube is comprehended: Of 4. such angles none may be comprehended for the same cause. Of three angles of an ordinate quinquangle, is made the angle of a Dodecahedrum. Of 4. none may possibly be made; For every such angle: For every one of them severally doe countervaile one right angle and 1/5 of the same, Therefore they would be foure, and three fifths. Of more therefore much lesse may it be possible.

This demonstration doth indeed very accurately and manifestly appeare, Although there may be an innumerable sort of ordinate plaines, yet of the kindes of angles five onely ordinate bodies may be made; From whence the Tetrahedrum, Octahedrum, and Icosahedrum are made upon a triangular base: the Cube upon a quadrangular: And the Dodecahedrum, upon a quinquangular.

Of Geometry *the twenty sixth Booke; Of a* Spheare.

1 *An imbossed solid is that which is comprehended of an imbossed surface.*

2. *And it is either a spheare or a Mingled forme.*

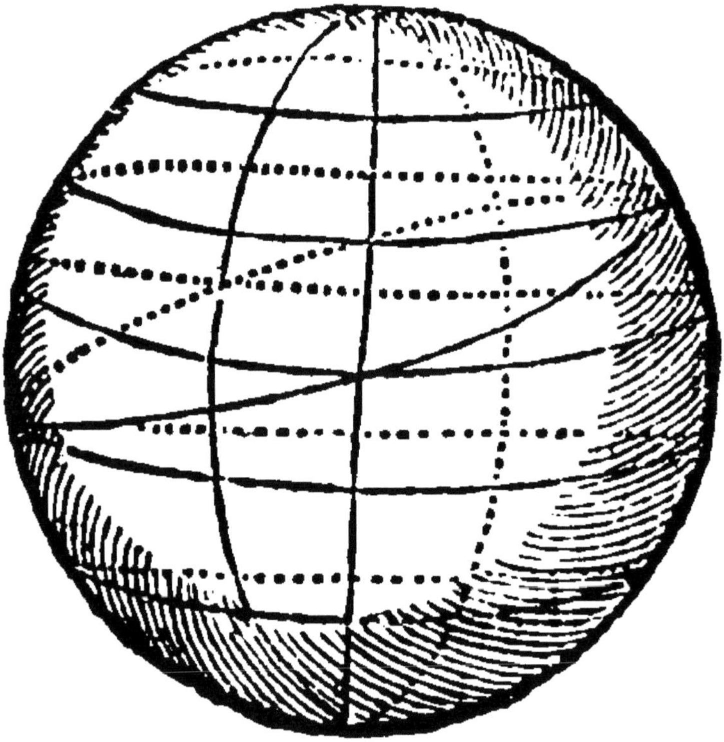

3. *A speare is a round imbossement.*

It may also be defined to be that which is comprehended of a sphearical surface. A sphearicall body in Greeke is called *Sphæra*, in Latine *Globus*, a Globe.

Therefore

4. *A Spheare is made by the conversion of a semicircle, the diameter standing still. 14 d xj.*

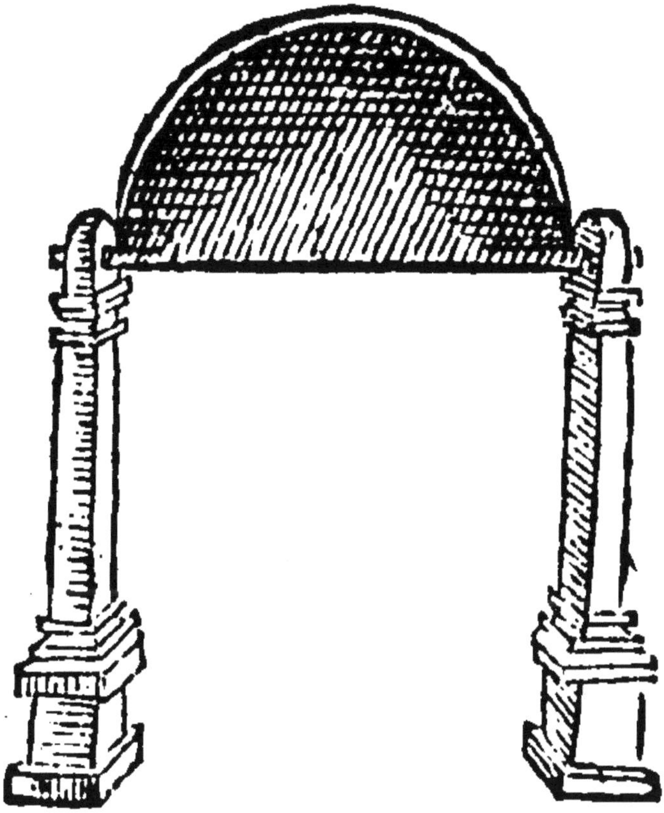

As here thou seest.

5. *The greatest circle of a spheare, is that which cutteth the spheare into two equall parts.*

Therefore

6. *That circle which is neerest to the greatest, is greater than that which is farther off.*

And

7. *Those which are equally distant from the greatest are equall.*

As in the example above written.

8. *The plaine of the diameter and sixth part of the sphearicall is the solidity of the spheare.*

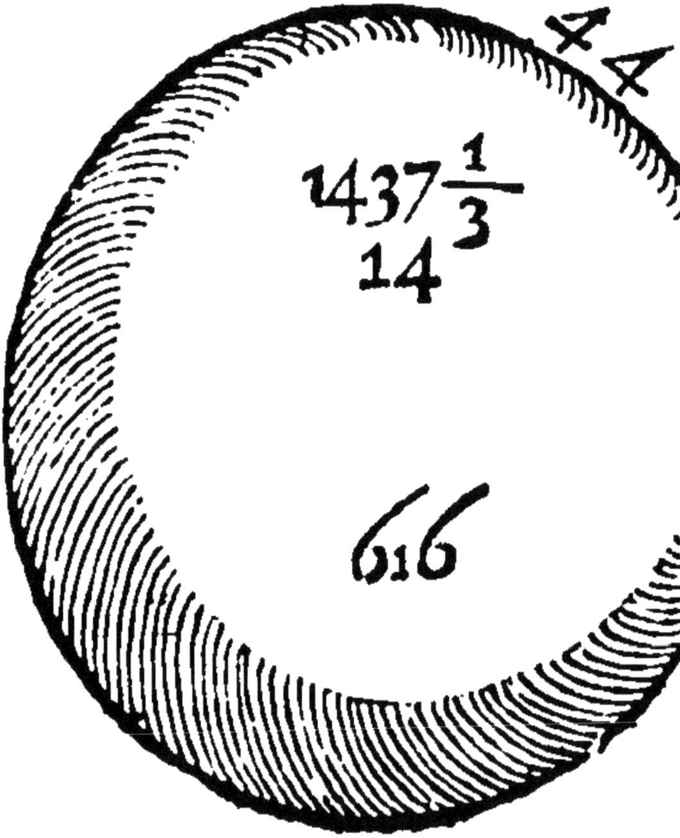

As before there was an analogy betweene a Circle and a Sphericall: so now is there betweene a Cube and a spheare. A cubicall surface is comprehended of sixe quadrate or square and equall bases: And a spheare in like manner is comprehended of sixe equall sphearicall bases compassing the cubicall bases. A cube is made by the multiplication of the sixth part of the base, by the side: And a spheare likewise is made by multiplying the sixth part of the sphearicall by the diameter, as it were by the side: so the plaine of 616/6 and 14, the diameter is 1437.1/3 for the solidity of the spheare.

Therefore

9. *As 21 is unto 11, so is the cube of the diameter unto the spheare.*

As here the Cube of 14 is 2744. For it was an easy matter for him that will compare the cube 2744, with the spheare, to finde that 2744 to be to 1437.1/3 in the least boundes of the same reason, as 21 is unto 11.

Thus much therefore of the Geodesy of the spheare: The geodesy of the Sectour and section of the spheare shall follow in the next place.

And

10. *The plaine of the ray, and of the sixth part of the sphearicall is the hemispheare.*

But it is more accurate and preciser cause to take the halfe of the spheare.

11. *Spheares have a trebled reason of their diameters.*

So before it was told you; That circles were one to another, as the squares of their diameters were one to another, because they were like plaines: And the diameters in circles were, as now they are in spheares, the homologall sides. Therefore seeing that spheres are figures alike, and of treble dimension, they have a trebled reason of their diameters.

12. *The five ordinate bodies are inscribed into the same spheare, by the conversion of a semicicle having for the diameter, in a tetrahedrum, a right line of value sesquialter unto the side of the said tetrahedrum; in the other foure ordinate bodies, the diagony of the same orginate.*

The Adscription of ordinate plaine bodies is unto a spheare, as before the Adscription plaine surfaces was into a circle; of a triangle, I meane, and ordinate triangulate, as Quadrangle, Quinquangle, Sexangle, Decangle, and Quindecangle. But indeed the Geometer hath both inscribed and circumscribed those plaine figures within a circle. But these five ordinate bodies, and over and above the Polyhedrum the Stereometer hath onely inscribed within the spheare. The Polyhedrum we have passed over, and we purpose onely to touch the other ordinate bodies.

13 *Out of the reason of the axeltree of the sphearicall the sides of the tetraedrum, cube, octahedrum and dodecahedrum are found out.*

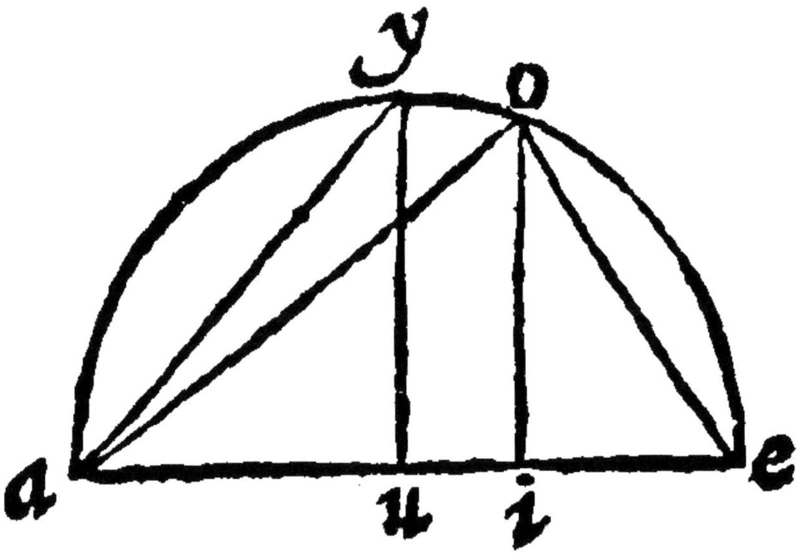

The axeltree in the three first bodies is rationall unto the side, as was manifested in the former. For it is of the sesquialter valew unto the side of the tetrahedrum; of treble, to the side of the cube: Of double, to the side of the Octahedrum. Therefore if the axis *ae*, be cut by a double reason in *i*: And the perpendicular *io*, be knit to *a*, and *e*, shall be the side of the tetrahedrum; and *oe*, of the cube, as was manifest by the 10 e viij, and 25 iiij: And the greater segment of the side of the cube proportionally cut, is by the 24 e, xxv.

If the same axis be cut into two halfes, as in *u*: And the perpendicular *uy*, be erected: And *y*, and *a*, be knit together, the same *ya*, thus knitting them, shall be the side of the Octahedrum, as is manifest in like manner, by the said 10 e, viij, and 25 e iiij.

The side of the Icosahedrum is had by this meanes.

14. *If a right line equall to the axis of the sphearicall, and to it from the end of the perpendicular be knit unto the center, a right line drawne from the cutting of the periphery unto the said end shall be the side of the Icosahedrum.*

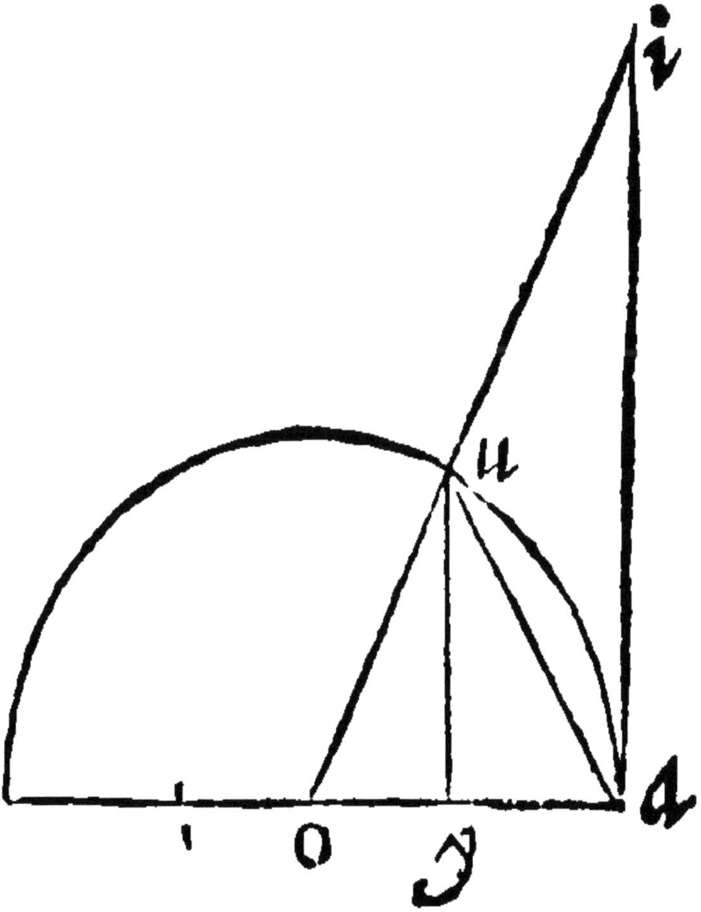

As here let the Axis *ae*; be the diameter of the circle *aue*, and *ai*, equall to the same axis, and perpendicular from the end, be knit unto the center, by the right line *io*: A right drawne from the section *u*, unto *a*, shall be the side of the

Icosahedrum. From *u*, let the perpendicular *uy*, be drawne: Here the two triangles *iao*, & *uyo*, are equiangles by the 13 e, vij. Therfore by the 12 e, vij. as *ia*, is unto *ao*: so is *uy*, unto *yo*. But *ia*, is the double of the said *ao*: Therefore *uy*, is the double of the same *yo*: Therefore by the 14 e, xij, it is of quadruple power unto it: And therefore also *uy*, and *yo*, that is, by the 9 e xij, *uo*, that is againe by the 28 e, iiij, *ao*, is of quintuple power to *yo*. But *yo*, is lesser than *ao*, that is, than *oe*: Let therefore *os*, be cut off equall to it. Now as the halfe of *ao*, is of quintuple valew to the halfe of *yo*: so the double *ae*, is of quintuple power to the double *ys*. Therefore, by the 18 e xxv. seeing that the diagony *ae*, is of quintuple power to *ys*; the said *ys*, shall be the side of the sexangle inscribed into a circle, circumscribing the quinquangle of the Icosahedrum. But the perpendicular *uy*, is equall to *ys*; because each of them is the double of *yo*. Wherefore *uy*, is the side of the sexangle. But *ay*, is the side of the Decangle: For it is equall to *se*: Because if from equall rayes *ao*, and *oe*, you take equall portions *oy*, and *os*: There shall remaine equall, *ya*, and *se*. And the Diagony of an Icosahedrum by the 16 e xxv, is compounded of the side of the sexangle, continued at each end with the side of the decangle. Wherefore *ay*, is the side of the decangle. Lastly, *ua*, whose power is as much as the sides of the sexangle and decangle, by the 15. e, xviij, shall be the side of an Icosahedrum.

15 *Of the five ordinate bodies inscribed into the same spheare, the tetrahedrum in respect of the greatnesse of his side is first, the Octahedrum, the second; the Cube, the third; the Icosahedrum, the fourth; and the Dodecahedrum, the fifth.*

As it will plainely appeare, if all of them be gathered into one, thus. For *ai*, the side of the Tetrahedrum, subtendeth a greater periphery than *ao*, the side of the Octahedrum; And *ao*, a greater than *ie*, the side of the Cube; because it subtendeth but the halfe: And *ie*, greater than *ue*, the side of the Icosahedrum: And *ue*, greater than *ye*, the side of Dodecahedrum.

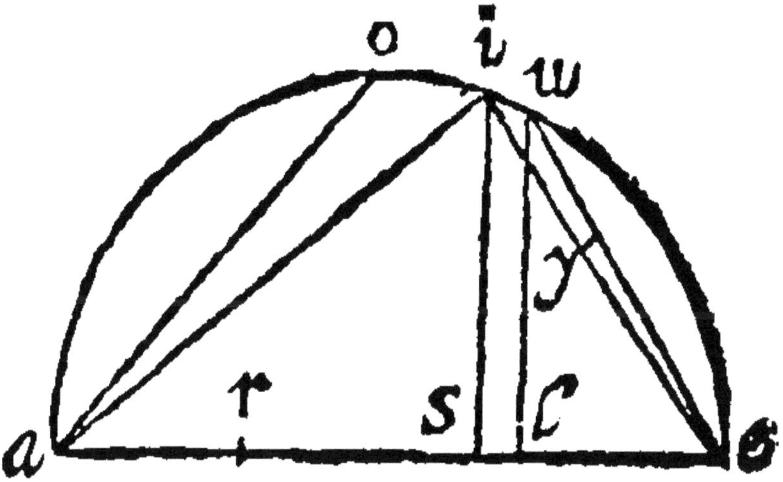

The latter, *Euclide* doth demonstrate with a greater circumstance. Therefore out of the former figures and demonstrations, let here be repeated, The sections of the axis first into a double reason in *s*: And the side of the sexangle *rl*: And the side of the Decangle *ar*, inscribed into the same circle, circumscribing the quinquangle of an icosahedrum: And the perpendiculars *is*, and *ul*.

Here the two triangles *aie*, and *ies*, are by the 8 e, viij. alike; And as *se*, is unto *ei*: So is *ie*, unto *ea*: And by 25 e, iiij, as *se*, is to *ea*: so is the quadrate of *se*, to the quadrate of *ei*: And inversly or backward, as *ae*, is to *se*: so is the quadrate of *ie*, to the quadrate of *se*. But *ae*, is the triple of *se*. Therefore the quadrate of *ie*, is the triple of *se*. But the quadrate of *as*, by the grant, and 14 e xij, the quadruple of the quadrate of *se*. Therefore also it is greater than the quadrate of *ie*: And the right line *as*, is greater than *ie*, and *al*, therefore is much greater. But *al*, is by the grant compounded of the sides of the sexangle and decangle *rl*, and *ar*. Therefore by the 1 c. 5 e, 18. it is cut proportionally: And the greater segment is the side of the sexangle, to wit, *rl*: And the greater segment of *ie*, proportionally also cut, is *ye*. Therefore the said *rl*, is greeter than *ye*: And even now it was shewed *ul*, was equall to *rl*. Therefore *ul*, is greater than *ye*: But *ue*, the side of the Icosahedrum, by 22. e vj. is greater than *ul*. Therefore the side of the Icosahedrum is much greater, then the side of the dodecahedrum.

Of Geometry *the twenty seventh Book; Of the Cone and Cylinder.*

1 *A mingled solid is that which is comprehended of a variable surface and of a base.*

For here the base is to be added to the variable surface.

2 *If variable solids have their axes proportionall to their bases, they are alike. 24. d xj.*

It is a Consectary out of the 19 e, iiij. For here the axes and diameters are, as it were, the shankes of equall angles, to wit, of right angles in the base, and perpendicular axis.

3 *A mingled body is a Cone or a Cylinder.*

The cause of this division of a varied or mingled body, is to be conceived from the division of surfaces.

4 *A Cone is that which is comprehended of a conicall and a base.*

Here the base is a circle.

Therefore

5 *It is made by the turning about of a right angled triangle, the one shanke standing still.*

As it appeareth out of the definition of a variable body.

And

6 *A Cone is rightangled, if the shanke standing still be equall to that turned about: It is Obtusangeld, if it be lesse: and acutangled, if it be greater. ê 18 d xj.*

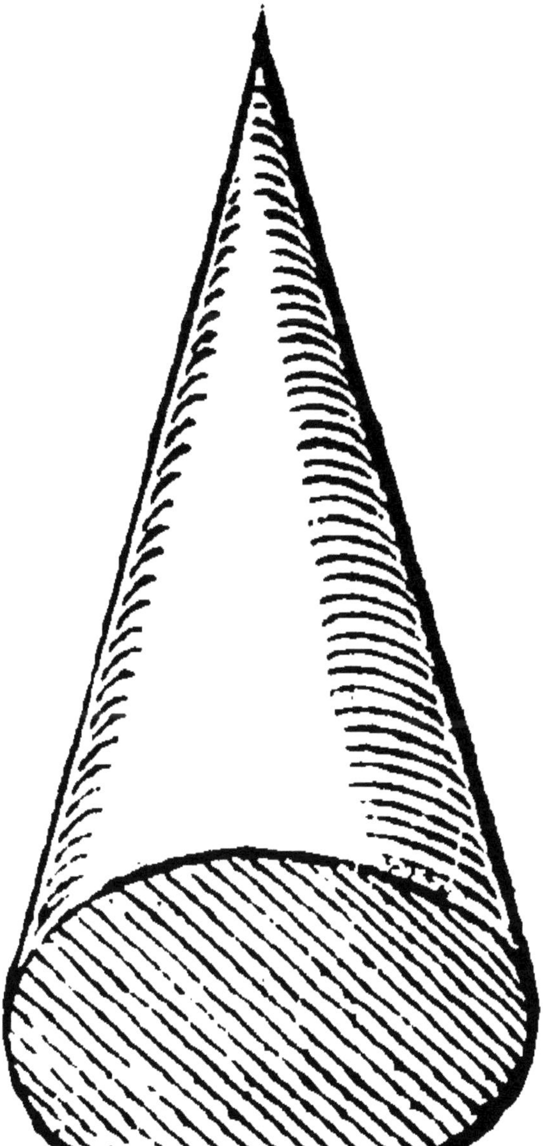

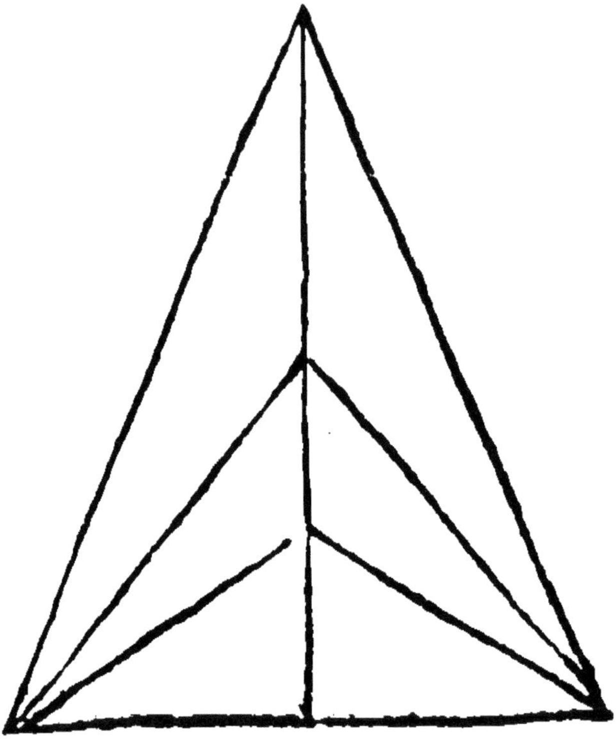

Here a threefold difference of the heighth of a Cone is professed, out of the threefold difference of the angles, whereby the toppe of the halfed cone is distinguished: Notwithstanding this consideration belongeth rather to the Optickes, than to Geometry. For a Cone a farre off seeme like triangle. Therefore according to the difference of the heighth, it appeareth with a right angled, or obtusangled or acutangled toppe: As here the least Cone is obtusangled: the middle one rightangled: and the highest acutangled. But the cause of this threefold difference in the angles from of the difference of the shankes, is out of the consectaries of the threefold triangle of a right line cutting the base into two equall parts, as appeareth at the end of the viij Booke.

And

7 *A Cone is the first of all variable.*

For a Cone is so the first in variable solids, as a triangle is in rectilineall plaines: As a Pyramis is in solid plaines: For neither may it indeed be divided into any other variable solids more simple.

And

8 *Cones of equall heighth are as their bases are 11. p xij.*

As here you see.

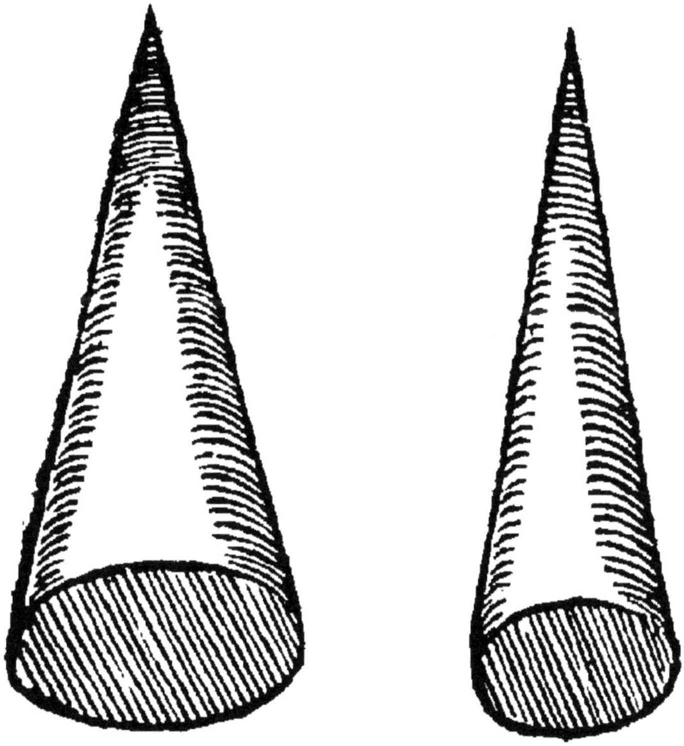

And

9 *They which are reciprocall in base and heighth are equall, 15 p xij.*

These are consectaries drawne out of the 16 and 18 e. iiij. As here you see.

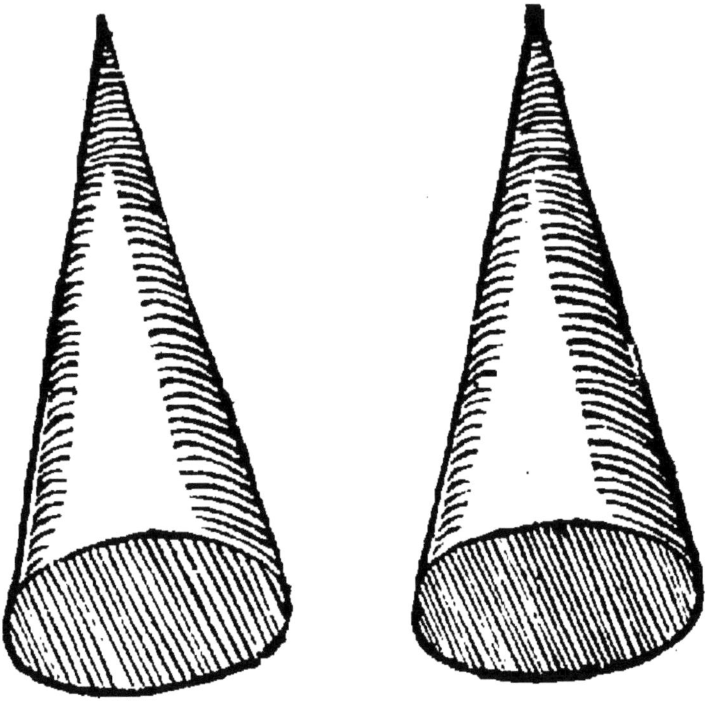

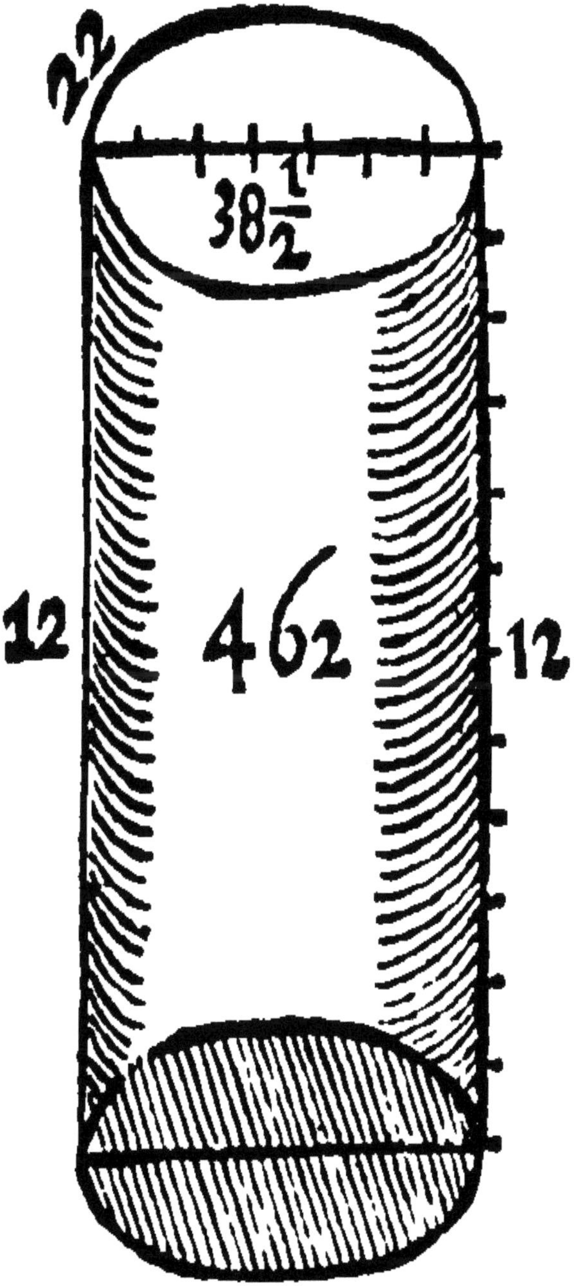
22
$38\frac{1}{2}$
12
462
12

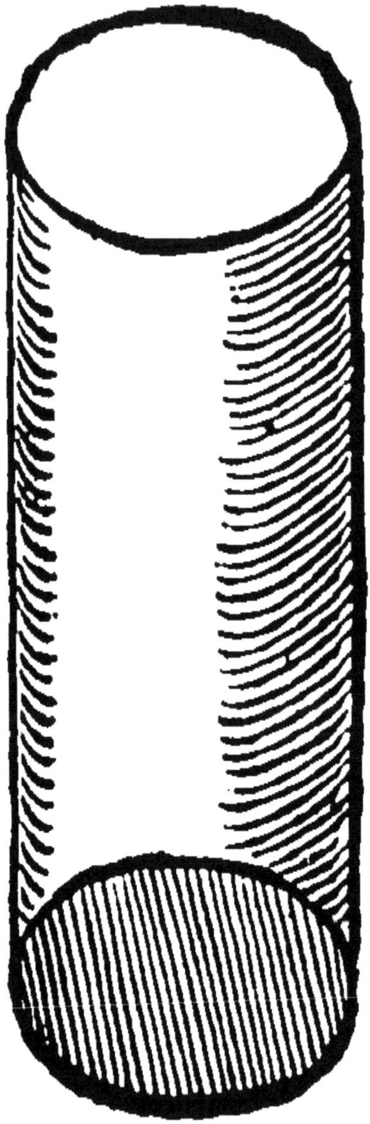

10 *A Cylinder is that which is comprehended of a cylindricall surface and the opposite bases.*

For here two circles, parallell one to another are the bases of a Cylinder.

Therefore

11 *It is made by the turning about of a right angled parallelogramme, the one side standing still. 21. d xj.*

As is apparant out the same definition of a varium.

12. *A plaine made of the base and heighth is the solidity of a Cylinder.*

The geodesy here is fetch'd from the prisma: As if the base of the cylinder be 38.1/2: Of it and the heighth 12, the solidity of the cylinder is 462.

This manner of measuring doth answeare, I say, to the manner of measuring of a prisma, and in all respects to the geodesy of a right angled parallelogramme.

If the cylinder in the opposite bases be oblique, then if what thou cuttest off from one base thou doest adde unto the other, thou shalt have the measure of the whole; as here thou seest in these cylinders, *a* and *b*.

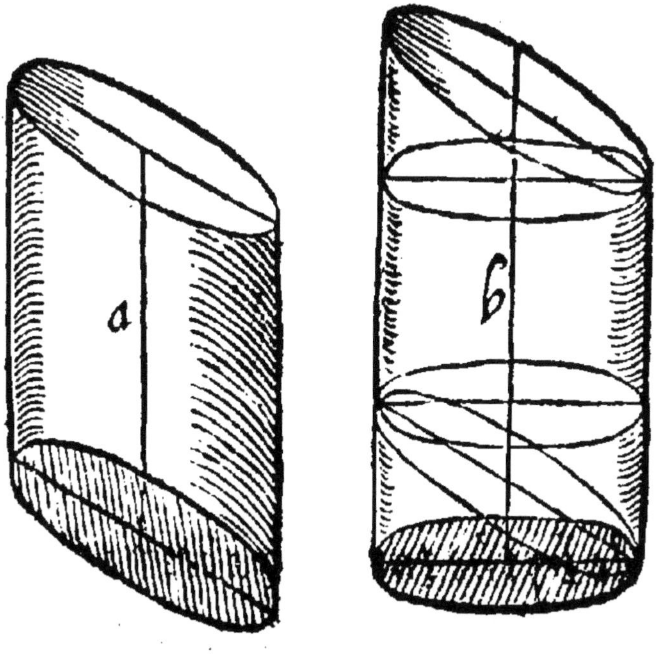

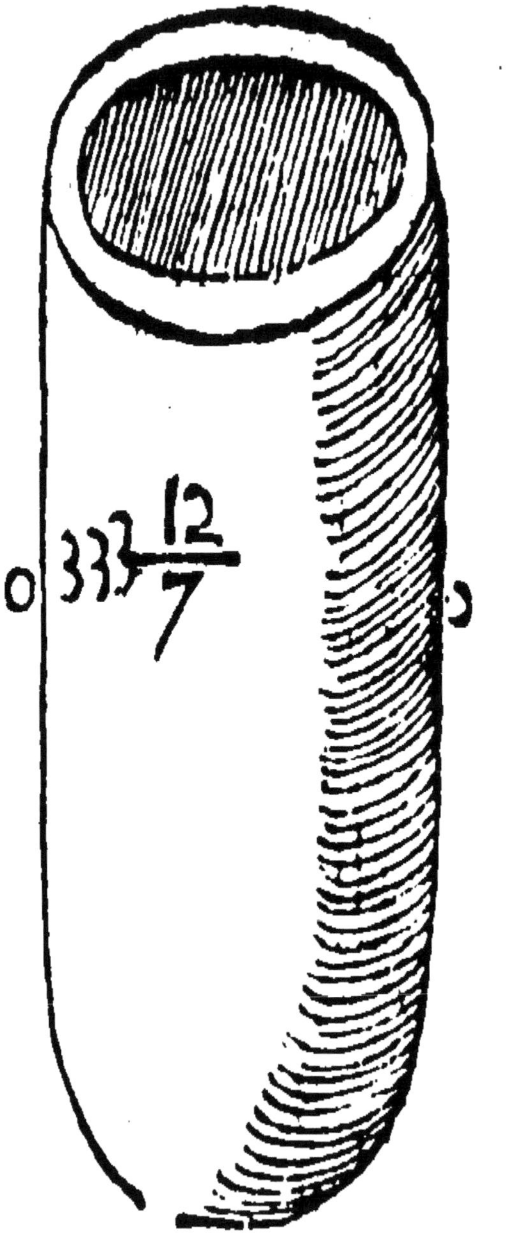

From hence the capacity or content of cylinder-like vessell or measure is esteemed and judged of. For the hollow or empty place is to be measured as if it were a solid body.

As here the diameter of the inner Circle is 6 foote: The periphery is 18.6/7: Therefore the plot or content of the circle is 28.2/7. Of which, and the heighth 10, the plaine is 282.6/7 for the capacity of the vessell. Thus therefore shalt thou judge, as afore, how much liquor or any thing else conteined, a cubicall foote may hold.

13. *A Cylinder is the triple of a cone equall to it in base and heighth. 10 p xij.*

The demonstration of this proposition hath much troubled the interpreters. The reason of a Cylinder unto a Cone, may more easily be assumed from the reason of a Prisme unto a Pyramis: For a Cylinder doth as much resemble a Prisme, as the Cone doth a Pyramis: Yea and within the same sides may a Prisme and a Cylinder, a Pyramis and a Cone be conteined: And if a Prisme and a Pyramis have a very multangled base, the Prisme and Cylinder, as also the Pyramis and Cone, do seeme to be the same figure. Lastly within the same sides, as the Cones and Cylinders, so the Prisma and Pyramides, from their axeltrees and diameters may have the similitude of their bases. And with as great reason may the Geometer demand to have it granted him, That the Cylinder is the treble of a Cone: As it was demanded and granted him, That Cylinders and Cones are alike, whose axletees are proportionall to the diameters of their bases.

Therefore

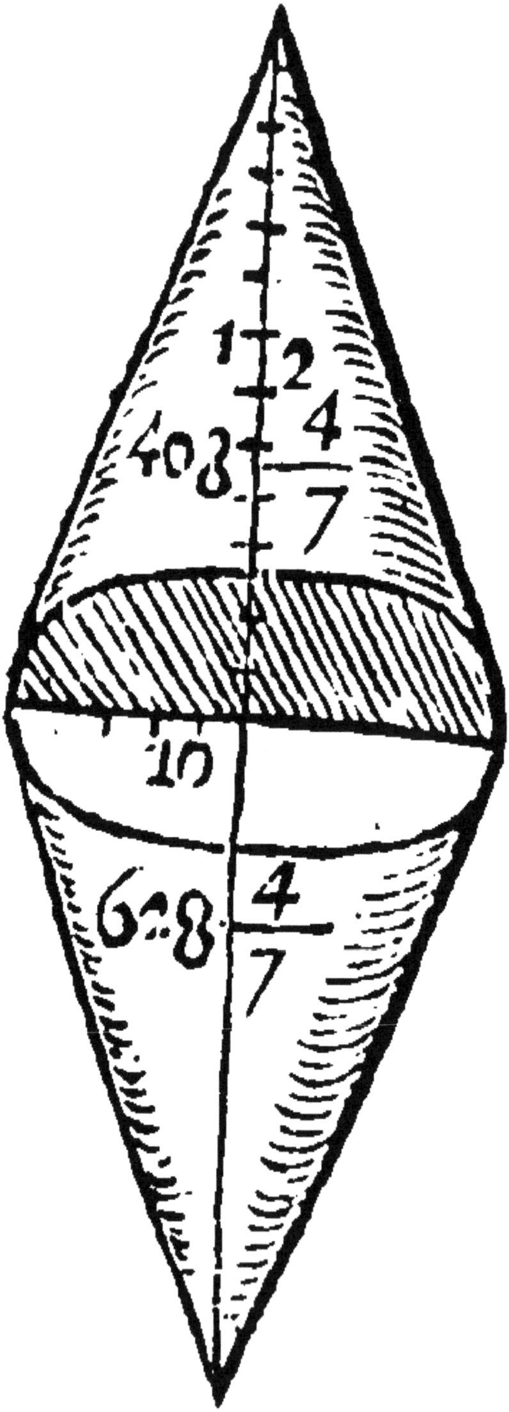
1
2
408
4
7
10
4
7

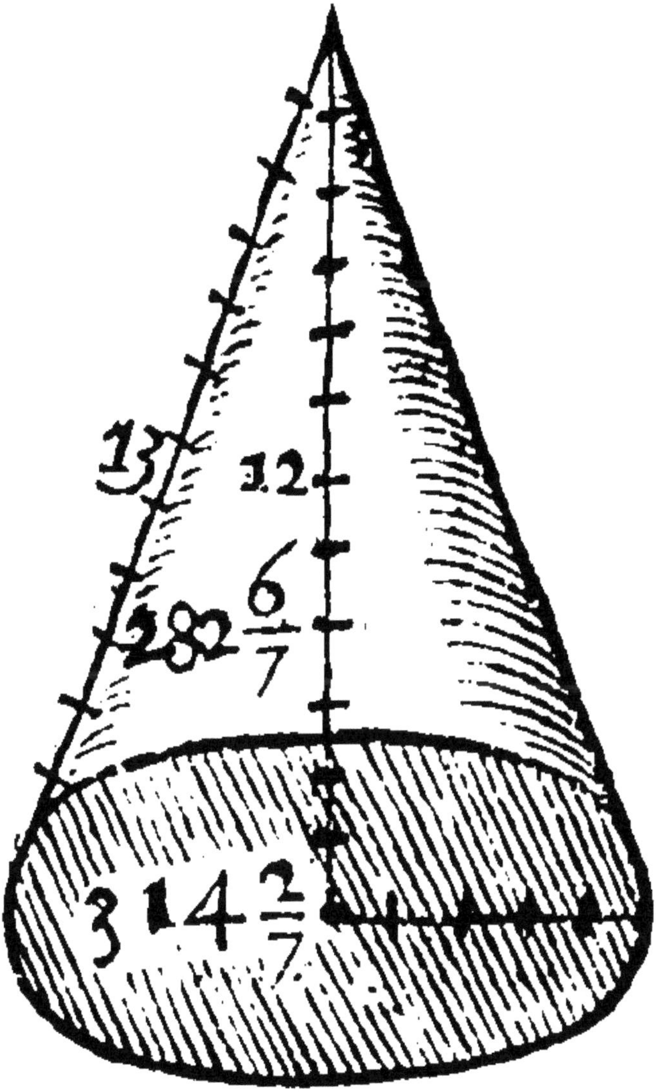

14. *A plaine made of the base and thid part of the height, is the solidity of the cone of equall base & height;*

The heighth is thus had. If the square of the ray of the base, be taken out of the square of the side, the side of the remainder shall bee the heighth, as is manifest by the 9 e xij. Here therefore the square of the ray 5, is 25. The square of 13, the side is 169. And 169 - 25, are 144; whose side is 12 for the heighth: The third part of which is 4. Now the circular base is 78.4/7: And the plaine of these is 314.2/7 for the solidity of the Cone.

But the analogie of a conicall unto a Cylinder like surface doth not answeare, that the Conicall should be the subtriple of the Cylindricall, as the Cone is the subtriple of the Cylinder.

Of two cones of one common base is made *Archimede's Rhombus*, as here, whose geodæsy shall be cut of two cones.

And

15. *Cylinder of equall heighth are as their bases are. 11 p xij.*

Sackes in which they carry corne, are for the most part of a cylinderlike forme. If an husbandman therefore shall lend unto his neighbour a sacke full or corne, and the base of the sacke be 4 foote over. And the neighbour afterward for that one sacke, shall pay him 4 sacke fulls, every sacke being as long as that was, yet but one foote over in the diameter, he may be thought peradventure to have repayed that which he borrowed in equall measure, to wit in heighth and base. But it shall be indeed farre otherwise: For there is a great difference betweene the quadrate of the foure severall diameters, 1. 1. 1. 1. that is 4: and 16, the quadrate of 4, the diameter of that sacke by which it was lent. For Circles are one unto another as the quadrates of their diameters are one to another, by the 2 e xv. Therefore he payd him but one fourth part of that which he borrowed of him.

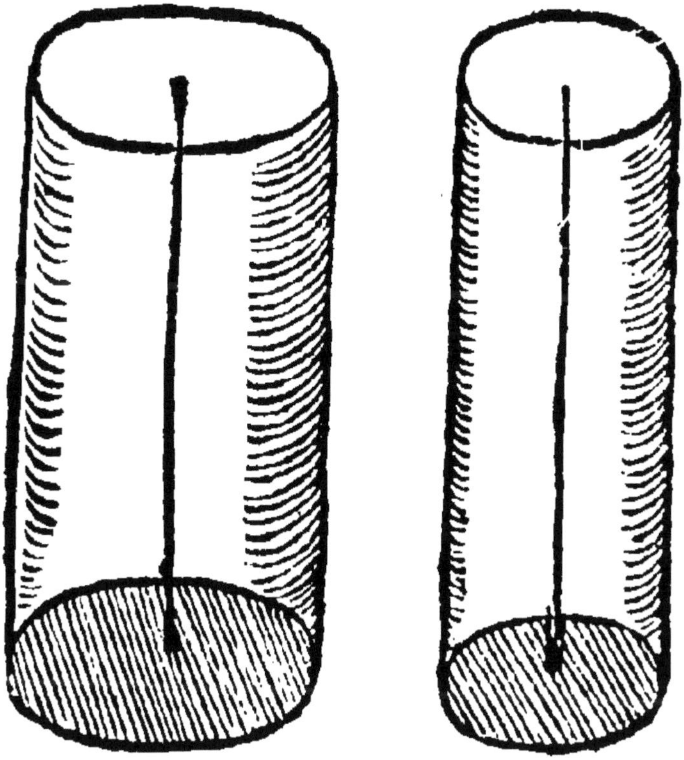

And

16 *Cylinders reciprocall in base and heighth are equall. 15 p xij.*

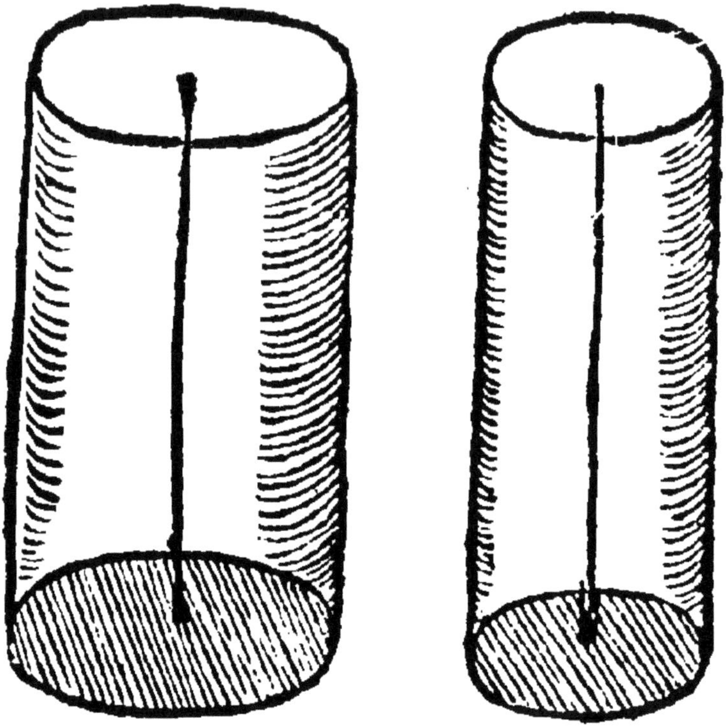

Both these affections are in common attributed to the equally manifold of first figures.

And

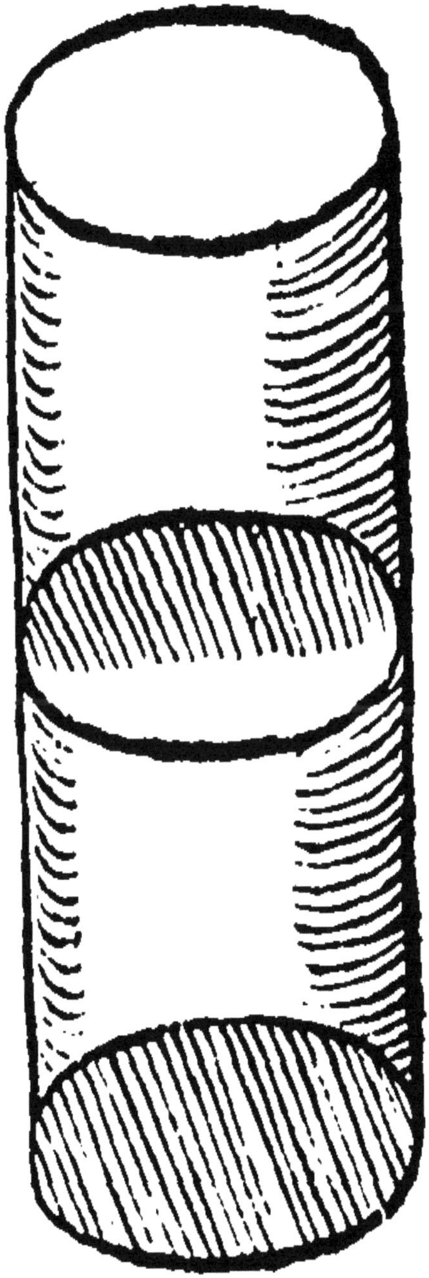

17. *If a cylinder be cut with a plaine surface parallell to his opposite bases, the segments are, as their axes are 13 p xij.*

As here thou seest. For the axes are the altitudes or heights. It is likwise a consectary following upon that generall theoreme of first figure, but somewhat varyed from it. It doth answere unto the 10 e 23.

The unequall sections of a spheare we have reserved for this place: Because they are comprehended of a surface both sphearicall and conicall, as is the sectour. As also of a plaine and sphearicall, as is the section: And in both like as in a Circle, there is but a greater and lesser segment. And the sectour, as before, is considered in the center.

18. *The sectour of a spheare is a segment of a spheare, which without is comprehended of a sphearicall within of a conicall bounded in the center, the greater of a concave, the lesser of a convex.*

Archimedes, maketh mention of such kinde of Sectours, in his 1 booke of the Spheare. From hence also is the geodesy following drawne. And here also is there a certaine analogy with a circular sectour.

19. *A plaine made of the diameter, and sixth part of the greater, or lesser sphearicall, is the greater or lesser sector.*

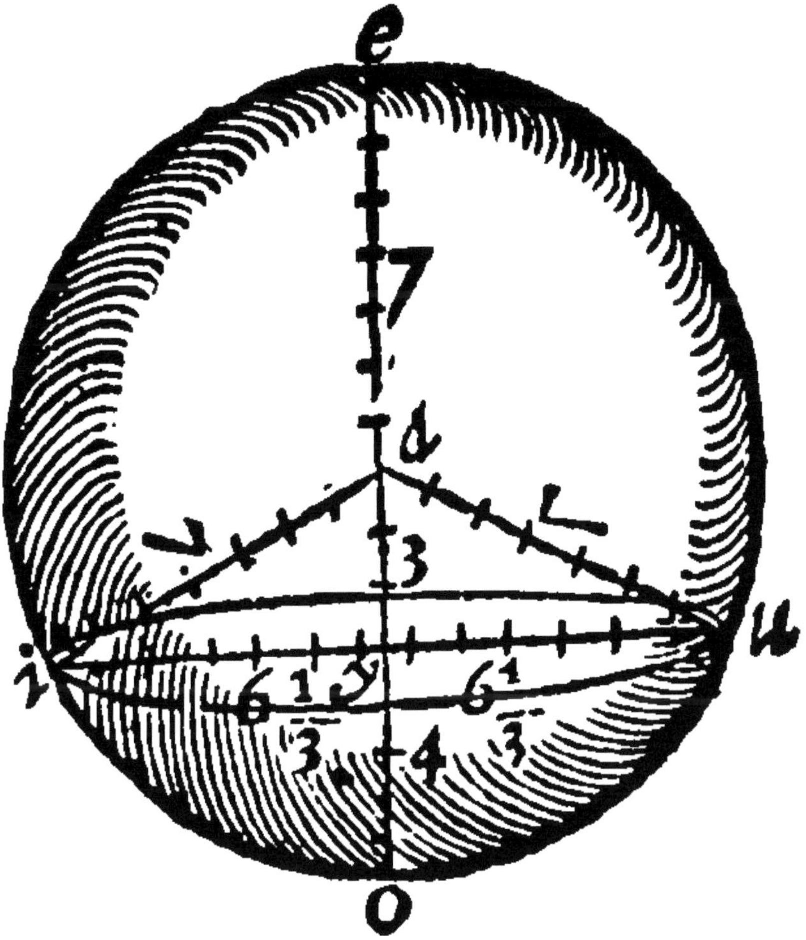

As here of the Diameter 14, and of 73.1/3 and 4.2/3 (which is the one sixth part of the greater sphearicall) the plaine is 1026.2/3 for the solidity of the greater sectour, so of the same diameter 14, and 29.1/3 which is the 1/6 part of 176, the lesser sphæricall, the plaine is 410.2/3 for the solidity of the lesser sectour.

And from hence lastly doth arise the solidity of the section, by addition and subduction.

20. *If the greater sectour be increased with the internall cone, the whole shall be the greater section: If the lesser be diminished by it, the remaine shall be the lesser section.*

As here the inner cone measured is 126.4/63. The greater sectour, by the former was 1026.2/3. And 1026.2/3 + 126.4/63 doe make 1152.46/63.

Againe the lesser sectour, by the next precedent, was 410.2/3: And here the inner cone is 126.4/63 And therefore 410.2/3 - 126.4/63 that is 284.38/63 is the lesser section.

FINISH.

*** END ***

This book has been produced in line with the EU GPSR guidelines about the safety of products.

The General Product Safety Regulation is the European Union's updated framework for ensuring that all consumer products, including books, are safe for consumers.

This book has been printed by Libri Plureos GmbH. The printer has issued safety certificates for the materials - like ink, paper and glue - being used.

The product identifier is: 9786253870027

The author is responsible for the content of the book, is the publisher of the works and bears full responsibility for it.

The book has been produced via Bookmundo. Bookmundo enables any author to share their stories with the rest of the world via printed books and ebooks and a broad distribution network.

Bookmundo will act as an intermediary in regard to questions about safety and will address them to the printer / author. Should there be any question in regard to the safety of the product, please contact us.

Bookmundo
Delftsestraat 33
3013AE Rotterdam
The Netherlands
info@bookmundo.com